FINITARY PROBABILISTIC METHODS IN ECONOPHYSICS

Econophysics applies the methodology of physics to the study of economics. However, whilst physicists have a good understanding of statistical physics, they are often unfamiliar with recent advances in statistics, including Bayesian and predictive methods. Equally, economists with knowledge of probabilities do not have a background in statistical physics and agent-based models. Proposing a unified view for a dynamic probabilistic approach, this book is useful for advanced undergraduate and graduate students as well as researchers in physics, economics and finance.

The book takes a finitary approach to the subject. It discusses the essentials of applied probability, and covers finite Markov chain theory and its applications to real systems. Each chapter ends with a summary, suggestions for further reading and exercises with solutions at the end of the book.

UBALDO GARIBALDI is First Researcher at the IMEM-CNR, Italy, where he researches the foundations of probability, statistics and statistical mechanics, and the application of finite Markov chains to complex systems.

ENRICO SCALAS is Assistant Professor of Physics at the University of Eastern Piedmont, Italy. His research interests are anomalous diffusion and its applications to complex systems, the foundations of statistical mechanics and agent-based simulations in physics, finance and economics.

FINITARY PROBABILISTIC METHODS IN ECONOPHYSICS

UBALDO GARIBALDI
IMEM-CNR, Italy

ENRICO SCALAS
University of Eastern Piedmont, Italy

CAMBRIDGE
UNIVERSITY PRESS

University Printing House, Cambridge CB2 8BS, United Kingdom

One Liberty Plaza, 20th Floor, New York, NY 10006, USA

477 Williamstown Road, Port Melbourne, VIC 3207, Australia

314-321, 3rd Floor, Plot 3, Splendor Forum, Jasola District Centre, New Delhi - 110025, India

79 Anson Road, #06-04/06, Singapore 079906

Cambridge University Press is part of the University of Cambridge.

It furthers the University's mission by disseminating knowledge in the pursuit of education, learning and research at the highest international levels of excellence.

www.cambridge.org
Information on this title: www.cambridge.org/9780521515597

First published 2010

A catalogue record for this publication is available from the British Library

ISBN 978-0-521-51559-7 Hardback

Contents

Foreword

What is this book about?

The theme of this book is the allocation of n objects (or elements) into g categories (or classes), discussed from several viewpoints. This approach can be traced back to the early work of 24-year-old Ludwig Boltzmann in his first attempt to derive Maxwell's distribution of velocity for a perfect gas in probabilistic terms.

Chapter 2 explains how to describe the state of affairs in which for every object listed 'alphabetically' or in a sampling order, its category is given. We can consider the descriptions of Chapter 2 as facts (taking place or not), and events as propositions (true or not) about facts (taking place or not). Not everything in the world is known, and what remains is a set of possibilities. For this reason, in Chapter 3, we show how events can be probabilized and we present the basic probability axioms and their consequences. In Chapter 4, the results of the previous two chapters are rephrased in the powerful language of random variables and stochastic processes.

Even if the problem of allocating n objects into g categories may seem trivial, it turns out that many important problems in statistical physics and some problems in economics and finance can be formulated and solved using the methods described in Chapters 2, 3 and 4. Indeed, the allocation problem is far from trivial. In fact, in the language of the logical approach to probability, traced back to Johnson and, mainly, to Carnap, the individual descriptions and the statistical descriptions are an essential tool to represent possible worlds. A neglected paper written by Brillouin showed that the celebrated Bose–Einstein, Fermi–Dirac and Maxwell–Boltzmann distributions of statistical physics are nothing other than particular cases of the one-parameter multivariate accommodation process, where bosons share the same predictive probability as in Laplace's succession rule. These are particular instances of the generalized multivariate Pólya distribution studied in Chapter 5 as a sampling distribution of the Pólya process. The Pólya process is an n-exchangeable stochastic process, whose characterization leads us to discuss Pearson's fundamental problem of practical statistics, that is, given that a result of an experiment has been observed

r times out of m trials, what is the probability of observing the result in the next (the $(m+1)$th) trial?

Up to Chapter 5, we study the allocation of n objects into g categories either as an accommodation process or as a sampling process. The index set of the finite individual stochastic processes studied in Chapters 4 and 5 may represent either the name of the objects or their order of appearance in a sampling procedure. In other words, objects are classified once for all. If time enters and change becomes possible, the simplest probabilistic description of n objects jumping within g categories as time goes by is given in terms of finite Markov chains, which are an extension in several directions of the famous Ehrenfest urn model. This is the subject of Chapter 6. For the class of irreducible Markov chains, those chains in which each state can be reached from any other state, there exists a unique invariant distribution, that is a probability distribution over the state space which is not affected by the time evolution. If the Markov chain is also aperiodic, meaning that the initial conditions are forgotten forever during time evolution, then the invariant distribution is also the equilibrium distribution. This means that, irrespective of the initial conditions, in the long run, the chain is described by its invariant distribution, which summarizes all the relevant statistical properties. Moreover, the ergodic problem of statistical physics finds a complete solution in the theory of finite Markov chains. For finite, aperiodic and irreducible Markov chains, time averages converge to ensemble averages computed according to the invariant (and equilibrium) distribution. Incidentally, in this framework, many out-of-equilibrium properties can be derived. The Ehrenfest–Brillouin Markov chain of Chapter 7 encompasses many useful models of physics, finance and economics. This is shown not only in Chapter 7, but also in Chapter 8, where a detailed review of some recent results of ours is presented. These models describe a change of category as a sequence of an Ehrenfest destruction (where an element is selected at random and removed from the system) followed by a Brillouin creation (where an element re-accommodates in the system according to a Pólya weight).

The Pólya distribution is the invariant and equilibrium distribution for the Ehrenfest–Brillouin Markov chain. In Chapter 9, we study what happens to the Pólya distribution and to the Ehrenfest–Brillouin process when the number of categories is not known in advance, or, what is equivalent, it is infinite. This naturally leads to a discussion of models of herding (an element joins an existing cluster) and innovation (an element decides to go alone and create a new cluster). If the probability of innovation still depends on the number of elements present in the system, the herding–innovation mechanism leads to the Ewens distribution for the size of clusters. On the contrary, in Chapter 10, the more difficult case is considered in which the innovation probability is a constant, independent of the number of elements in the system. The latter case leads to the Yule distribution of sizes displaying a power-law behaviour.

One can always move from the probabilistic description of individuals to the probabilistic description of occupation vectors to the probabilistic description of the frequency of occupation vectors. This is particularly simple if the descriptions have suitable symmetries. In Chapter 5, we show how to do that, for individual descriptions whose sampling distribution is the Pólya distribution. Two methods can be used, and both of them can be traced back to Boltzmann's work: the exact marginalization method and the approximate method of Lagrange multipliers leading to the most probable occupation vector. These methods are then used in the subsequent chapters as they give quantities that can be compared either with the results of Monte–Carlo simulations or with empirical data.

Throughout the book, we have worked with a *finitary* approach and we have tried to avoid the use of the infinite additivity axiom of probability theory. Of course, this is not always possible and the reader is warned that not all our derivations are fully rigorous even if we hope they are essentially correct. In principle, this book could be read with profit by anyone who masters elementary calculus. Integrals and derivatives are seldom used, but limits do appear very often and always at the end of calculations on finite systems. This procedure follows Boltzmann's conviction that the continuum is just a useful device to describe a world which is always finite and discrete. This conviction is shared by E.T. Jaynes, according to whom, if one wishes to use continuous quantities, one must provide the explicit limiting procedure from the finite assumptions.

A final *caveat* is necessary. This is not a textbook on Monte Carlo simulations. However, we do often use them and they are also included in solved exercises. The reader already acquainted with Monte Carlo simulations should not find any major difficulty in using the simulation programs written in *R* and listed in the book, and even in improving them or spotting bugs and mistakes! For absolute beginners, our advice is to try to run the code we have given in the book, while, at the same time, trying to understand it. Moreover, this book is mainly concerned with theoretical developments and it is not a textbook on statistics or econometrics, even if, occasionally, the reader will find estimates of relevant quantities based on empirical data. Much work and another volume would be necessary in order to empirically corroborate or reject some of the models discussed below.

Cambridge legacy

We are particularly proud to publish this book with Cambridge University Press, not only because we have been working with one of the best scientific publishers, but also because Cambridge University has an important role in the flow of ideas that led to the concepts dealt with in this book.

The interested reader will find more details in Chapter 1, and in the various notes and remarks around Chapter 5 and Section 5.6 containing a proof, in the spirit of Costantini's original one, of what, perhaps too pompously, we called the fundamental representation theorem for the Pólya process. As far as we know, both W.E. Johnson and J.M. Keynes were fellows at King's College in Cambridge and both of them worked on problems strictly related to the subject of this book.

Johnson was among Keynes' teachers and introduced the so-called postulate of *sufficientness*, according to which the answer to the fundamental problem of practical statistics, that is, given r successes over m trials, the probability of observing a success in the next experiment depends only on the number of previous successes r and the size of the sample m.

How to use this book?

This book contains a collection of unpublished and published material including lecture notes and scholarly papers. We have tried to arrange it in a textbook form. The book is intended for advanced undergraduate students or for graduate students in quantitative natural and social sciences; in particular, we had in mind physics students as well as economics students. Finally, applied mathematicians working in the fields of probability and statistics could perhaps find something useful and could easily spot mathematical mistakes. Incidentally, we deeply appreciate feedback from readers. Ubaldo Garibaldi's email address is `garibaldi@fisica.unige.it` and Enrico Scalas can be reached at `enrico.scalas@mfn.unipmn.it`.

Enrico Scalas trialled the material in Chapter 1, in Section 3.6, and parts of Chapter 4 for an introductory short course (20 hours) on probability for economists held for the first year of the International Doctoral Programme in Economics at Sant'Anna School of Advanced Studies, Pisa, Italy in fall 2008. The feedback was not negative. Section 3.6 contains the standard material used by Enrico for the probability part of a short (24 hours) introduction to probability and statistics for first-year material science students of the University of Eastern Piedmont in Novara, Italy. Finally, most of Chapters 4, 5 and 8 were used for a three-month course on stochastic processes in physics and economics taught by Enrico to advanced undergraduate physics students of the University of Eastern Piedmont in Alessandria, Italy. This means that the material in this book can be adapted to several different courses.

However, a typical one-semester econophysics course for advanced undergraduate students of physics who have already been exposed to elementary probability theory could include Chapters 2, 4, 5, 6, 7 and selected topics from the last chapters.

Acknowledgements

Let us begin with a disclaimer. The people we are mentioning below are neither responsible for any error in this book, nor do they necessarily endorse the points of view of the authors.

Domenico Costantini, with whom Ubaldo has been honoured to work closely for thirty years, is responsible for the introduction into Italy of the Cambridge legacy, mainly mediated by the work of Carnap. His deep interest in a probabilistic view of the world led him to extend this approach to statistical physics. Paolo Viarengo has taken part in this endeavour for the last ten years. We met them many times during the period in which this book was written, we discussed many topics of common interest with them, and most of the already published material contained in this book was joint work with them! Both Ubaldo and Enrico like to recall the periods in which they studied under the supervision of A.C. Levi many years ago. Finally, we thank Lindsay Barnes and Joanna Endell-Cooper at Cambridge University Press as well as Richard Smith, who carefully edited our text, for their excellent work on the manuscript.

Enrico developed his ever increasing interest in probability theory and stochastic processes while working on continuous-time random walks with Rudolf Gorenflo and Francesco Mainardi and, later, with Guido Germano and René L. Schilling.

While working on this book Enrico enjoyed several discussions on related problems with Pier Luigi Ferrari, Fabio Rapallo and John Angle. Michele Manzini and Mauro Politi were very helpful in some technical aspects.

For Enrico, the idea of writing this book was born between 2005 and 2006. In 2005, he discussed the problem of allocating n objects into g boxes with Tomaso Aste and Tiziana di Matteo while visiting the Australian National University. In the same year, he attended a Thematic Institute (TI) sponsored by the EU EXYSTENCE network in Ancona, from May 2 to May 21. Mauro Gallegati and Edoardo Gaffeo (among the TI organizers) independently developed an interest in the possibility of applying the concept of statistical equilibrium to economics. The TI was devoted to *Complexity, Heterogeneity and Interactions in Economics and Finance*. A group of students, among whom were Marco Raberto, Eric Guerci, Alessandra Tedeschi and Giulia De Masi, followed Enrico's lectures on *Statistical equilibrium in physics and economics* where, among other things, the ideas of Costantini and Garibaldi were presented and critically discussed. During the Econophysics Colloquium in November 2006 in Tokyo, the idea of writing a textbook on these topics was suggested by Thomas Lux and Taisei Kaizoji, who encouraged Enrico to pursue this enterprise. Later, during the Econophysics Colloquium 2007 (again in Ancona), Joseph McCauley was kind enough to suggest Enrico as a potential author to Cambridge University Press. Out of this contact, the proposal for this book was written jointly by Ubaldo and Enrico and the joint proposal was later accepted. In October 2008, Giulio Bottazzi (who has some nice papers on the application of the Pólya distribution to the spatial structure of economic activities) invited Enrico for a short introductory course on probability for economists where some of the material in this book was tested on the field.

Last, but not least, Enrico wishes to acknowledge his wife, Tiziana Gaggino, for her patient support during the months of hard work. The formal deadline for this book (31 December 2009) was also the day of our marriage.

1
Introductory remarks

This chapter contains a short outline of the history of probability and a brief account of the debate on the meaning of probability. The two issues are interwoven. Note that the material covered here already informally uses concepts belonging to the common cultural background and that will be further discussed below. After reading and studying this chapter you should be able to:

- gain a first idea on some basic aspects of the history of probability;
- understand the main interpretations of probability (classical, frequentist, subjectivist);
- compute probabilities of events based on the fundamental counting principle and combinatorial formulae;
- relate the interpretations to the history of human thought (especially if you already know something about philosophy);
- discuss some of the early applications of probability to economics.

1.1 Early accounts and the birth of mathematical probability

We know that, in the seventeenth century, probability theory began with the analysis of games of chance (a.k.a. gambling). However, dice were already in use in ancient civilizations. Just to limit ourselves to the Mediterranean area, due to the somewhat Eurocentric culture of these authors, dice are found in archaeological sites in Egypt. According to Svetonius, a Roman historian, in the first century, Emperor Claudius wrote a book on gambling, but unfortunately nothing of it remains today.

It is 'however' true that chance has been a part of the life of our ancestors. Always, and this is true also today, individuals and societies have been faced with unpredictable events and it is not surprising that this unpredictability has been the subject of many discussions and speculations, especially when compared with better predictable events such as the astronomical ones.

It is perhaps harder to understand why there had been no mathematical formalizations of probability theory until the seventeenth century. There is a remote possibility that, in ancient times, other treatises like the one of Claudius were available and also were lost, but, if this were the case, mathematical probability was perhaps a fringe subject.

A natural place for the development of probabilistic thought could have been physics. Real measurements are never exactly in agreement with theory, and the measured values often fluctuate. Outcomes of physical experiments are a natural candidate for the development of a theory of random variables. However, the early developments of physics did not take chance into account and the tradition of physics remains far from probability. Even today, formal education in physics virtually neglects probability theory.

Ian Hacking has discussed this problem in his book *The Emergence of Probability*. In his first chapter, he writes:

> A philosophical history must not only record what happened around 1660,[1] but must also speculate on how such a fundamental concept as probability could emerge so suddenly.
>
> Probability has two aspects. It is connected with the degree of belief warranted by evidence, and it is connected with the tendency, displayed by some chance devices, to produce stable relative frequencies. Neither of these aspects was self-consciously and deliberately apprehended by any substantial body of thinkers before the times of Pascal.

The official birth of probability theory was triggered by a question asked of Blaise Pascal by his friend Chevalier de Méré on a dice gambling problem. In the seventeenth century in Europe, there were people rich enough to travel the continent and waste their money gambling. de Méré was one of them. In the summer of 1654, Pascal wrote to Pierre de Fermat in order to solve de Méré's problem, and out of their correspondence mathematical probability theory was born.

Soon after this correspondence, the young Christian Huygens wrote a first exhaustive book on mathematical probability, based on the ideas of Pascal and Fermat, *De ratiociniis in ludo aleae* [1], published in 1657 and then re-published with comments in the book by Jakob Bernoulli, *Ars conjectandi*, which appeared in 1713, seven years after the death of Bernoulli [2]. It is not well known that Huygens' treatise is based on the concept of expectation for a random situation rather than on probability; expectation was defined as the fair price for a game of chance reproducing the same random situation. The economic flavour of such a point of view anticipates the writings by Ramsey and de Finetti in the twentieth century. However, for Huygens, the underlying probability is based on *equally possible cases*, and the fair price of the equivalent game is just a trick for calculations. Jakob Bernoulli's contribution to probability theory is important, not only for the celebrated theorem of *Pars IV*, relating the probability of an event to the

[1] This is when the correspondence between Pascal and Fermat took place.

relative frequency of successes observed iterating independent trials many times. Indeed, Bernoulli observes that if we are unable to set up the set of all *equally possible cases* (which is impossible in the social sciences, as for causes of mortality, number of diseases, etc.), we can estimate unknown probabilities by means of observed frequencies. The 'rough' inversion of Bernoulli's theorem, meaning that the probability of an event is identified with its relative frequency in a long series of trials, is at the basis of the ordinary frequency interpretation of probability. On the other side, for what concerns foundational problems, Bernoulli is important also for the *principle of non-sufficient reason* also known as *principle of indifference.* According to this principle, if there are $g > 1$ mutually exclusive and exhaustive possibilities, indistinguishable except for their labels or names, to each possibility one has to attribute a probability $1/g$. This principle is at the basis of the logical definition of probability.

A probabilistic inversion of Bernoulli's theorem was offered by Thomas Bayes in 1763, reprinted in Biometrika [3]. Bayes became very popular in the twentieth century due to the *neo-Bayesian* movement in statistics.

It was, indeed, in the eighteenth century, namely in 1733, that Daniel Bernoulli published the first paper where probability theory was applied to economics, *Specimen theoriae novae de mensura sortis*, translated into English in Econometrica [4].

1.2 Laplace and the classical definition of probability

In 1812, Pierre Simon de Laplace published his celebrated book *Théorie analytique des probabilités* containing developments and advancements, and representing a summary of what had been achieved up to the beginning of the nineteenth century (see further reading in Chapter 3). Laplace's book eventually contains a clear definition of probability. One can argue that Laplace's definition was the one currently used in the eighteenth century, and for almost all the nineteenth century. It is now called the *classical definition.*

In order to illustrate the classical definition of probability, consider a *dichotomous variable*, a variable only assuming two values in an experiment. This is the case when tossing a coin. Now, if you toss the coin you have two possible outcomes: H (for head) and T (for tails). The probability $\mathbb{P}(H)$ of getting H is given by the number of favourable outcomes, 1 here, divided by the total number of possible outcomes, all considered *equipossible*, 2 here, so that:

$$\mathbb{P}(H) = \frac{\text{\# of favourable outcomes}}{\text{\# of possible outcomes}} = \frac{1}{2}, \tag{1.1}$$

where # means *the number*. The classical definition of probability is still a good guideline for the solution of probability problems and for getting correct results in many cases. The task of finding the probability of an event is reduced to a combinatorial problem. One must enumerate and count all the favourable cases as well as all the possible cases and assume that the latter cases have the same probability, based on the principle of indifference.

1.2.1 The classical definition at work

In order to use the classical definition, one should be able to list favourable outcomes as well as the total number of possible outcomes of an experiment. In general, some calculations are necessary to apply this definition. Suppose you want to know the probability of getting exactly two heads in three tosses of a coin. There are 8 possible cases: $(TTT, TTH, THT, HTT, HHT, HTH, THH, HHH)$ of which 3 exactly contain 2 heads. Then, based on the classical definition, the required probability is $3/8$. If you consider 10 tosses of a coin, there are already 1024 possible cases and listing them all has already become boring. The *fundamental counting principle* comes to the rescue.

According to the commonsensical principle, for a finite sequence of decisions, their total number is the product of the number of choices for each decision. The next examples show how this principle works in practice.

Example In the case discussed above there are 3 decisions in a sequence (choosing H or T three times) and there are 2 choices for every decision (H or T). Thus, the total number of decisions is $2^3 = 8$.

Based on the fundamental counting principle, one gets the number of dispositions, permutations, combinations and combinations without repetition for n objects.

Example (Dispositions with repetition) Suppose you want to choose an object k times out of n objects. The total number of possible choices is n each time and, based on the fundamental counting principle, one finds that there are n^k possible choices.

Example (Permutations) Now you want to pick an object out of n, remove it from the list of objects and go on until all the objects are selected. For the first decision you have n choices, for the second decision $n-1$ and so on until the nth decision where you just have 1 to take. As a consequence of the fundamental counting principle, the total number of possible decisions is $n!$.

Example (Sequences without repetition) This time you are interested in selecting k objects out of n with $k \leq n$, and you are also interested in the order of the selected items. The first time you have n choices, the second time $n-1$ and so on, until the

kth time where you have $n-k+1$ choices left. Then, the total number of possible decisions is $n(n-1)\cdots(n-k+1) = n!/(n-k)!$.

Example (Combinations) You have a list of n objects and you want to select k objects out of them with $k \leq n$, but you do not care about their order. There are $k!$ ordered lists containing the same elements. The number of ordered lists is $n!/(n-k)!$ (see above). Therefore, this time, the total number of possible decisions (possible ways of selecting k objects out of n irrespective of their order) is $n!/(k!(n-k)!)$. This is a very useful formula and there is a special symbol for the so-called *binomial coefficient*:

$$\binom{n}{k} = \frac{n!}{k!(n-k)!}. \tag{1.2}$$

Indeed, these coefficients appear in the expansion of the nth power of a binomial:

$$(p+q)^n = \sum_{k=0}^{n} \binom{n}{k} p^k q^{n-k}, \tag{1.3}$$

where

$$\binom{n}{0} = \binom{n}{n} = 1, \tag{1.4}$$

as a consequence of the definition $0! = 1$.

Example (Combinations with repetition) Suppose you are interested in finding the number of ways of allocating n objects into g boxes, irrespective of the names of the objects. Let the objects be represented by crosses, $\times$, and the boxes by vertical bars. For instance, the string of symbols $| \times \times | \times ||$ denotes two objects in the first box, one object in the second box and no object in the third one. Now, the total number of symbols is $n+g+1$ of which 2 are always fixed, as the first and the last symbols must be a $|$. Of the remaining $n+g+1-2 = n+g-1$ symbols, n can be arbitrarily chosen to be crosses. The number of possible choices is then given by the binomial factor $\binom{n+g-1}{n}$.

Example (Tossing coins revisited) Let us consider once again the problem presented at the beginning of this subsection. This was: what is the probability of finding exactly two heads out of three tosses of a coin? Now, the problem can be generalized: what is the probability of finding exactly n heads out of N tosses of a coin ($n \leq N$)? The total number of possible outcomes is 2^N as there are 2 choices the first time, two the second and so on until two choices for the Nth toss. The number of favourable outcomes is given by the number of ways of selecting n places out of N and putting a head there and a tail elsewhere. Therefore

$$\mathbb{P}(\text{exactly } n \text{ heads}) = \binom{N}{n} \frac{1}{2^N}. \tag{1.5}$$

1.2.2 Circularity of the classical definition

Even if very useful for practical purposes, the classical definition suffers from circularity. In order to justify this statement, let us re-write the classical definition: *the probability of an event is given by the number of favourable outcomes divided by the total number of possible outcomes considered equipossible.* Now consider a particular outcome. In this case, there is only 1 favourable case, and if r denotes the total number of outcomes, one has

$$\mathbb{P}(\text{outcome}) = \frac{1}{r}. \tag{1.6}$$

This equation is the same for any outcome and this means that all the outcomes have the same probability. Therefore, in the classical definition, there seems to be no way of considering *elementary* outcomes with different probabilities and the equiprobability of all the outcomes is a consequence of the definition. Essentially, equipossibility and equiprobability do coincide. A difficulty with equiprobability arises in cases in which the outcomes have different probabilities. What about an unbalanced coin or an unfair die?

If the hidden assumption *all possible outcomes being equiprobable* is made explicit, then one immediately sees the circularity as probability is used to define itself, in other words *the probability of an event is given by the number of favourable outcomes divided by the total number of possible outcomes assumed equiprobable.* In summary, if the equiprobability of outcomes is not mentioned, it becomes an immediate consequence of the definition and it becomes impossible to deal with non-equiprobable outcomes. If, on the contrary, the equiprobability is included in the definition as an assumption, then the definition becomes circular. In other words, the equiprobability is nothing else than *a hypothesis* which holds in all the cases where it holds.

A possible way out from circularity was suggested by J. Bernoulli and adopted by Laplace himself; it is the so-called indifference principle mentioned above. According to this principle, if one has no reason to assign different probabilities to a set of exhaustive and mutually exclusive events (called outcomes or possibilities so far), then these events must be considered as equiprobable. For instance, in the case of the coin, in the absence of further indication, one has the following set of equations

$$\mathbb{P}(H) = \mathbb{P}(T), \tag{1.7}$$

and

$$\mathbb{P}(H) + \mathbb{P}(T) = 1, \tag{1.8}$$

yielding $\mathbb{P}(H) = \mathbb{P}(T) = 1/2$, where the outcomes H and T are exhaustive (one of them must occur) and mutually exclusive (if one obtains H, one cannot get T at the same time).

The principle of indifference may seem a beautiful solution, but it leads to several problems and paradoxes identified by J. M. Keynes, by J. von Kries in *Die Prinzipien der Wahrscheinlichkeitsrechnung* published in 1886 [5] and by Bertrand in his *Calcul des probabilités* of 1907 [6]. Every economist knows the *General Theory* [7], but few are aware of *A Treatise on Probability*, a book published by Keynes in 1921 and including one of the first attempts to present a set of axioms for probability theory [8]. Let us now consider some of the paradoxes connected with the principle of indifference. Suppose that one does not know anything about a book. Therefore the probability of the statement *this book has a red cover* is the same as the probability of its negation *this book does not have a red cover*. Again here one has a set of exhaustive and mutually exclusive events, whose probability is 1/2 according to the principle of indifference. However, as nothing is known about the book, the same considerations can be repeated for the statements *this book has a green cover*, *this book has a blue cover*, etc. Thus, each of these events turns out to have probability 1/2, a paradoxical result. This paradox can be avoided if one further knows that the set of possible cover colours is finite and made up of, say, r elements. Then the probability of *this book has a red cover* becomes $1/r$ and the probability of *this book does not have a red cover* becomes $1 - 1/r$. Bertrand's paradoxes are subtler and they make use of the properties of real numbers. Already with integers, if the set of events is countable, the indifference principle leads to a distribution where every event has zero probability as $\lim_{r\to\infty} 1/r = 0$ but where the sum of these zero probabilities is 1, a puzzling result which can be dealt with using measure theory. The situation becomes worse if the set of events is infinite and non-countable. Following Bertrand, let us consider a circle and an equilateral triangle inscribed in the circle. What is the probability that a randomly selected chord is longer than the triangle side? Three possible answers are:

1. One of the extreme points of the chord can indifferently lie in any point of the circle. Let us then assume that it coincides with a vertex of the triangle, say vertex A. Now the chord direction can be selected by chance, that is we assume that the angle θ of the chord with the tangent to the circle in A is uniformly distributed. Now the chord is longer than the side of the triangle only if its $\pi/3 < \theta < 2\pi/3$.

The triangle defines three circle arcs of equal length, this means that the required probability is 1/3.
2. Fixing a direction, and considering all chords parallel to that direction, all chords whose distance from the centre is $< r/2$ are longer than the triangle side; then considering the distance from the centre of a chord parallel to the fixed direction uniformly distributed gives a probability equal to 1/2.
3. Random selection of a chord is equivalent to random selection of its central point. In order for the chord to be longer than the triangle side, the distance of its central point from the centre of the circle must be smaller than one-half of the circle radius. Then the area to which this point must belong is 1/4 of the circle area and the corresponding probability turns out to be 1/4 instead of 1/3.

An important vindication for the indifference principle has been given by E.T. Jaynes, who not only reverted Bertrand's so-called ill-posed problem into *The Well Posed Problem* [9], but also extended it to a more general Maximum Entropy Principle, one of major contemporary attempts to save Laplace's tradition against the frequentist point of view described below.

The history of thought has seen many ways for avoiding the substantial difficulties of the classical definition or for circumventing its circularity. One of these attempts is the frequentist approach, where probabilities are roughly identified with observed frequencies of outcomes in repeated independent experiments. Another solution is the subjectivist approach, particularly interesting for economists as probabilities are defined in terms of *rational bets*.

1.3 Frequentism

The principle of indifference introduces a logic or subjective element in the evaluation of probabilities. If, in the absence of any reason, one can assume equiprobable events, then if there are specific reasons one can make another assumption. Then probability assignments depend on one's state of knowledge of the investigated system. Empiricists opposed similar views and tried to focus on the outcomes of real experiments and to define probabilities in terms of frequencies. Roughly speaking, this viewpoint can be explained as follows, using the example of coin tossing. According to frequentists the probability of H can be approximated by repeatedly tossing a coin, by recording the sequence of outcomes $HHTHTTHHTTTH\cdots$, counting the number of Hs and dividing for the total number of trials

$$\mathbb{P}(H) \sim \frac{\#\text{ of } H}{\#\text{ of trials}}. \tag{1.9}$$

The ratio on the right-hand side of the equation is the *empirical relative frequency* of the outcome H, a useful quantity in descriptive statistics. Now, this ratio is hardly exactly equal to 1/2 and the frequentist idea is to extrapolate the sequence of trials to infinity and to define the probability as

$$\mathbb{P}(H) = \lim_{\#\text{ of trials}\to\infty} \frac{\#\text{ of } H}{\#\text{ of trials}}. \tag{1.10}$$

This is the preferred definition of probability in several textbooks introducing probability and statistics to natural scientists and in particular to physicists. Probability becomes a sort of measurable quantity that does not depend on one's state of knowledge, it becomes *objective* or, at least, based on empirical evidence. Kolmogorov, the founder of axiomatic probability theory, was himself a supporter of frequentism and his works on probability theory have been very influential in the twentieth century.

The naïve version of frequentism presented above cannot be a solution to the problems discussed before. Indeed, the limit appearing in the definition of probability is not the usual limit defined in calculus for the convergence of a sequence. There is no analytic formula for the number of heads out of N trials and nobody can toss a coin for an infinite number of times. Having said that, one can notice that similar difficulties are present when one wants to define real numbers as limits of Cauchy sequences of rational numbers. Following this idea, a solution to the objection presented above has been proposed by Richard von Mises; starting from 1919, he tried to develop a rigorous frequentist theory of probability based on the notion of *collective*. A collective is an infinite sequence of outcomes where each attribute (e.g. H in the case of coin-tossing dichotomous variables) has a limiting relative frequency not depending on place selection. In other words, a collective is an infinite sequence whose frequencies have not only a precise limit in the sense of (1.10), but, in addition, the same limit must hold for *any* subsequence chosen in advance. The rationale for this second and most important condition is as follows: whatever strategy one chooses, or game system one uses, the probability of an outcome is the same. von Mises' first memoir was *Grundlagen der Wahrscheinlichkeitsrechnung*, which appeared in the fifth volume of *Mathematische Zeitschrift* [10]. Subsequently, he published a book, *Wahrscheinlichkeit, Statistik und Wahrheit. Einführung in die neue Wahrscheinlichkeitslehre und ihre Anwendungen* (Probability, statistics and truth. Introduction to the new probability theory and its applications) [11]. There are several difficulties in von Mises' theory, undoubtedly the deepest attempt to define randomness as the natural basis for a frequentist and objective view of probability.

The main objection to frequentism is that most events are not repeatable, and in this case it is impossible, even in principle, to apply a frequentist definition of probability based on frequencies simply because these frequencies cannot be measured at all. von Mises explicitly excluded these cases from his theory. In other words, given an event that is not repeatable such as *tomorrow it will rain*, it is a nonsense to ask for its probability. Notice that most of economics would fall outside the realm of repeatability. If one were to fully accept this point of view, most applications of probability and statistics to economics (including most of econometrics) would become meaningless. Incidentally, the success of frequentism could explain why there are so few probabilistic models in theoretical economics. The late Kolmogorov explored a method to avoid problems due to infinite sequences of trials by developing a *finitary* frequentist theory of probability connected with his theory of information and computational complexity.

Note that the use of observed frequencies to calculate probabilities can be traced back to Huygens, who used mortality data in order to calculate survival tables, as well as bets on survival. The success of frequentism is related to the social success among statisticians and natural scientists of the methods developed by R.A. Fisher who was a strong supporter of the frequentist viewpoint. Namely, Fisher tried to systematically exclude any non-frequency notion from statistics. These methods deeply influenced the birth of econometrics, the only branch of economics (except for mathematical finance) making extensive use of probability theory. The frequentism adopted by a large majority of orthodox[2] statisticians amounts to refusing applications of probability outside the realm of repeatable events. The main effect of this attitude is not using a priori or initial probabilities within Bayes' theorem (*la probabilité de causes par l'evenements* in the Laplacean formulation). This attitude is just setting limits to the realm of probability, and has nothing to do with the definition of the concept. A sequence of n trials such as $HHT \ldots H$ should be more correctly written as $H_1 H_2 T_3 \ldots H_n$, and almost everybody would agree that $\forall i, \mathbb{P}(H_i) = (\#\text{ of } H)/n$, if the only available knowledge is given by the number of occurrences of H and by the number of trials n. However, usually, one is mainly interested in $\mathbb{P}(H_{n+1})$: the probability of an event not contained in the available evidence. There is no logical constraint in assuming that at the next trial the coin will behave as before, but after Hume's criticism of the Principle of Uniformity of Nature, this is a practical issue. Therefore, assuming (1.10) turns out to be nothing else than a *working hypothesis*.

1.4 Subjectivism, neo-Bayesianism and logicism

Frequentism wants to eliminate the subjective element present in the indifference principle. On the contrary, subjectivism accepts this element and amplifies it by defining probability as the degree of belief that each individual assigns to an event. This event need not refer to the future and it is not necessary that the

[2] The adjective orthodox is used by the neo-Bayesian E.T. Jaynes when referring to statisticians following the tradition of R.A. Fisher, J. Neyman and E. Pearson.

event is repeatable. Being *subjective*, the evaluation of probability may differ from individual to individual. However, any individual must assign his/her probabilities in a coherent way, so that, for instance, a set of exhaustive and mutually exclusive events has probabilities summing up to 1. Indeed, probabilities are related to bets. The evaluation of probabilities in terms of bets was independently proposed by Frank Plumpton Ramsey and by the Italian mathematician Bruno de Finetti. This is a rare case where an Italian scientist who published his results after an Anglo-Saxon counterpart is better known within the scientific community. Ramsey died when he was 26 years old in 1930, whereas de Finetti died in 1985 at the age of 79. Ramsey published his results in 1926 in his notes on *Truth and Probability* [12] and de Finetti wrote his essay on the logical foundations of probabilistic reasoning (*Fondamenti logici del ragionamento probabilistico*) in 1930 [13]. From 1930 to 1985, de Finetti had a lot of time to further develop and publicize his views and he also published many important papers on statistics and probability, based on the notion of exchangeability. The most important theorem on exchangeable sequences is known as de Finetti's theorem in the probabilistic literature.

In particular, de Finetti presented an operational procedure to define probabilities. Assume that, in return for an amount x, a rational individual is given 1 if the event A occurs and 0 if the event A does not occur. Now, further assume that the loss function L of the individual is $(1-x)^2$ if A takes place and $(0-x)^2 = x^2$ if A does not take place. The value x the rational individual has to choose is such that the expected loss attains its minimum. Let $p = \mathbb{P}(A)$ be the (subjective) probability of occurrence for the event A and let $1-p$ be the probability of A not taking place, then the expected loss is

$$\mathbb{E}[L] = p(1-x)^2 + (1-p)x^2, \tag{1.11}$$

so that

$$\mathbb{E}[L] = p - 2px + x^2 = [p(1-p)^2 + (1-p)p^2] + (p-x)^2. \tag{1.12}$$

The analysis of this quadratic function of x shows that the expected loss is minimal for $x = p = \mathrm{P}(A)$ and our rational agent must choose $x = p$ in order to minimize his/her loss. Hence, one can elicit the (subjective) probability of a person by inviting him/her to bet. The extensions of this analysis to a finite set of mutually exclusive and exhaustive events is straightforward. For years, in Rome, Bruno de Finetti organized a lottery based on these ideas to forecast Sunday football (soccer for US readers) match outcomes among his students and colleagues.

In de Finetti's view the core of probability theory coincides with Bayes' Theorem, allowing a continuous updating of probabilities after empirical observations or measurements. The bridge with the classical Laplacean tradition is rebuilt, and many shortcuts invented by orthodox statisticians to sidestep the gap, such as estimates, significance tests and hypothesis tests become useless. The work by de Finetti was almost ignored by Italian and European statisticians, but it became popular in the USA due to Savage's book *The Foundations of Statistics* published in 1954 [14]. With this book, the so-called Bayesian renaissance began.

An objection to this line of thought can be based on the results of empirical economics and psychology. It turns out that human beings are not able to correctly evaluate probabilities even in simple cases. By the way, this also happens to many students and scholars of probability trying to solve elementary exercises even after many years of practice. Probability is highly counterintuitive and even if one can conceive a rational agent who can base his/her choices in the presence of uncertainty on perfect probabilistic calculations, this is not a human being.

Against the extreme subjectivism, an objective non-frequentist view can be based on the following viewpoint: if two individuals share the same knowledge, their probability assignments must coincide. We have already mentioned Jaynes who tries to give an objective (or, better, intersubjective) measure of the available knowledge *via* Shannon's entropy and information theory. Another very important line of thought can be traced back to W.E. Johnson (Keynes' guide and adviser), H. Jeffreys (Jaynes' guide) and R. Carnap. Jeffreys was a physicist and his book *Theory of Probability* was the milestone of the Bayesian renaissance [15]. Carnap was a logician, and he developed axiomatic methods for predictive inferences, mixing prior information and empirical frequencies without using Bayes' formula. All these points of view are particularly appealing for theoretical economists as individuals are assumed to be rational agents, whose probability assignments follow what is known as *normative theory* in the theory of choice.

1.5 From definitions to interpretations

If one is able to define the concept of an entity, then one is able to derive all its properties from this definition. In the case of probability all the definitions discussed above fortunately lead to the same set of properties. For instance, given two mutually exclusive events (not necessarily exhaustive), the probability of the occurrence of one of them or the other is the sum of the two separate probabilities, because (for Laplace) favourable cases add, or (for a frequentist) the number of past occurrences add, or (for a subjectivist) the fair prices of the bets add. In 1933, Kolmogorov axiomatized the properties of probability and the great success of his work changed the logical status of the foundations of probability. Now the classical, frequentist or subjective notions of probability can be considered as possible interpretations of the axioms. An axiomatic theory can be applied to facts only in a hypothetical way, and the agreement between theory and experiments is the only conceivable signal of correctness. Hence equiprobable cases in games of chance, long series of statistical data in economics and physics or rational expectations in social sciences are all ways for introducing those probability values needed to apply an axiomatic system to reality.

Further reading

A set of books and papers has already been quoted in this chapter. Many of them are available online and can be retrieved by properly using appropriate search engines.

The history of probability is discussed in many books. Keith Devlin devoted a popular science book to the correspondence between Pascal and Fermat.

K. Devlin, *The Unfinished Game: Pascal, Fermat, and the Seventeenth-Century Letter That Made the World Modern*, Basic Books, (2009).

Philosophically inclined readers will find Hacking's books of great interest.

I. Hacking, *The Emergence of Probability*, Cambridge University Press, Cambridge UK (1975).

I. Hacking, *The Taming of Chance*, Cambridge University Press, Cambridge UK (1990).

An account of von Mises' ideas is available in English.

R. von Mises and H. Geiringer, *Mathematical Theory of Probability and Statistics*, Academic Press, New York (1964).

E.T. Jaynes wrote an introduction to probability theory from the neo-Bayesian viewpoint. His book has been published posthumously.

E.T. Jaynes, *Probability Theory: The Logic of Science*, Cambridge University Press, Cambridge UK (2003).

A classical introductory book for probability theory was written by Parzen. It is a rigorous introductory book including many insightful examples.

E. Parzen, *Modern Probability Theory and its Applications*, Wiley, New York (1960).

References

[1] C. Huygens, *Libellus de ratiociniis in ludo aleae or The Value of all Chances in Games of Fortune; Cards, Dice, Wagers, Lotteries, etc. Mathematically Demonstrated*, printed by S. Keimer for T. Woodward, near the Inner-Temple-Gate in Fleet Street, London (1714). English translation of the original Latin version.

[2] J. Bernoulli, *Ars conjectandi, opus posthumum. Accedit Tractatus de seriebus infinitis, et epistola gallice scripta de ludo pilae reticularis. The Art of Conjecturing, together*

with Letter to a Friend on Sets in Court Tennis, The Johns Hopkins University Press, Baltimore MA (2006). English translation of the original Latin version.

[3] T. Bayes, *An Essay towards Solving a Problem in the Doctrine of Chances*, Philosophical Transactions, **53**, 370–418 (1763). Reprinted in Biometrika, **45**, 296–315 (1958).

[4] D. Bernoulli, *Specimen Theoriae Novae de Mensura Sortis. Exposition of a New Theory on the Measurement of Risk*, Econometrica, **22**, 23–36 (1954). English translation of the original Latin version.

[5] J. von Kries, *Die Prinzipien der Wahrscheinlichkeitsrechnung*, J.C.B. Mohr, Tübingen (1886).

[6] J.L. Bertrand, *Calcul des probabilités*, Gauthier-Villars, Paris (1907).

[7] J.M. Keynes, *The General Theory of Employment, Interest and Money*, Macmillan & Co., London (1936).

[8] J.M. Keynes, *A Treatise on Probability*, Macmillan & Co., London (1921).

[9] E.T. Jaynes, *The Well Posed Problem*, Foundations of Physics, **3**, 477–491 (1973).

[10] R. von Mises, *Grundlagen der Wahrscheinlichkeitsrechnung*, Mathematische Zeitschrift, **5**, 52–99 (1919).

[11] R. von Mises, *Wahrscheinlichkeit, Statistik und Wahrheit. Einfürung in die neue Wahrscheinlichkeitslehre und ihre Anwendungen*, Springer, Vienna (1928).

[12] F.P. Ramsey, *Truth and Probability*, in R.B. Braithwaite (ed.), *The Foundations of Mathematics and Other Logical Essays*, Chapter 7, pp. 156–198, Kegan, Paul, Trench, Trubner & Co., London, and Harcourt, Brace and Company, New York, NY (1931).

[13] B. de Finetti, *Fondamenti logici del ragionamento probabilistico*, Bollettino dell'Unione Matematica Italiana, **9** (Serie A), 258–261 (1930).

[14] L.J. Savage, *The Foundations of Statistics*, Wiley, New York, NY (1954).

[15] H. Jeffreys, *Theory of Probability*, Oxford University Press, Oxford, UK (1961).

2
Individual and statistical descriptions

In this chapter, the problem of allocating n *objects* or *elements* into g *categories* or *classes* is discussed. Probability deals with events and events are made of descriptions of facts. Given that the main subject of this book concerns populations of objects allocated into categories, the formal concept of description is introduced below.

After studying this chapter, you should be able to:

- define α- and s-descriptions, frequency occupation vectors and partition vectors;
- count the number of α-descriptions and the number of frequency occupation vectors;
- describe the above concepts by means of elementary examples;
- solve simple combinatorial problems related to the allocation of n objects into g categories.

2.1 Joint α-descriptions and s-descriptions

Let us consider a set of n elements $U = \{u_1, u_2, \ldots, u_n\}$ representing a finite population of n physical entities, or physical elements. The symbol # is used for the *cardinality* of a set, that is $\#U = n$ denotes that the set U contains n distinct elements. A *sample* of U is any subset $S \subseteq U$, i.e. $S = \{u_{i_1}, u_{i_2}, \ldots, u_{i_m}\}$. The size of S is its cardinality, i.e. $\#S = \#\{u_{i_1}, u_{i_2}, \ldots, u_{i_m}\} = m \leq n$.

2.1.1 Example: five firms in a small town I

Let U represent the set composed of five firms in a small town. How many distinct samples of size 2 can be formed? Being a sample of a set, two samples are different if they differ in the presence of at least an element. The first element can be chosen in 5 ways, whereas for the second only 4 choices are possible. According to the *fundamental counting principle*, in this case, there are $5 \times 4 = 20$ ordered couples (see Section 1.2.1). However, the sample is a set and not a sequence, and each sample is represented by two couples. Hence, the number of distinct samples is just 10, that is $\frac{5!}{2!3!} = \binom{5}{2}$. The general form is $\binom{n}{m}$, if $\#U = n$ and $\#S = m$.

2.1.2 Joint alphabetic descriptions or lists

Suppose now that the objects belonging to U are grouped into g categories or classes in which a variate X is partitioned. The *joint α-description of all the elements* of the population U (the joint alphabetic description or list) with respect to these categories is the sequence $\boldsymbol{\omega} = (\omega_1, \omega_2, \ldots, \omega_n)$, with $\omega_i \in \{1, \ldots, g\}$. More generally we can represent the set of the categories[1] $X = \{x, \ldots, x^{(g)}\}$, whose cardinality is $\#X = g$, as an alphabet, and each description as a word composed by n symbols taken from X. In the case of words, each word is a set function mapping each position $i \in \{1, \ldots, n\}$ into an element of the alphabet X. In the jargon of statisticians, X is some property of the elements, partitioned into g categories. Categories should be clearly defined, mutually exclusive and collectively exhaustive. In this way, any entity of the given population unequivocally belongs to one, and only one, of the proposed categories. Returning to our population, each description is a set function $\omega: U \to X$ mapping each element $u_i \in U$ into an element of the set of categories X. $\omega(u_i) = x^{(k)}$ *denotes* that the ith element belongs to the kth category, in short $\omega_i = k$. The set of all distinct set functions, denoted by X^U, is then the space of all possible descriptions $\boldsymbol{\omega} = (\omega_i, i \in \{1, \ldots, n\})$, with $\omega_i \in X$, or $\boldsymbol{\omega} = (\omega(u_i), i \in \{1, \ldots, n\})$ with $\omega(u_i) \in X$.

Using the fundamental counting principle, it is possible to prove that the number of possible distinct joint α-descriptions is $\#X^U = g^n$, or

$$\#X^U = W(\mathbf{X}|n, g) = g^n. \tag{2.1}$$

Here and in the following the symbol of conditioning | is extended to the meaning of 'conditioned on', 'consistent (compatible) with'. Moreover, the term *description* should always denote the variate with respect to which elements are classified. Strictly speaking, one should always use the notation 'X-description' where the variate is explicitly written down.

2.1.3 Example: five firms in a small town II

Let U represent the above-mentioned set made up of five firms, and let X denote three industrial sectors, directly labelled by the integer numbers 1, 2 and 3. A

[1] Whenever it is possible we use $\{\ldots\}$ for sets, containing different elements by definition, while we use $(\ldots)$ for ordered sets, or sequences, or vectors, which may contain repeated values. Indeed, sequences are set functions mapping natural numbers into some set.

possible joint description of all the elements of the population is $\omega_1 = 1, \omega_2 = 1, \omega_3 = 2, \omega_4 = 3, \omega_5 = 2$ and it is possible to write $\boldsymbol{\omega} = (1,1,2,3,2)$. In this case the total number of possible joint α-descriptions is $W(\boldsymbol{\omega}|n=5, g=3) = 3^5 = 243$ as firm u_1 may have 3 possible labels, u_2 may have 3 possible labels, and so on.

2.1.4 A remark on sample descriptions

More formally and generally, one can interpret the sequence $\boldsymbol{\omega} = (\omega_1, \ldots, \omega_n)$, with $\omega_i \in X$, where the symbol X refers to the specific variate one is considering (income, strategy, sector, etc.) as follows; the subscript i refers to the order of observation (according to the planning of an experiment), and ω_i stands for the ith observed value (the ith outcome) of the variate X. Following Kolmogorov's original ideas, an *experiment* is nothing other than a partition of the possible outcomes or results into non-overlapping classes; performing an experiment amounts to classifying the observed objects into some class without any doubt. The joint α-description of all the elements corresponds to a sequence following a specific order (the alphabetical order), so that ω_i coincides with the value given to the ith element. An alternative[2] and, for many purposes, equivalent description of the population can be obtained by means of complete sampling without replacement. If U is sampled without replacement, the drawing order of the n elements is any permutation of the ordered sequence $(u_1, u_2, \ldots, u_n)$; a *sample description* of the variate X means that for each drawn element, one observes not the name, rather the value of X. In order to avoid confusion, it is useful to introduce a different symbol, say ς, in order to denote such a description. Then ς_i represents the category of the element drawn at the ith step. Sample descriptions will be denoted in short as *s-descriptions*.

2.1.5 Example: five firms in a small town III

Referring to the example in Section 2.1.3, given that the joint α-description is $\boldsymbol{\omega} = (1,1,2,3,2)$, a possible s-description is $\boldsymbol{\varsigma} = (3,1,2,1,2)$, which is compatible with the fact that u_4 has been observed first. Notice that this s-description is compatible with the sequence of draws u_4, u_1, u_3, u_2, u_5, but also with u_4, u_2, u_3, u_1, u_5 and so on. In the next paragraph, the number of s-descriptions will be discussed. Assume that one observer, say A, prefers to call objects in alphabetic order, whereas a second observer, say B, prefers to sample objects from an urn or following any other possible procedure, but not using alphabetical ordering. The descriptions given by A and B will differ. What is invariant?

[2] It is usually forbidden to attach permanent labels like 'names' to the so-called 'indistinguishable' particles in quantum mechanics, whereas nothing prevents a set of elementary particles being analyzed sequentially with respect to some individual property.

2.2 Frequency vector and individual descriptions

Given that an α-description is a set function $\omega : U \to X$, all elements that are mapped into the same category belong to the same class, and the inverse image of ω partitions U into disjoint classes. Returning to the previous example, denoting by $\omega^{-1}(x^{(j)})$ the set of all elements mapped into the same value $x^{(j)}$, we have:

$$\omega^{-1}(1) = \{u_1, u_2\}, \omega^{-1}(2) = \{u_3, u_5\}, \omega^{-1}(3) = \{u_4\}. \tag{2.2}$$

Now the cardinality of these classes is the *the occupation-number vector of the categories*, or the *frequency vector of the elements* associated with the α-description $\boldsymbol{\omega}$. Hence, $\mathbf{n}(\boldsymbol{\omega}) = (n_1(\boldsymbol{\omega}), \ldots, n_g(\boldsymbol{\omega}))$, where $n_k(\boldsymbol{\omega}) = \#\{\omega_i = k, i = 1, \ldots, n\}$ counts how many elements belong to the kth category, for $k = 1, \ldots, g$. Equivalently, one can say that $n_k(\boldsymbol{\omega}) = \#\omega^{-1}(k)$.

Let $\boldsymbol{\varsigma}$ be an s-description compatible with a given α-description $\boldsymbol{\omega}$, in short $\boldsymbol{\varsigma}|\boldsymbol{\omega}$. Hence if $\boldsymbol{\omega} = (\omega_1, \omega_2, \ldots, \omega_n)$, and $\boldsymbol{\varsigma} = (\varsigma_1, \varsigma_2, \ldots, \varsigma_n)$, given that $\boldsymbol{\varsigma}$ is nothing else than a permutation of $\boldsymbol{\omega}$, it turns out that $\mathbf{n}(\boldsymbol{\varsigma}) = \mathbf{n}(\boldsymbol{\omega})$. The set of all s-descriptions $\boldsymbol{\varsigma}$ compatible with a given α-description $\boldsymbol{\omega}$ provides an alternative equivalent definition of $\mathbf{n}(\boldsymbol{\omega})$. The number of distinct s-descriptions $\boldsymbol{\varsigma}$ compatible with a given α-description $\boldsymbol{\omega}$ is a function of the occupation number vector of $\boldsymbol{\omega}$, $\mathbf{n}(\boldsymbol{\omega}) = (n_1, \ldots, n_g)$

$$W(\boldsymbol{\varsigma}|\boldsymbol{\omega}) = W(\boldsymbol{\varsigma}|\mathbf{n}(\boldsymbol{\omega})) = \frac{n!}{n_1! \cdots n_g!}. \tag{2.3}$$

This result is further discussed in Exercise 2.1.

The occupation-number vector can be conceived as an individual description of the g categories; on the other side, seen as the frequency vector, it also gives a statistical description of the elements, as it does not take into account either the individual names of the elements or their order within the sample. Hence, if the frequency vector is known, but the individual α-description is unknown, the number of α-descriptions compatible with the given frequency vector is

$$W(\boldsymbol{\omega}|\mathbf{n}) = \frac{n!}{n_1! \cdots n_g!}, \tag{2.4}$$

that is the same number as (2.3). Let Y denote the description of the *frequency vector;* Y assumes values within the set of integer non-negative g-ples summing up to n, that is $Y^U = \{\mathbf{n} = (n_1, \ldots, n_g) : \sum_1^g n_j = n\}$. It can be shown (see Exercise 2.4) that the number of possible distinct occupation number vectors in a system of n elements and g classes is

$$\#Y^U = W(\mathbf{n}|n, g) = \binom{n+g-1}{n}. \tag{2.5}$$

2.2.1 Example: five firms in a small town IV

With reference to the Example of Section 2.1.3, the system of five firms divided into three sectors is described by the occupation number vector $\mathbf{n} = (n_1 = 2, n_2 = 2, n_3 = 1)$. This means that two firms belong to sector 1, two firms belong to sector 2 and one firm belongs to sector 3. In this example, the α-description, ω, is known, however the s-description ς is not specified. Therefore, if one observes all the firms in a sequence drawn without replacement in any order, there are $5!/(2!2!1!) = 30$ possible resulting sequences. If we knew only the frequency vector, the number of compatible α-descriptions would be the same: $5!/(2!2!1!) = 30$. A priori, the number of possible α-descriptions would have been $g^n = 3^5 = 243$.

2.3 Partitions

The description introduced in Section 2.1 concerns elements or physical entities, sorted by their names or randomly; this is a first level of description; the second level of description, introduced in Section 2.2 concerns classes or categories, and can be seen both as an individual description of classes, and as a statistical description of elements. At each new level, the focus is on more abstract objects, and their individual description can be considered as a statistical description of the objects of the previous level. This hierarchy can be traced at least back to Boltzmann (see the Example in the next Section 2.3.1 and Exercise 2.3).

The less complete description is in terms of the *frequency vectors of occupation numbers* or *partition vectors*: $\mathbf{z} = (z_0 \ldots z_n)$, where $z_k = \#\{n_j = k, j = 1, \ldots, g\}$, for any $k = 0, 1, \ldots, n$, that is the number (not the names or labels) of categories with k elements. This is the frequency distribution of categories; therefore, it is the frequency of a frequency of elements. The constraints for $\mathbf{z}$ are the following: $\sum_0^n z_i = g$, and $\sum_0^n i z_i = n$. The number of distinct category descriptions $\mathbf{n}$ compatible with a given partition vector $\mathbf{z}$ is:

$$W(\mathbf{n}|\mathbf{z}; n, g) = \frac{g!}{z_0! z_1! \cdots z_n!}, \tag{2.6}$$

similar to Equation (2.3). In both cases, it amounts to the number of permutations of all 'individuals' divided by all permutations of individuals belonging to the same class. In the case of Equation (2.3), individuals are elements, whereas in the case of Equation (2.6), they are categories. The number of distinct partition vectors given n and g is denoted by $W(\mathbf{z}|n, g)$ and is the analogue of Equation (2.5); it is related to the number of partitions of an integer, but no general closed formula is available.

2.3.1 Example: The three levels of description in physics

Ludwig Boltzmann introduced the three levels of description in his first paper devoted to kinetic theory, published in 1868 (see further reading below), where he tried to derive Maxwell's distribution of velocities. Boltzmann's problem was distributing a given amount of energy E to a fixed number of molecules g. According to classical mechanics, energy is a continuous variable; therefore it seems impossible to solve the problem by means of combinatorics, that is by discrete counting. Boltzmann's idea was to divide energy into n 'energy elements' of value ε with the total energy E given by $E = n\varepsilon$, to be then allocated to the g molecules. A complete description $\boldsymbol{\omega}$ of the n elements on the g categories (molecules) has no immediate physical meaning; as for the second level, a description $\mathbf{n}$ of the g molecules has a clear physical meaning: it gives the energy of each individual molecule. But the aim of physicists is understanding how many molecules have energy $0, \varepsilon, 2\varepsilon, \ldots, n\varepsilon$ that is obtaining the partition vector $\mathbf{z}$, representing the *energy distribution*. Boltzmann gave complicated names to these descriptions, such as 'state descriptions'. These names are not so important as they are often incomplete, but they are very popular in statistical mechanics. To avoid confusion, one must always declare the 'state of what with respect to what'. This example taken from Boltzmann will be often used below. The reader should be careful not to mix up Boltzmann's 'quantization' with an anticipation of quantum mechanics, even if some historians of science did so. In Boltzmann's method, his final discrete results were subject to a limiting procedure where $n \to \infty$, $\varepsilon \to 0$ with $n\varepsilon = E$ so that continuity is restored.

2.3.2 Example: n coins, distributed over g agents

Consider a system of n coins, distributed over g agents. Supposing that each coin is labelled, for a given coin, the α-description $\boldsymbol{\omega} = (j_1, j_2, \ldots, j_n)$, with $j_i \in \{1, \ldots, g\}$ tells one to which agent the ith coin belongs; the frequency vector of the elements, $\mathbf{n} = (n_1, \ldots, n_g)$, is the agent description and gives the number of coins (the wealth) of each agent; finally, the partition vector $\mathbf{z} = (z_0, \ldots, z_n)$ describes the number of agents with $0, 1, \ldots, n$ coins and is commonly referred to as the *wealth distribution*; however, it is a description (an event) and it should not be confused with a probability distribution.

2.4 Partial and marginal descriptions

An experiment resulting in some alphabetic description $\boldsymbol{\omega}$ can be considered as a set of n partial experiments or individual observations, each of them limited to an individual in the population. So far, global descriptions of the system have been discussed, that is joint descriptions of all the elements of a system. Because the allocation of objects into categories may change in time, it is often useful to focus on a specific element, or a specific category, and to follow it as time goes by.

The *marginal description* of the ith element is given by the ith term of $\boldsymbol{\omega} = (\omega_1,\ldots,\omega_i,\ldots,\omega_n)$. Assume that $\omega_i = j$: writing that $\omega_i = j$ amounts to claiming that $\boldsymbol{\omega}$ belongs to the set of all the joint descriptions $\{\boldsymbol{\omega} : \boldsymbol{\omega} = (\omega_1 = \cdot,\ldots,\omega_i = j,\ldots,\omega_n = \cdot)\}$, with $\omega_i = j$ and any value for $\omega_{k\neq i}$.

The marginal description of the jth category $Y_j = n_j$ is the set of all distinct descriptions $(Y_1 = \cdot,\ldots,Y_j = n_j,\ldots,Y_g = \cdot)$ with $Y_j = n_j$ and any value for $Y_{k\neq j}$, with the usual constraint $\sum_{k=1}^{g} n_k = n$. This amounts to moving from a classification into g categories to a classification into only 2 classes: the jth category and the other classes, sometimes collectively called the *thermostat* by physicists.

When sampling from a population whose frequency vector is $\mathbf{n}$, it would be interesting to know what happens if 1, 2 or any number $m \leq n$ of elements are drawn and observed. In this case, the reduced sequence $(\varsigma_1, \varsigma_2,\ldots,\varsigma_m)$ defines the set of all the joint s-descriptions $(\varsigma_1, \varsigma_2,\ldots,\varsigma_m, \varsigma_{m+1}\ldots,\varsigma_n)$ with the first m terms fixed and any possible value for the remaining $n-m$ terms. The frequency vector of the sample is denoted by $\mathbf{m} = (m_1,\ldots,m_g)$. How many individual sequences ς pass through a given $\mathbf{m}$? The number of s-descriptions of the population with occupation numbers $\mathbf{m}$ given $\mathbf{n}$, $m_i \leq n_i$, can be computed using the fundamental counting principle. It is equal to the number of sequences to the frequency vector $\mathbf{m}$ in the first m observations, that is $m!/(m_1!\cdots m_g!)$ multiplied by the number of sequences connecting $\mathbf{m}$ to the frequency vector $\mathbf{n}-\mathbf{m}$ in the remaining $n-m$ observations, that is $(n-m)/((n_1-m_1)!\cdots(n_g-m_g)!)$ so that:

$$W(\varsigma|\mathbf{m},\mathbf{n}) = \frac{m!}{m_1!\ldots m_g!}\frac{(n-m)!}{(n_1-m_1)!\ldots(n_g-m_g)!}. \tag{2.7}$$

Considering that $W(\varsigma|\mathbf{n}) = \dfrac{n!}{n_1!\ldots n_g!}$, the ratio $W(\varsigma|\mathbf{m},\mathbf{n})/W(\varsigma|\mathbf{n})$ is

$$\frac{W(\varsigma|\mathbf{m},\mathbf{n})}{W(\varsigma|\mathbf{n})} = \frac{m!}{m_1!\ldots m_g!}\frac{(n-m)!}{(n_1-m_1)!\ldots(n_g-m_g)!}\frac{n_1!\ldots n_g!}{n!}. \tag{2.8}$$

An alternative form of Equation (2.8) is

$$\frac{W(\varsigma|\mathbf{m},\mathbf{n})}{W(\varsigma|\mathbf{n})} = \frac{\prod_{i=1}^{g}\binom{n_i}{m_i}}{\binom{n}{m}}. \tag{2.9}$$

In Fig. 2.1, the case $g = 2$ is shown, for which a graphical representation is possible. The set of sequences producing $\mathbf{n}$ are delimited by thick lines, while those passing through $\mathbf{m}$ are delimited by two thin sets of lines.

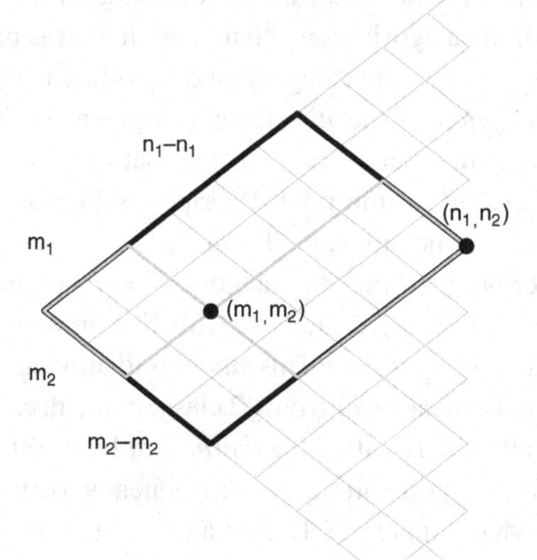

Figure 2.1. Basic representation for sampling with $g = 2$ categories. This figure visually illustrates the concepts discussed in Section 2.4. The vector $\mathbf{m} = (m_1, m_2)$ and the composition vector $\mathbf{n} = (n_1, n_2)$ are represented by the highlighted nodes in the graph. All the paths (histories) leading to $\mathbf{n}$ are contained in the external parallelogram. All the paths compatible with the observation of $\mathbf{m} = (m_1, m_2)$ are contained in the two smaller parallelograms with a common vertex in the node representing $\mathbf{m} = (m_1, m_2)$.

The ratio $W(\varsigma|\mathbf{m},\mathbf{n})/W(\varsigma|\mathbf{n})$ given by (2.8) can be rearranged to give

$$\frac{m!}{m_1!\dots m_g!}\frac{(n-m)!}{(n_1-m_1)!\dots(n_g-m_g)!}\frac{n_1!\dots n_g!}{n!}$$
$$= \frac{m!}{m_1!\dots m_g!}\frac{(n-m)!}{n!}\prod_{i=1}^{g}\frac{n_i!}{(n_i-m_i)!}.$$

If $n_i >> m_i$, that is if the population is large and the sample size is small, one has $(n-m)!/n! \simeq n^{-m}$, $n_i!/(n_i-m_i)! \simeq n_i^{m_i}$ and

$$\frac{W(\varsigma|\mathbf{m},\mathbf{n})}{W(\varsigma|\mathbf{n})} \simeq \frac{m!}{m_1!\dots m_g!}\prod_{i=1}^{g}\left(\frac{n_i}{n}\right)^{m_i}. \tag{2.10}$$

As will be discussed in Chapter 3, under the hypothesis of equal probability of all $\varsigma|\mathbf{n}$, Equation (2.9) represents the *multivariate hypergeometric distribution*, whereas Equation (2.10) is the *multinomial* approximation to the hypergeometric distribution for small sample sizes. In other words, for small samples of large populations, sampling without replacement and sampling with replacement virtually coincide.

2.4.1 Example: Two agents selected out of $n = 4$ agents belonging to $g = 3$ categories

Consider $n = 4$ agents and $g = 3$ strategies (e.g. 1 = 'bull', optimistic; 2 = 'bear', pessimistic; 3 = 'neutral', normal), and assume that the joint α-description is $\omega = (1,2,3,2)$. Now two agents are drawn without replacement (for a pizza with a friend):

1. List all sample descriptions $(\varsigma_i)_{i=1,\ldots,4}$.
   They are
$$\{(1,2,2,3),(1,2,3,2),(1,3,2,2),(2,1,2,3),(2,1,3,2),\\ (2,2,1,3),(2,2,3,1),(2,3,1,2),(2,3,2,1),(3,1,2,2),\\ (3,2,1,2),(3,2,2,1)\}.$$
   Their number is $W(\varsigma|\mathbf{n}) = \dfrac{4!}{1!2!1!} = 12.$
2. Write all possible descriptions of the reduced sample $(\varsigma_i)_{i=1,2}$.
   If the rule is applied and the first two digits of each permutation are taken, one gets some repetitions: {(1, 2), (1, 2), (1, 3), (2, 1), (2, 1), (2, 2), (2, 2), (2, 3), (2, 3), (3, 1), (3, 2), (3, 2)}, so that the possible descriptions are {(1, 2), (1, 3), (2, 1), (2, 2), (2, 3), (3, 1), (3, 2)}, and this is due to the fact that only 'bears' can appear together, as the draw is without replacement. It is useful to note that (1, 2), (2, 1), (2, 2), (2, 3), (3, 2) can appear twice as the initial part of the complete sequences, while (1, 3), (3, 1) can appear only once.
3. Write all possible frequency vectors for the reduced sample and count their number.
   $\mathbf{n} = (1,2,1)$ as there are 1 bull, 2 bears and 1 neutral. The possible sample descriptions are (1, 2), (1, 3), (2, 1), (2, 2), (2, 3), (3, 1) and (3, 2).
   Here (1, 2), (2, 1) belong to $\mathbf{m} = (1,1,0)$, and
$$W(\varsigma|\mathbf{m},\mathbf{n}) = \frac{m!}{m_1!\cdots m_g!}\frac{(n-m)!}{(n_1-m_1)!\cdots(n_g-m_g)!} = \frac{2!}{1!1!0!}\frac{2!}{1!1!1!} = 4; \tag{2.11}$$
   (1, 3), (3, 1) belong to $\mathbf{m} = (1,0,1)$, and $W(\varsigma|\mathbf{m},\mathbf{n}) = \dfrac{2!}{1!0!1!}\dfrac{2!}{0!2!0!} = 2;$
   (2, 2) belongs to $\mathbf{m} = (0,2,0)$, and $W(\varsigma|\mathbf{m},\mathbf{n}) = \dfrac{2!}{0!2!0!}\dfrac{2!}{1!0!1!} = 2;$

(2, 3), (3, 2) belong to $\mathbf{m} = (0,1,1)$, and $W(\varsigma|\mathbf{m},\mathbf{n}) = \frac{2!}{0!1!1!}\frac{2!}{1!1!0!} = 4$. Note that $\sum_{\mathbf{m}} W(\varsigma|\mathbf{m},\mathbf{n}) = W(\varsigma|\mathbf{n}) = 12$, where the sum is on the set $\{\mathbf{m} : m_i \leq n_i, \sum m_i = 2\}$.

2.5 Exercises

2.1 Prove that a set of n objects, divided into g classes with respect to the variate X, with frequency vector $(n_1,\ldots,n_g)$, can generate $n!/(n_1!\ldots n_g!)$ distinct s-descriptions.

2.2 How many different samples can be drawn out of a population of n elements?

2.3 Consider a system of five 'energy elements' (whose value is ε) allocated into three molecules.

1. Is it consistent to assume that the partition vector is $(z_0 = 1, z_1 = 0, z_2 = 1, z_3 = 1)$?
2. What is its physical meaning?

2.4 Prove that the number of possible distinct occupation number vectors in a system of n elements and g classes is

$$W(\mathbf{n}|n,g) = \binom{n+g-1}{n}. \tag{2.12}$$

2.5 Repeat the analysis of the Example in Section 2.4.1 for $n = 3$ and $g = 3$, the following joint α-description, $\boldsymbol{\omega} = (1,2,3)$ and a sample of 1 element.

2.6 Summary

In this chapter, the following descriptions were introduced for a set of n objects or elements partitioned into g classes with respect to a variate X.

The complete joint α-description gives the most complete characterization. For instance, for $n = 4$ and $g = 3$, the vector $\boldsymbol{\omega} = (1,2,2,3)$ means that the first object belongs to class or category 1, the second and third objects to class or category 2 and the fourth object to class or category 3.

By using the tool of sampling without replacement, one can pass from the joint α-description to the joint s-description. A joint s-description will be a permutation of the vector giving the joint α-description resulting from the process of sampling the n objects without replacement. A possible joint s-description corresponding to the joint α-description given above is $\boldsymbol{\varsigma} = (2,1,2,3)$.

The occupation number vector gives the number of objects belonging to the various classes. For instance, in the case discussed above, one has $\mathbf{n} = (1,2,1)$ meaning that there is one object belonging to class 1, two objects belonging to class 2 and one object belonging to class 3.

The partition vector gives the number of classes with 0 up to n objects. In the example, one has $\mathbf{z} = (0,2,1,0,0)$ and this means that there are zero classes with 0 objects, two classes with 1 object, one class with 2 objects, zero classes with 3 objects and zero classes with 4 objects.

Further reading

Boltzmann's paper of 1868 is available in a collection of his works edited by F. Hasenöhrl.

L. Boltzmann, *Studien über das Gleichgewicht der lebendigen Kraft zwischen bewegten materiellen Punkten*, in Wissenschaftliche Abhandlungen, Volume 1, edited by F. Hasenöhrl, Barth, Leipzig (1868).

A contemporary account of Boltzmann's method can be found in a paper by Bach.

A. Bach, *Boltzmann's probability distribution of 1877*, Archive for History of Exact Science, **41**, 1–40 (1990).

Bach also discusses the basic descriptions on which this chapter is based. The reader can also refer to Ya.G. Sinai's book for a simple approach to the description of finite sets.

Ya.G. Sinai, *Probability Theory. An Introductory Course*, Springer-Verlag, Berlin, Heidelberg (1992).

For basic statistical notions one can see Kendall and Stuart's book.

M.G. Kendall and A. Stuart, *The Advanced Theory of Statistics*, 2nd edition, Volume 1, Griffin & Co., London (1953).

Much of this chapter is based on elementary combinatorics. Merris' textbook contains a clear statement of the fundamental counting principle. The first two chapters of his book contain most of the proofs and material needed to read this chapter with profit.

R. Merris, *Combinatorics*, Wiley, Hoboken, NJ (2003).

A standard reference for combinatorics, available in virtually all maths libraries, is the first volume of Feller's monograph on probability. Probably influenced by quantum mechanical problems, Feller uses the concepts of *distinguishable* and *indistinguishable* objects, which to the authors of this book appear redundant and perhaps misleading.

W. Feller, *An Introduction to Probability Theory and its Applications*, 2nd edition, Volume 1, Wiley, New York (1957).

In Chapter 7, quantum statistics is derived in the framework of a classical model developed by Brillouin and without using indistinguishability.

3

Probability and events

In this chapter, the space of all individual descriptions compatible with what is known on a finite population is introduced. This is an instance of what probabilists call either *sample space* or *set of logical possibilities* or *set of elementary descriptions*; it is usually represented by the symbol Ω. The subsets of Ω can be composed by means of logical connectives, and the resulting algebra is an algebra of subsets of the sample space. Each element of this algebra is called an *event*. Each event can be assigned a non-negative real number in the interval [0,1]: its *probability*. Given a set of mutually disjoint and exhaustive events, their probabilities sum up to 1. Individual decompositions will be defined, whose meaning is the probability distribution for the ith agent to be in each category.

If you already know elementary probability theory, you can read the chapter directly and skip the appendix. For those who are not familiar with elementary probability theory, an appendix explains the basic concepts needed to read the chapter with profit.

After studying this chapter, you should be able to:

- understand the notion of an event, as a collection of joint descriptions;
- understand the composition of two events in terms of union and intersection, and the complement of an event;
- define the notion of probability space;
- discuss some deep hypotheses usually added to the basic axioms of probability;
- use simple urn models, in order to compute predictive probabilities.

3.1 Elementary descriptions and events

As discussed in Chapter 2, a system of n elements classified into g categories can be described by some alphabetic list $\boldsymbol{\omega} = (\omega_i)_{i=1,\dots,n}$ or any equivalent sample description, or by the occupation-number vector $\mathbf{n} = (n_j)_{j=1,\dots,g}$, or by frequency vectors of occupation numbers a.k.a. partition vectors: $\mathbf{z} = (z_k)_{k=0,\dots,n}$. The most fundamental descriptions of the system are the individual sequences $\boldsymbol{\omega}$; setting up a probability space for these descriptions is equivalent to considering the space Ω including all possible individual sequences. Here, it is not necessary to distinguish between α- and s-descriptions. An element (an outcome, a point) of Ω is denoted by $\boldsymbol{\omega} = (\omega_1, \ldots, \omega_n)$, with $\omega_i \in X = \{1, \ldots, g\}$, and its cardinality is $\#\Omega = g^n$. An event is a subset of Ω.

As Ω is a finite set with g^n elements, the family $\mathcal{P}(\Omega)$ of all subsets of Ω (its power set), including Ω itself and the empty set $\varnothing$, is finite, as well, and it has 2^{g^n} elements. In this finite case we can state that every set A belonging to $\mathcal{P}(\Omega)$, in symbols $A \in \mathcal{P}(\Omega)$, is an event. $\mathcal{P}(\Omega)$ is closed under the binary operations of union $\cup$ and intersection $\cap$ and the unary operation of complementation. In other words, if $A \in \mathcal{P}(\Omega)$ and $B \in \mathcal{P}(\Omega)$, then $A \cup B \in \mathcal{P}(\Omega)$, $A \cap B \in \mathcal{P}(\Omega)$ and $A^c \in \mathcal{P}(\Omega)$.

3.1.1 Example: dichotomous wealth distribution

In order to illustrate the above considerations with an example, consider a system of three agents, Alice, u_1, Bob, u_2, and Charles, u_3, and two qualities, poor (not rich), $P = 1$, and rich, $R = 2$. The fundamental description of this system is assumed to be one of the $2^3 = 8$ individual sequences $\boldsymbol{\omega} = (\omega_1, \omega_2, \omega_3)$ with $\omega_i \in \{1, 2\}$. Before observing the system, what can one expect regarding the wealth distribution of this small population?

The set of all individual descriptions is $\Omega = \{\boldsymbol{\omega}_i, i = 1, \ldots, 8\}$, in detail $\boldsymbol{\omega}_1 = (1,1,1)$, $\boldsymbol{\omega}_2 = (1,1,2)$, $\boldsymbol{\omega}_3 = (1,2,1)$, $\boldsymbol{\omega}_4 = (1,2,2)$, $\boldsymbol{\omega}_5 = (2,1,1)$, $\boldsymbol{\omega}_6 = (2,1,2)$, $\boldsymbol{\omega}_7 = (2,2,1)$, $\boldsymbol{\omega}_8 = (2,2,2)$.

All subsets of Ω are events. The simplest events in Ω are the singletons, such as $\{\boldsymbol{\omega}_5\}$, containing a single joint description. Observe that the vector $\boldsymbol{\omega}_5$ describes a world where Alice is rich while Bob and Charles are poor, whereas the set $\{\boldsymbol{\omega}_5\}$ is an event true if $\boldsymbol{\omega}_5$ occurs. One should consider descriptions as facts (taking place or not), and events as propositions (true or not) about facts (taking place or not). This useful distinction is borrowed from Wittgenstein's *Tractatus Logico-Philosophicus* [1].

The event $\{\boldsymbol{\omega}_1, \boldsymbol{\omega}_2\}$ is true when both Alice and Bob are poor, irrespective of Charles' wealth. Note that $\{\boldsymbol{\omega}_1, \boldsymbol{\omega}_2\} = \{\boldsymbol{\omega}_1\} \cup \{\boldsymbol{\omega}_2\}$.

Some events are worthy of special attention. Let $A_j^{(i)}$ denote the event in which the ith individual is in state j: $\omega_i = j$ (see Section 2.4). For instance $A_1^{(2)}$ denotes the event $A_1^{(2)} = \{\boldsymbol{\omega}_1, \boldsymbol{\omega}_2, \boldsymbol{\omega}_5, \boldsymbol{\omega}_6\}$; $A_1^{(2)}$ is true when Bob is rich, irrespective of Alice's and Charles' status. A singleton, $\{\boldsymbol{\omega}_i\}$, can be written as the intersection of three marginal events. As an example, one has $\{\boldsymbol{\omega}_8\} = A_2^{(1)} \cap A_2^{(2)} \cap A_2^{(3)}$; this event is true if Alice, Bob and Charles are rich.

Moreover, the complement (negation) of an event is still an event: $\{\omega_8\}^c = \{\omega_1, \omega_2, \ldots, \omega_7\}$ is the event that is true if at least one agent is poor. It can be directly checked that every set operation $\cup$, $\cap$ and ^{c} on subsets of Ω yields some set in $\mathcal{P}(\Omega)$, including Ω and the empty set $\varnothing$.

Suppose one is interested in checking whether Alice is rich. The event $A_1^{(1)} = A_1$ (for ease of notation) is true if Alice is rich. In order to plan an empirical observation, one must consider all the possibilities, and in order to complete a Boolean lattice (see Section 3.6), one has to introduce at the least the complement $A_1^c = A_2$ as well, the event true if Alice is not rich. If the options were g instead of 2, here one would further consider all the marginal events A_j, $j = 2, \ldots, g$. Starting from A_1 and A_2 and combining them by intersection, union and complementation, one finds only two new events, namely $\Omega = A_1 \cup A_2$, whose meaning is 'Alice is rich or not rich' – a truism – and $\varnothing = A_1 \cap A_2$, whose meaning is 'Alice is both rich and not rich' – a contradiction. Hence, the lattice (algebra) generated by A_1 is $\sigma(A_1) = \{\varnothing, A_1, A_2, \Omega\}$, and $\{A_1, A_2\}$ are the constituents (a.k.a. atoms) of the lattice.

If Bob is taken into account as well, starting from the two events A_1 ('Alice is rich') and $A_1^{(2)} = B_1$ ('Bob is rich') whose complement is B_2 and considering all the intersections $C_{ij} = A_i \cap B_j$; $i, j = 1, 2$, any event concerning the wealth of Alice and Bob is obtained as a union of the sets C_{ij}.

The simplest probability assignment for the system discussed above is the uniform distribution on all the individual descriptions. If one introduces a non-negative weight function such as $p(\omega_1) = \ldots = p(\omega_8)$ and $\sum_{i=1}^{8} p(\omega_i) = 1$, then, for any i, $p(\omega_i) = 1/8$. In this case, given that $\mathbb{P}(C) = \sum_{\omega \in C} p(\omega)$, it turns out that the probability of any event is proportional to the number of individual descriptions it contains. In other words, one gets

$$\mathbb{P}(C) = \frac{\#C}{\#\Omega}; \tag{3.1}$$

this is nothing else but the number of favourable cases divided by the number of possible cases, coinciding with the so-called classical definition of probability given by Equation (1.1). Each agent has four possibilities for being rich, then $\mathbb{P}(A_1^{(i)}) = 4/8 = 1/2$, for $i = 1, 2, 3$. The same applies to being poor, as $A_2^{(i)} = A_1^{(i)c}$, and as a consequence of the axioms of probability, one has that $\mathbb{P}(A_1^{(i)c}) = 1 - \mathbb{P}(A_1^{(i)}) = 1/2$. Note that for any singleton $\mathbb{P}(\{\omega\}) = \mathbb{P}(A_{\omega_1}^{(1)} \cap A_{\omega_2}^{(2)} \cap A_{\omega_3}^{(3)}) = 1/8$ and that $\mathbb{P}(A_{\omega_1}^{(1)})\mathbb{P}(A_{\omega_2}^{(2)})\mathbb{P}(A_{\omega_3}^{(3)}) = 1/2^3 = 1/8$. Hence

$$\mathbb{P}\left(A_{\omega_1}^{(1)} \cap A_{\omega_2}^{(2)} \cap A_{\omega_3}^{(3)}\right) = \mathbb{P}\left(A_{\omega_1}^{(1)}\right)\mathbb{P}\left(A_{\omega_2}^{(2)}\right)\mathbb{P}\left(A_{\omega_3}^{(3)}\right). \tag{3.2}$$

This means that the n sub-experiments $\mathcal{A}^{(i)} = \{A_1^{(i)},\dots,A_g^{(i)}\}$, each of them describing the observation of the ith individual, are independent. In other words, an important consequence of assigning the uniform distribution on the set of individual descriptions is the *independence* of the marginal events. Moreover, $\mathbb{P}(A_{\omega_i}^{(i)}) = 1/2$ does not depend on i, the label of the agent, and agents are *equidistributed*, in other words the probability of belonging to a category is the same for all the g agents. Finally, it does not depend on ω_i, meaning that the distribution is *uniform* for each agent.

3.2 Decompositions of the sample space

As discussed in the appendix, a decomposition of the sample space Ω is a family of events $\mathcal{E} = \{E_i : E_i \cap E_j = \varnothing, \cup_{i=1}^n E_i = \Omega\}$. Note that, for ease of notation and following common usage, from now on, the intersection symbol $\cap$ between sets is sometimes omitted when no confusion is possible.

Given a decomposition $\mathcal{E} = E_1,\dots,E_n$ of Ω, the probabilities $\mathbb{P}(E_i)$ are the *probability distribution* associated with $\mathcal{E}$ and $\sum_{i=1}^n \mathbb{P}(E_i) = 1$.

The collection of all singletons is the thinnest decomposition of the description space Ω, or $\Omega = \mathcal{E}_\omega = \{\{\omega\} : \omega \in \Omega\}$; the singletons are mutually disjoint and their union is Ω. For the sample space of individual descriptions $\#\mathcal{E}_\omega = g^n$. However, this is not the only possible decomposition of Ω. Note that both $\{\mathbf{n}\}$ and $\{\mathbf{z}\}$ are events (subsets of Ω); they were discussed in Chapter 1 and, given their properties, it is possible to verify that

- the set of distinct occupation vectors $\mathcal{E}_{\mathbf{n}} = \{\mathbf{n} = (n_1,\dots,n_g) : \sum_1^g n_j = n\}$ is a decomposition of Ω, with $\#\mathcal{E}_{\mathbf{n}} = \binom{n+g-1}{n}$;
- the same holds true for the set of the distinct partition vectors $\mathcal{E}_{\mathbf{z}} = \{\mathbf{z} = (z_0 \dots z_n) : \sum_0^n z_i = g, \sum_0^n i z_i = n\}$;
- it follows that $\sum_{\mathbf{n} \in \mathcal{E}_{\mathbf{n}}} \mathbb{P}(\mathbf{n}) = \sum_{\mathbf{z} \in \mathcal{E}_{\mathbf{z}}} \mathbb{P}(\mathbf{z}) = 1$.

All the above decompositions are associated with experiments involving all individuals and categories.

- If the attention is fixed on the ith sub-experiment, $\{A_1^{(i)},\dots,A_g^{(i)}\}$ is a decomposition of Ω and its meaning is that the ith observed agent will result in one and only one value of $\{1,\dots,g\}$. Therefore $(\mathbb{P}(A_j^{(i)}))_{j=1,\dots,g}$ is the probability distribution associated with the ith observed agent, and $\sum_{j=1}^g \mathbb{P}(A_j^{(i)}) = 1$.
- If the attention is fixed on a specific category j, and all individuals are classified as belonging or not to this category, one has a dichotomous partition of Ω of the kind $\mathbf{n} = (n_j, n - n_j)$ (see Section 2.4).

3.2.1 Example: the symmetric binomial distribution

Considering the example of Section 3.1.1, one can group the individual descriptions according to their occupation number vectors $n_1 = \#$ poor, $n_2 = \#$ rich. In this example, each vector $\mathbf{n} = (n_1, n_2)$ represents a wealth distribution, whereas in the Example of Section 2.3.2, the wealth distribution was described by a partition vector $\mathbf{z}$. Indeed, here, the objects to be allocated into categories are agents, and in Section 2.3.2 the objects were coins, to be distributed into categories there represented by agents. Also, here, the total wealth is not fixed, but in Section 2.3.2 it was fixed. The possible occupation vectors as well as the corresponding individual descriptions are listed below. $\omega_1 = (1,1,1)$ corresponds to $(n_1 = 3, n_2 = 0)$: Alice, Bob and Charles are all poor. $(n_1 = 2, n_2 = 1)$ contains three descriptions, $\omega_2 = (1,1,2)$, $\omega_3 = (1,2,1)$, $\omega_5 = (2,1,1)$; these are the cases with two poor agents and a rich one. $(n_1 = 1, n_2 = 2)$ contains three descriptions too, $\omega_4 = (1,2,2)$, $\omega_6 = (2,1,2)$, $\omega_7 = (2,2,1)$; these are the cases with a rich agent and two poor ones. Finally, $\omega_8 = (2,2,2)$ is the happy situation where all the agents are rich, $(n_1 = 0, n_2 = 3)$.

All the possible occupation number vectors are a decomposition of Ω. If the individual joint descriptions are equiprobable, one gets $\mathbb{P}(n_1 = 3, n_2 = 0) = \mathbb{P}(n_1 = 0, n_2 = 3) = 1/8$, and $\mathbb{P}(n_1 = 2, n_2 = 1) = \mathbb{P}(n_1 = 1, n_2 = 2) = 3/8$. Given the constraint $n_2 = 3 - n_1$, the probability distribution on the statistical description can be written as

$$\mathbb{P}(n_1 = k) = \binom{3}{k}\left(\frac{1}{2}\right)^3, \; k = 0, \ldots, 3. \tag{3.3}$$

Therefore, the probability of the occupation $\mathbf{n}$ can be written as the probability of a given sequence $(1/2)^3$ multiplied by the number of sequences $W(\varsigma|\mathbf{n})$ given by Equation (2.3). For n agents then

$$\mathbb{P}(n_1 = k) = \binom{n}{k}\left(\frac{1}{2}\right)^n, \; k = 0, \ldots, n, \tag{3.4}$$

that is the symmetric binomial distribution.

As for the partition vectors, $(n_1 = 3, n_2 = 0)$ and $(n_1 = 0, n_2 = 3)$ belong to $(z_0 = 1, z_1 = 0, z_2 = 0, z_3 = 1)$, meaning that one category is empty and the other contains all the three agents; $(n_1 = 2, n_2 = 1)$ and $(n_1 = 1, n_2 = 2)$ belong to $(z_0 = 0, z_1 = 1, z_2 = 1, z_3 = 0)$. Note that the constraints allow only two values of $\mathbf{z}$. Finally $\mathbb{P}(z_0 = 1, z_1 = 0, z_2 = 0, z_3 = 1) = \mathbb{P}(n_1 = 0, n_2 = 3) + \mathbb{P}(n_1 = 3, n_2 = 0) = 2/8 = 1/4$ and $\mathbb{P}(z_0 = 0, z_1 = 1, z_2 = 1, z_3 = 0) = \mathbb{P}(n_1 = 1, n_2 = 2) + \mathbb{P}(n_1 = 2, n_2 = 1) = 3/4$.

One could bet 3 : 1 to find elements divided into two clusters, and 1 : 3 to find them all in the same state of wealth.

3.2.2 Example: the Ehrenfest urn

Let us consider a box, divided into two symmetric communicating parts, and n molecules free to move from one part to the other. If all the individual descriptions of n molecules on $g = 2$ states are equiprobable then each molecule has probability $1/2$ to be found either on the left or on the right part of the box, and this probability is independent from the state of other molecules. Considering the description $\mathbf{n} = (n_1, n_2 = n - n_1)$, due to the constraint $n_1 + n_2 = n$, as above one has

$$\mathbb{P}(n_1 = k) = \binom{n}{k}\left(\frac{1}{2}\right)^n, k = 0, 1, \ldots, n. \tag{3.5}$$

Note that due to symmetry $\mathbb{P}(n_1 = k) = \mathbb{P}(n_1 = n - k) = \mathbb{P}(n_2 = k)$, and given that the symmetric binomial distribution is bell-shaped the most probable occupation vector is close to $k = n/2$. Conversely the probability of finding all molecules on the same side is $2(1/2)^n$. The tendency to a fifty–fifty allocation of molecules increases very rapidly as a function of the number n of molecules. This looks quite reasonable in the case of gas diffusion.

3.2.3 Example: energy distribution of two molecules

Coming back to Boltzmann's ideas (see the Example in Section 2.3.1), let us assume that all the individual descriptions of n energy elements $\varepsilon = E/n$ on $g = 2$ molecules are equiprobable: each element has probability $1/2$ of belonging to one molecule or to the other; this probability does not depend on the behaviour of any other energy element. By using the previous examples, one can see that the probability that the first molecule has energy equal to $E_1 = k\varepsilon$, or, equivalently, that it contains a fraction of the total energy $k(E/n)$, is

$$\mathbb{P}(E_1 = k\varepsilon) = \binom{n}{k}\left(\frac{1}{2}\right)^n, k = 0, 1, \ldots, n. \tag{3.6}$$

If n is even, the most probable value is $n/2$, that is $E_1 = E/2$. In the case of g molecules, the most probable value for any molecule would have been E/g. This looks reasonable, but a bit of inspection is enough to show that this cannot be the case. Boltzmann observed that in the continuous limit $\varepsilon \to 0$, and $n \to \infty$ with $E = n\varepsilon$, each molecule would have energy E/g with probability one, in disagreement with Maxwell's energy distribution. This means that a uniform distribution of energy elements on molecules cannot be the right one.

3.3 Remarks on distributions

The uniform distribution is often presented as a 'milestone' of probability theory. As already discussed in the introductory chapter, this is due to historical reasons, and to the widespread application of this approach to gambling. Methods due to Laplace were discussed, criticized and finally abandoned during the nineteenth century, when probability started to be applied to the social sciences, to physics, to statistics as well as to genetics. The difficulty related to equiprobable 'possible cases' from which 'favourable ones' are taken based on counting can be traced back to J. Bernoulli's *Ars Conjectandi* and to Bernoulli's theorem (see references of Chapter 1). The problem of the so-called 'inversion' of Bernoulli's theorem (that is estimating probabilities *via* observed frequencies) is the core subject of Bayes' memoir *Essay Towards Solving a Problem in the Doctrine of Chances* published posthumously in 1763 in the *Philosophical Transactions of the Royal Society* (see references of Chapter 1). As a matter of fact, for any probability space there are as many possible uniform probability distributions as decompositions of the sample space. If you wish to follow the 'indifference principle' and then use a uniform distribution somewhere, which decomposition of the sample space should you prefer?

3.3.1 Independence and identical distribution of events: the multinomial distribution

From the uniform distribution on all individual descriptions[1]

$$\mathbb{P}(\{\omega\}) = \frac{1}{\#\Omega} = \frac{1}{g^n}, \tag{3.7}$$

one derives that the individual events $A^{(i)}_{\omega_i}$ are independent, have the same probability distribution and this distribution is uniform. Note that (3.7) does not contain any parameter added to n and g. Considering that any function $\mathbb{P}(\cdot)$ satisfying the axioms of probability is worthy of study, it is now time to move from (3.7) to a $\mathbb{P}(\cdot)$ containing a parameter vector $(p_i \geq 0)_{i=1,\dots,g}$, $\sum_{j=1}^{g} p_j = 1$ by setting

1. $\mathbb{P}(\{\omega\}) = \mathbb{P}(A^{(1)}_{\omega_1} \dots A^{(n)}_{\omega_n}) = \mathbb{P}(A^{(1)}_{\omega_1}) \dots \mathbb{P}(A^{(n)}_{\omega_n})$,
2. $\mathbb{P}(A^{(i)}_{\omega_i}) = p_j$ if $\omega_i = j$. In this case $\mathbb{P}(\{\omega\}) = \prod_{j=1}^{g} p_j^{n_j}$,

where n_j is the number of $\omega_i = j$, that is the number of agents in the jth category.

The meaning is as follows: the notation $A^{(i)}_j$ denotes all the joint descriptions for which $\omega_i = j$, that is the ith element of the description belongs to the jth category;

[1] In the following, when no confusion is possible, both $\{\omega\}$ and ω will be called individual descriptions.

this notation was introduced in Section 3.1.1; therefore $\mathbb{P}(A^{(i)}_{\omega_i})$ is the probability that the ith element belongs to the category denoted by ω_i; for instance, $\mathbb{P}(A^{(2)}_1)$ is the probability that the second element is in the first category. If $\mathbb{P}(A^{(i)}_j) = p_j$, then the probability of belonging to the jth category coincides for all the elements. If $\mathbb{P}(\{\boldsymbol{\omega}\}) = \prod_{j=1}^{g} p_j^{n_j}$ holds, the events $\{A^{(i)}_{\omega_i}\}_{i=1,\dots,n}$ are still independent and identically distributed (i.i.d.), but this distribution is no longer uniform, except in the case $p_j = 1/g$, bringing us back to the uniform distribution on the joint individual descriptions. At odds with (3.7), all individual descriptions are no longer equiprobable, as their probabilities depend on the occupation vector $\mathbf{n}$. However, it is still true that all the descriptions belonging to the same occupation vector $\mathbf{n}$ have the same probability. The probability of an occupation vector $\mathbf{n}$ is the probability of a given individual description multiplied by the number of possible sequences belonging to the same occupation vector $\mathbf{n}$, which is given by Equation (2.3). In other words, one gets the multinomial distribution of parameters n and $\mathbf{p} = (p_1, \dots, p_g)$:

$$\mathbb{P}(\mathbf{n}) = \mathrm{MD}(\mathbf{n};\mathbf{p}) = \frac{n!}{n_1! \cdots n_g!} \prod_{j=1}^{g} p_j^{n_j}, \tag{3.8}$$

where $\mathrm{MD}(\mathbf{n};\mathbf{p})$ is a shortcut for denoting the last term in Equation (3.8).

3.3.2 Example: casting a die with skewed prize

If a fair die gives ‘1’ the prize is 0, if it gives ‘2 or 3’ the prize is 1, if it gives ‘4 or 5 or 6’ the prize is 2. What is the probability of a prize 0 after throwing the die three times? In order to solve this problem, consider the Ω-space of a fair die cast three times using these descriptions $\boldsymbol{\omega} = (\omega_1, \omega_2, \omega_3)$, $\omega_i \in \{1,2,3,4,5,6\}$, and introducing the individual events $A^{(i)}_{\omega_i}$ which occur if the ith throw gives ω_i. The prize associated with the ith throw is the event $G^{(i)}_{\xi_i}$, $\xi_i \in \{0,1,2\}$, and $G^{(i)}_0 = A^{(i)}_1$, $G^{(i)}_1 = A^{(i)}_2 \cup A^{(i)}_3$, $G^{(i)}_2 = A^{(i)}_4 \cup A^{(i)}_5 \cup A^{(i)}_6$. If all the descriptions in Ω have the same probability, one can set $p_i = \mathbb{P}(G^{(i)}_{\xi_i})$ and obtain $p_0 = 1/6$, $p_1 = 2/6 = 1/3$, $p_2 = 3/6 = 1/2$, with $\sum_{i=0}^{2} p_i = 1$ as for any throw $(G^{(i)}_{\xi_i})_{\xi_i \in \{0,1,2\}}$ is a partition of the sample space. One gets the prize 0 if and only if $\boldsymbol{\omega} = (1,1,1)$, and $\mathbb{P}(1,1,1) = \mathbb{P}(A^{(1)}_1)\mathbb{P}(A^{(2)}_1)\mathbb{P}(A^{(3)}_1) = p_0^3 = (1/6)^3 = 1/216$.

Alternatively, one can use the new sample space $\Omega^{(\xi)}$ composed of all the descriptions $\boldsymbol{\xi} = (\xi_1, \xi_2, \xi_3)$, with $\xi_i \in \{0,1,2\}$ which directly give the prize associated with each throw. Note that $\Omega^{(\xi)}$ only has 3^3 elements, whereas the previous sample space contained 6^3 elements. A probability assignment on $\Omega^{(\xi)}$ consistent with the previous one is $p_{\boldsymbol{\xi}} = p_{\xi_1} p_{\xi_2} p_{\xi_3}$, where $p_{\xi_i} = 1/6$ if $\xi_i = 0$, $p_{\xi_i} = 1/3$ if $\xi_i = 1$ and

$p_{\xi_i} = 1/2$ if $\xi_i = 2$. In this way

$$\mathbb{P}(\{\boldsymbol{\xi}\}) = \prod_{i=0}^{2} p_i^{n_i}, \tag{3.9}$$

where n_i is the number of $\xi_j = i$, that is the number of throws with the ith prize. Again, one gets $\mathbb{P}(0,0,0) = (1/6)^3 = 1/216$.

As mentioned at the end of Section 3.3.1, the probability of getting n_0 times the prize 0, n_1 times the prize 1 and n_2 times the prize 2 is

$$\mathbb{P}(\mathbf{n}) = \frac{3!}{n_0!n_1!n_2!}\prod_{i=0}^{2} p_i^{n_i}. \tag{3.10}$$

This is an instance of the multinomial distribution $\text{MD}(\mathbf{n};\mathbf{p})$ with $n = 3$, $\mathbf{p} = (1/6, 1/3, 1/2)$.

This example shows that the events needed to represent the problem can be built out of any suitable sample space Ω. Moreover, recalling the example in Section 3.1.1, one is not obliged to represent a lattice (algebra) by means of all its sets. What is essential is that probability values are assigned to all the atoms (constituents) of the lattice. In the present case, this means fixing $\mathbb{P}(G_{\xi_1}^{(1)} \cap G_{\xi_2}^{(2)} \cap G_{\xi_3}^{(3)})$ for any value ξ_1, ξ_2, ξ_3 representing the thinnest decomposition of Ω relevant for the problem. Instead of $3 \cdot 3 - 1 = 8$ free probability values, the assumed independence and equidistribution of events belonging to different trials reduces the free probability values to $3 - 1 = 2$.

3.3.3 Sampling without replacement: the multivariate hypergeometric distribution

As in Section 2.1.3, suppose one knows that the set of five firms is described by $\boldsymbol{\omega} = (1,1,2,3,2)$. If a random sample of size 2 is drawn, what is the probability of obtaining two firms belonging to the sector labelled by 1? In order to answer this question, one first notices that a complete random sampling without replacement is a sequence $(\varsigma_1, \varsigma_2, \varsigma_3, \varsigma_4, \varsigma_5)$ given by a permutation of $(1,1,2,3,2)$. The number of distinct permutations can be obtained from Equation (2.3) with $n = 5$, $n_1 = n_2 = 2$ and $n_3 = 1$:

$$\frac{n!}{n_1!n_2!n_3!} = \frac{5!}{2!2!1!} = 30. \tag{3.11}$$

The conditioned sample space can be denoted by the symbol

$$\Omega_{\mathbf{n}}^{(\varsigma)} = \{(\varsigma_1, \varsigma_2, \varsigma_3, \varsigma_4, \varsigma_5) | \mathbf{n} = (2,2,1)\}. \tag{3.12}$$

It is usual to assume that all the possible samples ς have the same probability; in the present case this assumption leads to:

$$\mathbb{P}(\varsigma|\mathbf{n}) = \left(\frac{n!}{n_1!n_2!n_3!}\right)^{-1} = \frac{1}{30}. \tag{3.13}$$

Indeed, the assumption

$$\mathbb{P}(\varsigma|\mathbf{n}) = \left(\frac{n!}{n_1!\dots n_g!}\right)^{-1}, \tag{3.14}$$

can be regarded as a definition of *random sampling*, and can be justified both from an objective and from a subjective point of view. With this assumption, it is possible to relate all the combinatorial formulae in Section 2.4 to probabilities. In particular, Equation (2.9) becomes

$$\mathbb{P}(\mathbf{m}) = \mathrm{HD}(\mathbf{m};\mathbf{n}) = \frac{\prod_{i=1}^{g}\binom{n_i}{m_i}}{\binom{n}{m}}, \tag{3.15}$$

which is the multivariate hypergeometric distribution of parameters n, $\mathbf{n}$.

In the example discussed above, there are $g = 3$ categories, and one has $n = 5$, $\mathbf{n} = (2,2,1)$ and $m = 2$, $\mathbf{m} = (2,0,0)$. Therefore, by means of the hypergeometric distribution, one gets

$$\mathbb{P}(\mathbf{m} = (2,0,0)) = \frac{\binom{2}{2}\binom{2}{0}\binom{1}{0}}{\binom{5}{2}} = \frac{2!3!}{5!} = \frac{2}{5\cdot 4} = \frac{1}{10}. \tag{3.16}$$

Note that the set of distinct $\mathbf{m} = (m_1, m_2, m_3)$, such that $m_1 + m_2 + m_3 = 2$, has 6 elements in agreement with Equation (2.5).

3.4 Probability assignments

Contrary to other approaches, Kolmogorov's axiomatic method does not provide any tool to determine event probabilities. Therefore, in order to assign probabilities, it is necessary to use other considerations and, as discussed in Chapter 1, the indifference principle is often useful. In this section, probability assignments to sample spaces related to the partition of n objects into g categories will be further discussed. The classical definition (3.1), when applied to the sample space of n elements and g categories, leads to Equation (3.7), meaning that all the individual descriptions are equiprobable, and the events $A^{(1)}_{\omega_1},\ldots,A^{(n)}_{\omega_n}$ are independent, identically and uniformly distributed. In particular, one has $\mathbb{P}(A^{(i)}_j) = 1/g$.

When applied to the conditional sample space $\Omega^{(\varsigma)}_{\mathbf{n}}$, the same assumption (3.1) assigns equal probability to all the descriptions belonging to **n**. Now, the situation is quite different: the events $\{A^{(i)}_{\omega_i}\}_{i=1,\ldots,n}$ are still identically distributed, with probability $\mathbb{P}(A^{(i)}_j) = n_j/n$ for all individuals, but they are no more independent.

3.4.1 Absolute and predictive probability assignments

There are two possible ways of assigning probabilities to events in the sample space Ω. As mentioned before, the thinnest partition of Ω is given by the events corresponding to the individual descriptions $\boldsymbol{\omega}$. On this partition, a probability function is obtained defining a function $p(\boldsymbol{\omega}) \geq 0$ for each description $\boldsymbol{\omega}$, with the condition $\sum_{\boldsymbol{\omega}} p(\boldsymbol{\omega}) = 1$. The singletons $\{\boldsymbol{\omega}\}$ can be written as the intersection of n marginal events, leading to

$$p(\boldsymbol{\omega}) = \mathbb{P}(\{\boldsymbol{\omega}\}) = \mathbb{P}\left(A^{(1)}_{\omega_1}\ldots A^{(n)}_{\omega_n}\right). \tag{3.17}$$

The *absolute* method consists in fixing the left-hand side of (3.17) and deriving $\mathbb{P}(E)$ for any event E by adding the probabilities of all the individual descriptions whose union gives E.

The alternative *predictive* method is based on the calculation of the right-hand side of (3.17) using the multiplicative rule based on conditional probabilities:

$$\mathbb{P}\left(A^{(1)}_{\omega_1}\ldots A^{(n)}_{\omega_n}\right) = \mathbb{P}\left(A^{(1)}_{\omega_1}\right)\mathbb{P}\left(A^{(2)}_{\omega_2}|A^{(1)}_{\omega_1}\right)\ldots\mathbb{P}\left(A^{(n)}_{\omega_n}|A^{(1)}_{\omega_1}\ldots A^{(n-1)}_{\omega_{n-1}}\right). \tag{3.18}$$

The probability assignment to the sequence is built step-by-step by giving probability values to the next subresult, conditioned on the previous subresults. In many interesting cases, predictive probabilities can be interpreted in terms of urn models. The knowledge of $\mathbb{P}(A^{(m+1)}_j|A^{(1)}_{\omega_1}\ldots A^{(m)}_{\omega_m})$ for any $m = 0,1,\ldots,n-1$ is equivalent to the knowledge of the probability of the singletons $\{\boldsymbol{\omega}\}$ from which the probability of $A^{(i)}_j$ can then be recovered using the absolute method (see Exercise 3.4). The predictive method is very intuitive, and it is valid also in the case $n \to \infty$, typical of stochastic processes (see Chapter 4). The quantities $\mathbb{P}(A^{(m+1)}_j|A^{(1)}_{\omega_1}\ldots A^{(m)}_{\omega_m})$ are called *predictive probabilities*. In the independent and identically distributed case they are

$$\mathbb{P}\left(A^{(m+1)}_j|A^{(1)}_{\omega_1}\ldots A^{(m)}_{\omega_m}\right) = p_j, \tag{3.19}$$

where, indeed, the probability of being in the jth class does not depend on the index of the individual (identical distribution) and does not depend on the allocation of previous observed individuals as well (independence).

In the case of equiprobability of all descriptions belonging to $\mathbf{n}$, the predictive probabilities are:

$$\mathbb{P}\left(A_j^{(m+1)}|A_{\omega_1}^{(1)}\dots A_{\omega_m}^{(m)}\right)=\frac{n_j-m_j}{n-m}, \tag{3.20}$$

where $m_j \leq n_j$ is the number of observations belonging to the jth category up to step m. The meaning of this formula can be clarified by the usual model for such a situation: an urn of known composition $\mathbf{n}=(n_1,\dots,n_g)$, with $\sum_{i=1}^{g} n_i = n$, where the g categories represent g colours. Balls are drawn without replacement from the urn. If at each draw all the balls remaining in the urn have the same probability of being drawn, one gets $\mathbb{P}(A_j^{(1)}) = n_j/n$, $\mathbb{P}(A_j^{(2)}|A_i^{(1)}) = (n_j-\delta_{j,i})/(n-1),\dots,$ and, in general, all the equations summarized by (3.20) have an interpretation in terms of event probabilities within the urn model. In this case events are negatively correlated, as drawn balls of given colours reduce the probabilities of drawing the same colours (see Exercise 3.4). As a final remark, in Chapters 4 and 5, it will be shown that (3.19) leads to the multinomial distribution (3.8) whereas (3.20) leads to the multivariate hypergeometric distribution (3.15).

3.4.2 Degrees of freedom for a probability assignment

In general, in order to define a probability function on the sample space of n elements and g categories we need to fix $g^n - 1$ probability values. The uniform distribution assignment is the simplest, as everything follows from the knowledge of only two parameters: g and n. The independent and identically distributed case is slightly more complicated, as it requires a set of $g-1$ parameters $p_j \geq 0$, such that $\sum_{j=1}^{g} p_j = 1$, representing the marginal probability that an object belongs to the jth class. The case of a sample space conditioned on a given occupation number also has $g-1$ degrees of freedom, as it requires a set of $g-1$ parameters $n_j \geq 0$, such that $\sum_{j=1}^{g} n_j = n$, representing the distribution of the n objects into the g classes.

As discussed above, the case of the $\mathbf{n}$-conditional sample space has an interpretation in terms of an urn model. Indeed, the independent case also can be described in terms of an urn model: an urn where balls of g different colours are drawn with replacement and where p_j is the probability of drawing a ball of the jth colour. Recall that in the other case the parameter n_j represents the *initial* number of balls of colour j.

There is a more difficult case when one considers that all occupation vectors are possible, and $\mathbb{P}(\mathbf{n})$ is the probability distribution on them. In this case, the number of degrees of freedom d_f is

$$d_f = \binom{n+g-1}{n} - 1. \tag{3.21}$$

If also in this case it is assumed that, conditional on **n**, all the compatible descriptions are equiprobable, according to Equation (3.14), this probability space can be interpreted in terms of draws from an urn with uncertain composition, and $\mathbb{P}(\mathbf{n})$ describes this uncertainty.

3.4.3 *Example: again on molecules and energy elements*

Coming back to the example of Section 3.2.2, assume now that all occupation vectors of n energy elements on $g = 2$ molecules are equiprobable; this means that all the descriptions $(n,0)$, $(n-1,1)$, ..., $(1,n-1)$, $(0,n)$ have the same probability. When the continuous limit is taken, this hypothesis leads to the fundamental hypothesis of classical statistical mechanics, that is equal a-priori probability of all molecular descriptions compatible with the total energy. Given that Equation (3.14) holds, a joint description $\boldsymbol{\omega}$ of the n elements has a probability depending on $(k, n-k)$. One finds

$$\mathbb{P}(\{\omega\}) = \sum_k \mathbb{P}(\{\omega\}|k)\mathbb{P}(k) = \mathbb{P}(\{\omega\}|k)\mathbb{P}(k), \tag{3.22}$$

given that there is only a value of k compatible with $\boldsymbol{\omega}$. Moreover, $\mathbb{P}(k)$ is uniform, that is $\mathbb{P}(k) = 1/(n+1)$, and as a consequence of (3.14):

$$\mathbb{P}(\{\omega\}|k) = \left(\frac{n!}{k!(n-k)!}\right)^{-1}, \tag{3.23}$$

leading to, as a consequence of (3.22):

$$\mathbb{P}(\{\omega\}) = \frac{1}{n+1}\frac{k!(n-k)!}{n!}. \tag{3.24}$$

Thus, the assumption of uniform probability on occupation vectors leads to a non-uniform distribution for the individual descriptions where the probability of each individual description only depends on k, the number of energy elements belonging to the first molecule. This probability is maximum for $k = 0$ and $k = n$, corresponding to the occupation vectors $(0,n)$, and $(n,0)$, where it reaches the value $1/(n+1)$, whereas, for even n, the minimum is in $(n/2, n/2)$. In other words, there are many descriptions corresponding to the vector $(n/2, n/2)$, and the sum of their weights coincides with the weight of $\omega = (1,\ldots,1)$, the only description corresponding to $(n,0)$. It can be shown that the predictive probability ruling such individual descriptions is

$$\mathbb{P}\left(A_j^{(m+1)}|A_{\omega_1}^{(1)}\ldots A_{\omega_m}^{(m)}\right) = \frac{1+m_j}{2+m}. \tag{3.25}$$

Indeed, applying Equation (3.25) to the fundamental sequence made up of k values equal to 1 followed by $n-k$ values equal to 2 leads to the following value of $\mathbb{P}(\{\omega\})$:

$$\frac{1}{2}\frac{1+1}{2+1}\cdots\frac{1+k-1}{2+k-1}\frac{1}{2+k}\frac{1+1}{2+k+1}\cdots\frac{1+n-k-1}{2+n-1} = \frac{1}{n+1}\frac{k!(n-k)!}{n!}, \tag{3.26}$$

corresponding to the value obtained above. This value does not depend on the order in which the k components equal to 1 appear in the sequence. As will be clear in Chapter 7, Equation (3.25) can be used to derive the so-called Bose–Einstein statistics using a purely probabilistic approach due to Brillouin. For this reason, some scholars claimed that the Bose–Einstein statistics was originally discovered by Boltzmann while trying to derive Maxwell's distribution of velocities in a gas. Note that the Boltzmann method considered n energy elements into 2 molecules, whereas the derivation of Bose–Einstein statistics would be based on n molecules into 2 states.

The probability of allocating the first energy element to the first molecule is $1/2$, but the probability of the mth element being assigned to the first molecule depends on the allocation of all the previous energy elements. Contrary to (3.20), this probability is an increasing function of the number of 'successes' in the previous trials. This process can be described by means of an urn model as well: the so-called Pólya urn. Consider an urn containing balls belonging to two categories, the first category of balls is labelled with 1 and the second one is labelled with 2. At the beginning, half of the balls are of type 1 and the other half are of type 2. Each time a ball is drawn, it is replaced into the urn, together with a new ball of the same type. If, at any draw, all the balls in the urn have the same probability to be selected, Equation (3.25) follows.

3.5 Remarks on urn models and predictive probabilities

3.5.1 Urn models, accommodation models and possible misunderstandings

Urn models are commonly used, but, often, their probabilistic meaning is unclear. In the case of equiprobability of all the individual descriptions corresponding to the occupation vector $\mathbf{n}$, the predictive probability is given by the ratio of $n_j - m_j$ over $n-m$, where n_j is the number of individuals or objects belonging to category j, m_j is the number of observed individuals belonging to j and $n_j - m_j$ is the number of individuals belonging to the jth category not yet observed; moreover, n denotes the total number of individuals and m is the number of observations, so that $n-m$ is the total number of individuals yet to be observed. Here, there is a perfect isomorphism between observing individuals from a given population and drawing balls from an auxiliary urn, whose initial composition is $n_1,\ldots,n_g$ and corresponds to the composition of the population. Indeed, even today, statisticians can use balls in one-to-one correspondence with real objects in order to randomly sample individuals from a finite population. In this case, one can imagine that the sample is created step by step, and when one has a sample described by $\mathbf{m}$, a description of the sampled population is given by the vector $(m_1,\ldots,m_g)$, with $m_i \le n_i$. When m equals n, sampling must stop and the auxiliary urn is empty. Now, all the population has been transferred to the sample, or better the sample coincides with the whole population.

Starting from this pertinent example, it is tempting to extend urn models to independent predictive probabilities. In such a case, one has predictive probabilities given by (3.19) with $p_j \in \mathbb{Q}$, rational number, and the auxiliary urn contains $n_j = Np_j$ balls of type j, where $N : Np_j \in \mathbb{N}$ for all js and the drawn ball is replaced in order to preserve independence and equidistribution for all draws. In this case, there is no real correspondence between balls and individuals, the fractions of balls of type j in the urn are a mere expedient to produce the correct sequences of drawn values. This mechanism is valid for any sample size $0 < m \le n$, and it could be extended to any number of trials.

Returning to the example of n energy elements to be allocated into g molecules, one can imagine having a system of g colours. The predictive probability (here with the meaning of an allocation probability)

$$\mathbb{P}\left(A_j^{(m+1)}|A_{\omega_1}^{(1)}\ldots A_{\omega_m}^{(m)}\right) = \frac{1+m_j}{g+m}, \tag{3.27}$$

can be represented by a Pólya urn model, whose initial content is g balls, one for each colour; then the urn is updated after each draw by replacing the drawn ball and by adding a ball whose colour is the same. Again, the Pólya urn model is nothing else than an expedient tool to assign probability values. At the end of the accommodation process, one still has n energy elements in the system, but $n+g$ balls in the Pólya urn.

While in the hypergeometric case, the urn and the population to be examined have the same cardinality, so that the drawing process necessarily ends after n steps, in the Bernoullian case, when the population to be examined decreases to zero, the composition of the auxiliary urn stays constant, and in the Pólya case the content of the urn increases.

3.5.2 A realistic interpretation of the predictive probability

So far, two kinds of individual description have been discussed: alphabetical descriptions and sampling descriptions. This difference refers to the meaning of the index i in the event $A^{(i)}_{\omega_i}$, that is the object which is examined in the ith observation. Assume that the Body-Mass-Index (BMI) of a population of size n is studied; one could believe that the allocation of n individuals into g BMI classes is a datum that observation simply discovers. In this sense an alphabetical description of individuals into g BMI classes admits a probabilistic description only if one considers probability as a measure of ignorance; even if each individual is indeed allocated to some BMI class, it is not known to which class the individual belongs. Even if the (frequency) vector of the population is known, an orthodox (frequentist) statistician would say that there is a single alphabetical description that is *true but unknown*, so that the set of possible individual descriptions cannot be a domain for probability.

It is a quite different situation if, given the frequency vector $\mathbf{n}$, one draws all individuals without replacement at random. Indeed, it is important to stress that the deep meaning of the term *at random* is far from clear, and it resisted research work by scholars such as von Mises, Church, Kolmogorov and so on. Instead of *at random* it is advisable to use the cautious expression *all elements being equiprobable with respect to drawing*. As mentioned above, the equiprobability of all sequences $\mathbb{P}(\{\varsigma\}|\mathbf{n}) = (W(\varsigma|\mathbf{n}))^{-1}$, that is Equation (3.14), is widely accepted as the most important axiom in mathematical statistics.

The above discussion concerns the common statistical context where the allocation of objects into categories is a datum preceding the statistical analysis. However, this is not the case that will be discussed in the following. First of all, one would like to describe individuals (or agents) changing category as time goes by; therefore, alphabetical descriptions will change in time as well. Incidentally, in this framework, the probability $\mathbb{P}(\{\omega\})$ has a natural frequency interpretation in terms of the time fraction in which the description $\boldsymbol{\omega}$ is true. Secondly, as mentioned at the beginning of Section 3.5.1, the predictive probability $\mathbb{P}(A_j^{(m+1)}|A_{\omega_1}^{(1)}\ldots A_{\omega_m}^{(m)})$ has two different meanings. It can be considered as a sampling formula, where one discovers which category the $(m+1)$th object belongs to, and it can also be considered as an *accommodation probability*, where the $(m+1)$th object enters the jth category, after the previous accommodation of m objects.

Returning to n elements to be allocated into g categories, Equation (3.25) generalizes to Equation (3.27), and it describes how the allocation of the first m energy elements influences the probability of accommodation of the $(m+1)$th element. The resulting sequence is neither an alphabetical one nor a sampling one; it is an accommodation sequence. The name of the energy element is not important, as it is not contained in Equation (3.25). The order is that of appearance into the system. If

Equation (3.25) holds true, all the elements are identical with respect to probability, but they are distinguished by their time of appearance. This remark explains why it is not meaningful to speak of *indistinguishable* objects as is often done in physics when referring to particles in the framework of quantum mechanics. In a typical detection experiment, one can refer to the first, the second, etc. particle detected in a certain region of space, even if the *names* (or labels) of these particles are not available.

3.6 Appendix: outline of elementary probability theory

This appendix is devoted to an introduction to elementary probability theory seen as a measure on Boolean lattices (a.k.a. Boolean algebras). In particular, only finite additivity is assumed for the union of disjoint sets in a Boolean algebra. Unfortunately, finite additivity does not imply infinite additivity which is necessary to build a complete theory. These issues will be further discussed in the next chapters. Readers already familiar with elementary probability theory can skip the appendix. On the contrary, readers who were never exposed to probability theory should read this appendix before reading other parts of this chapter.

3.6.1 Boolean lattices

Both propositional logic and the elementary algebra of sets share an algebraic structure known as Boolean lattice or Boolean algebra.

Definition (Boolean lattice) A Boolean lattice is an algebraic structure $\langle \mathcal{B}, 0, 1, ', +, \cdot \rangle$ where $\mathcal{B}$ is a set, $0 \in \mathcal{B}$, $1 \in \mathcal{B}$, $'$ is a unary operation on $\mathcal{B}$ (that is a function $\mathcal{B} \to \mathcal{B}$), and $+, \cdot$ are binary operations on $\mathcal{B}$ (that is functions $\mathcal{B}^2 \to \mathcal{B}$) satisfying the following axioms ($a, b, c \in \mathcal{B}$):

1. Associative property 1
$$a + (b + c) = (a + b) + c; \tag{3.28}$$
2. Associative property 2
$$a \cdot (b \cdot c) = (a \cdot b) \cdot c; \tag{3.29}$$
3. Commutative property 1
$$a + b = b + a; \tag{3.30}$$
4. Commutative property 2
$$a \cdot b = b \cdot a; \tag{3.31}$$

5. Distributive property 1

$$a \cdot (b+c) = (a \cdot b) + (a \cdot c); \tag{3.32}$$

6. Distributive property 2

$$a + (b \cdot c) = (a+b) \cdot (a+c); \tag{3.33}$$

7. Identity 1

$$a + 0 = a; \tag{3.34}$$

8. Identity 2

$$a \cdot 1 = a; \tag{3.35}$$

9. Property of the complement 1

$$a + a' = 1; \tag{3.36}$$

10. Property of the complement 2

$$a \cdot a' = 0. \tag{3.37}$$

A simple model of this abstract structure is given by the following set-theoretical interpretation. Given a finite set Ω and the set of its subsets $\mathcal{P}(\Omega)$, one can identify $\mathcal{B}$ with $\mathcal{P}(\Omega)$, 0 with $\varnothing$, 1 with Ω, $'$ with the complement ^{c}, $+$ with the union $\cup$ and $\cdot$ with the intersection $\cap$. Direct inspection shows that $\langle \mathcal{P}(\Omega), \varnothing, \Omega, ^c, \cup, \cap \rangle$ is a Boolean lattice. In order to find the interpretation in propositional logic, one can consider the classes of equivalent propositions. They are a Boolean lattice where 0 is the class of contradictions, 1 the class of tautologies, $'$ corresponds to the logical connective *NOT*, $+$ corresponds to *OR* and $\cdot$ corresponds to *AND*. The prescription that propositions/events are a lattice makes it possible to negate them, or connect them freely, resulting always in some meaningful proposition/event.

3.6.2 Example: Boolean lattice generated by subsets of Ω

The power set $\mathcal{P}(\Omega)$ is a good sample space as it shares the structure of a Boolean lattice. However, in general, it is neither desirable nor necessary to consider all the sets in $\mathcal{P}(\Omega)$ as events.

Definition (Boolean lattice generated by $\mathcal{E}$) Given a finite set Ω and a set of its subsets $\mathcal{E}$, the Boolean lattice $\sigma(\mathcal{E})$, generated by $\mathcal{E}$, is the smallest Boolean lattice on Ω containing all the sets in $\mathcal{E}$.

Now, consider the case where Ω is a finite set containing 5 objects: $\Omega = \{a,b,c,d,e\}$. Assume that the set $\mathcal{E}$ is made up of two sets $E_1 = \{a,b,c\}$ and $E_2 = \{b,d,e\}$. The Boolean lattice generated by $\mathcal{E}$ can be generated by a direct application of the axioms. It must include $\varnothing$ and Ω, $F_1 = E_1^c = \{d,e\}$ and $F_2 = E_2^c = \{a,c\}$ as well as all the possible unions and intersections of these sets and their complements, $F_3 = E_1 \cap E_2 = \{b\}$ and $F_4 = F_1 \cup F_2 = E_1^c \cup E_2^c = \{a,c,d,e\}$. Further unions and intersections create no new sets. In summary, one has the following Boolean lattice $\sigma(E_1,E_2)$:

$$\sigma(E_1,E_2) = \{\varnothing, F_1, F_2, F_3, F_2 \cup F_3, F_1 \cup F_3, F_1 \cup F_2, \Omega\}. \tag{3.38}$$

$F_1 = \{d,e\}$, $F_2 = \{a,c\}$ and $F_3 = \{b\}$ are mutually disjoint sets whose union is Ω, in formulae: for $i \neq j$, $F_i \cap F_j = \varnothing$, $\cup_{i=1}^n F_i = \Omega$, and all the other sets in $\sigma(E_1,E_2)$ are unions of these sets. They are called *atoms* of the Boolean lattice. What is the fastest route to find the *atoms* of the lattice?

3.6.3 Example: constituents

It turns out that, if the Boolean lattice of sets $\mathcal{F}$ is finite, it can be decomposed into constituents, that is subsets $A_1,\ldots,A_n$ of Ω such that $A_i \cap A_j = \varnothing$ if $i \neq j$, $\cup_{i=1}^n A_i = \Omega$, and any set $B \in \mathcal{F}$ is the union of A_is. In order to see that this is true, first label the sets in $\mathcal{F}$ in an arbitrary way:

$$\mathcal{F} = \{B_1,\ldots,B_r\}.$$

Then for any set B in $\mathcal{F}$ set $B^1 = B$ and $B^{-1} = B^c$. Consider now a sequence $\mathbf{b} = (b_1,\ldots,b_r)$, where each b_i is either 1 or -1 and define the set

$$B^{\mathbf{b}} = \cap_{i=1}^r B_i^{b_i}.$$

The set $B^{\mathbf{b}}$ still belongs to the Boolean lattice as this is closed with respect to a finite number of unions and intersections. (This is indeed a theorem and De Morgan's laws are necessary to prove it. They are discussed in the next example.) It can happen that $B^{\mathbf{b}} = \varnothing$, but for any $\omega \in \Omega$ there is a vector $\mathbf{b}$ for which $\omega \in B^{\mathbf{b}}$; then ω is in $B_i^{b_i}$ for one of the $b_i = \pm 1$. Therefore, not all the sets $B^{\mathbf{b}}$ are empty. Moreover, if $\mathbf{b} \neq \mathbf{c}$ then $B^{\mathbf{b}} \cap B^{\mathbf{c}} = \varnothing$. Indeed, if $\mathbf{b} \neq \mathbf{c}$ then $b_i \neq c_i$ for some i. Therefore, if B_i^1 is in the intersection defining $B^{\mathbf{b}}$, its complement is in the intersection defining $B^{\mathbf{b}}$, this means that $B^{\mathbf{c}} \subset (B^{\mathbf{b}})^c$ and that $B^{\mathbf{b}} \cap B^{\mathbf{c}} = \varnothing$. In summary, the sets $B^{\mathbf{b}}$ are either empty or non-empty and they are mutually disjoint. Moreover they are a finite number. It suffices to define the sets A_i as the non-empty $B^{\mathbf{b}}$s to get the constituents of Ω.

3.6.4 Example: De Morgan's laws

In elementary set theory, De Morgan's laws state that

$$(A \cup B)^c = A^c \cap B^c, \tag{3.39}$$

and

$$(A \cap B)^c = A^c \cup B^c. \tag{3.40}$$

In propositional calculus one has that

$$NOT\ (p\ OR\ q) == ((NOT\ p)\ AND\ (NOT\ q))$$

and

$$NOT\ (p\ AND\ q) == ((NOT\ p)\ OR\ (NOT\ q)),$$

where $==$ denotes logical equivalence. These laws are a direct consequence of the axioms defining a Boolean lattice. For $a, b \in \mathcal{B}$, one must prove the following theorem:

Theorem (De Morgan's laws)

$$(a+b)' = a' \cdot b', \tag{3.41}$$

and

$$(a \cdot b)' = a' + b'. \tag{3.42}$$

Only (3.41) will be proved here. The proof of (3.42) is left to the reader as an exercise. The following lemma is useful:

Lemma (Sum of an element of $\mathcal{B}$ with identity 2)

$$a + 1 = 1. \tag{3.43}$$

Proof For the sake of simplicity, the commutative properties of $+$ and $\cdot$ will be freely used in this proof. Given $a \in \mathcal{B}$, as a consequence of (3.34) one has

$$a' + 0 = a', \tag{3.44}$$

and using (3.35) and (3.37) this becomes

$$a' \cdot 1 + a' \cdot a = a'. \tag{3.45}$$

Now (3.32) yields

$$a' \cdot (a+1) = a', \tag{3.46}$$

so that the identification

$$a+1 = 1, \tag{3.47}$$

follows. □

It is now possible to prove (3.41).

Proof The following chain of equalities follows from the axioms and from (3.43):

$$a+b+a' \cdot b' \overset{(3.33)}{=} (a+b+a') \cdot (a+b+b') \overset{(3.36)}{=} (a+1) \cdot (b+1) \overset{(3.43)}{=} 1 \cdot 1 \overset{(3.34)}{=} 1, \tag{3.48}$$

so that $a' \cdot b'$ can be identified as $(a+b)'$ as a consequence of (3.36). □

3.6.5 Probability space and probability axioms

Let Ω be a finite set, $\mathcal{F}$ be a Boolean lattice of subsets of Ω and $\mathbb{P} : \Omega \to [0,1]$ be a non-negative real-valued function defined on $\mathcal{F}$ and with values in the interval $[0,1]$. The triplet $\langle \Omega, \mathcal{F}, \mathbb{P} \rangle$ is a probability space if $\mathbb{P}$ satisfies the following axioms due to Kolmogorov:

A1. $\mathbb{P}(A) \geq 0, \forall A \in \mathcal{F}$;
A2. $\mathbb{P}(\Omega) = 1$;
A3. $\mathbb{P}(A \cup B) = \mathbb{P}(A) + \mathbb{P}(B), \forall A, B \in \mathcal{F}$ and $A \cap B = \varnothing$.

In the framework of probability theory, the set Ω is called *sample space*. Sometimes it is referred to as the *total set*, and as it is an event in $\mathcal{F}$, it is also called the *sure event*. Moreover, it is not necessary that $\mathcal{F}$ coincides with the set $\mathcal{P}(\Omega)$ of all subsets of Ω.

3.6.6 Elementary consequences of the axioms and conditional probabilities

Elementary consequences of the axioms

The axioms immediately lead to some elementary consequences. In particular, from a theorem on the probability of the complement, the relationship $\mathbb{P}(\varnothing) = 0$ is derived. Moreover, one can obtain the probability of the union of two sets even if they are not disjoint. Then:

Theorem (Probability of complement) For each A in $\mathcal{F}$ one has for $A^c = \{\omega \in \Omega : \omega \notin A\}$

$$\mathbb{P}(A^c) = 1 - \mathbb{P}(A). \tag{3.49}$$

Proof For each subset, A, of Ω one has that $A \cup A^c = \Omega$ and $A \cap A_c = \varnothing$, therefore as a consequence of **A2** and **A3**, the following chain of equalities holds true

$$1 = \mathbb{P}(\Omega) = \mathbb{P}(A \cup A^c) = \mathbb{P}(A) + \mathbb{P}(A^c), \tag{3.50}$$

hence $\mathbb{P}(A^c) = 1 - \mathbb{P}(A)$. □

Corollary (Probability of empty set) $\mathbb{P}(\varnothing) = 0$.

Proof One has that $\Omega^c = \varnothing$ and $\mathbb{P}(\Omega) = 1$ according to **A2**. Therefore, one has $\mathbb{P}(\varnothing) = \mathbb{P}(\Omega^c) = 1 - \mathbb{P}(\Omega) = 1 - 1 = 0$. □

Theorem (Probability of union) For each couple of sets A and B in $\mathcal{F}$, one has

$$\mathbb{P}(A \cup B) = \mathbb{P}(A) + \mathbb{P}(B) - \mathbb{P}(A \cap B). \tag{3.51}$$

Proof $A \cup B$ can be written as the union of three disjoint sets:

$$A \cup B = (A \cap B^c) \cup (A^c \cap B) \cup (A \cap B), \tag{3.52}$$

so that

$$\mathbb{P}(A \cup B) = \mathbb{P}(A \cap B^c) + \mathbb{P}(A^c \cap B) + \mathbb{P}(A \cap B). \tag{3.53}$$

Moreover, one has that A and B can be written as a union of the following disjoint sets

$$A = (A \cap B^c) \cup (A \cap B), \tag{3.54}$$

and

$$B = (A^c \cap B) \cup (A \cap B), \tag{3.55}$$

which means that

$$\mathbb{P}(A \cap B^c) = \mathbb{P}(A) - \mathbb{P}(A \cap B), \tag{3.56}$$

and

$$\mathbb{P}(A^c \cap B) = \mathbb{P}(B) - \mathbb{P}(A \cap B). \tag{3.57}$$

Replacing these equations in (3.53) yields the thesis (3.51). □

A repeated application of **A3** and of the associative property (3.28) for Boolean lattices immediately leads to the

Theorem (Addition Theorem) Given n mutually disjoint events (that is $\forall i, j$ with $i \neq j$, $(A_i \cap A_j = \varnothing)$), then

$$\mathbb{P}(A_1 \cup A_2 \cup \ldots \cup A_n) = \mathbb{P}(A_1) + \mathbb{P}(A_2) + \ldots + \mathbb{P}(A_n). \tag{3.58}$$

Definition (Decomposition (or partition) of the sample space): a family of mutually disjoint sets $\mathcal{A} = \{A_i\}_{i=1,\ldots,n}$ whose union coincides with Ω is called a decomposition (or partition) of the sample space. In formulae: $\mathcal{A} = \{A_i : A_i \cap A_j = \varnothing, \cup_{i=1}^n A_i = \Omega\}$. One usually refers a decomposition as to a family of *mutually exclusive and exhaustive* events. If each of them has a probability value $\mathbb{P}(A_i) \geq 0$ (to satisfy **A1**), due to the addition theorem (3.58) $\mathbb{P}(\cup_{i=1}^n A_i) = \sum_{i=1}^n \mathbb{P}(A_i) = \mathbb{P}(\Omega) = 1$ (to satisfy **A2**). Let $\mathcal{F}$ be the lattice generated by $\mathcal{A}$, that is only events generated by $\mathcal{A} = \{A_i\}_{i=1,\ldots,n}$ are considered. Then the probability of each set in $\mathcal{F}$ can be obtained by using **A3**. A sample space can be decomposed in many different ways.

Conditional probabilities

As mentioned above, the probability of a generic union of two sets depends on the probability of the intersection. In Kolmogorov's theory, the probability of the intersection of two events is related to the conditional probability *via* the following:

Definition (Conditional probability) If $\mathbb{P}(B) > 0$, for each set A and B in $\mathcal{F}$, the conditional probability of A given B is defined as follows

$$\mathbb{P}(A|B) = \frac{\mathbb{P}(A \cap B)}{\mathbb{P}(B)}. \tag{3.59}$$

The conditioning event, B, is often called *evidence*, whereas the conditioned one, A, is called *hypothesis*. If $P(B) > 0$, one can show that the following properties hold true:

A1′. $\mathbb{P}(A|B) \geq 0$, $\forall A \in \mathcal{F}$;
A2′. $\mathbb{P}(\Omega|B) = 1$;
A3′. $\mathbb{P}(A \cup C|B) = \mathbb{P}(A|B) + \mathbb{P}(C|B)$, $\forall A, C \in \mathcal{F}$ and $A \cap C = \varnothing$. That is, defining the measure $\mathbb{P}_B(\cdot) = \mathbb{P}(\cdot|B)$, it turns out that $\mathbb{P}_B(\cdot)$ satisfies the same axioms as $\mathbb{P}(\cdot)$. Therefore, one can consider conditional probabilities as probabilities defined on

a restriction of the sample space where the event B is always true (always takes place). As an immediate consequence of the definition one gets

$$\mathbb{P}(A \cap B|C) = \mathbb{P}(A|B \cap C)\mathbb{P}(B|C). \tag{3.60}$$

Given that

$$\mathbb{P}(A \cap B) = \mathbb{P}(B)\mathbb{P}(A|B), \tag{3.61}$$

by induction, the

Theorem (Multiplication theorem, or product rule)

$$\mathbb{P}(A_1 \cap A_2 \cap \ldots \cap A_n) = \mathbb{P}(A_1)\mathbb{P}(A_2|A_1)\ldots\mathbb{P}(A_n|A_{n-1} \cap \ldots \cap A_1) \tag{3.62}$$

can be derived. This theorem is very important in the theory of stochastic processes and will be further discussed in Exercise 3.1.

In other theories the conditional probability is primitive; it is a biargumental function $\mathbb{P}(\cdot|\cdot)$ defined for any couple of events. In such a case (3.61) is considered as a definition of $\mathbb{P}(A \cap B)$ in terms of the conditional probability $\mathbb{P}(A|B)$. Conditional probabilities are difficult and elusive to deal with and they can generate confusion as in the following examples.

Example (Retrodiction) Suppose you have a deck of 52 cards. Two cards are drawn without replacement and in a sequence. If the first card is a K, what is the probability that the second card is a K? Almost everybody acquainted with elementary probabilities gives the correct answer $3/51$. Now, imagine you know that the second card is a K without having any information on the first card, then what is the probability of drawing a K at the first draw? The correct answer is again $3/51$. Why?

Example (Monty Hall quiz) You are in front of three closed doors, numbered from 1 to 3. Two of them are empty and the third one contains a valuable prize, say a luxury car. You do not know where the car is. The quiz master (who knows where the prize is) asks you to choose one door. You choose door number 1. Now the quiz master opens door 3 (it is empty). Does this information change your probability of winning? The correct answer is no, but a lot of discussions have taken place about it. And what if the quiz master gives you the possibility of changing your choice? What is better for you? Changing your mind and selecting door number 2 or keeping your first decision? In our opinion this part of the game is simpler to understand, because changing makes you a winner if your initial choice was wrong (2 times out of 3), and a loser if your initial choice was right (1 time out of 3), so reversing the probability of winning without change. This game has become very important and popular in experimental psychology and experimental economics. It is rather difficult to give the correct answer at the first attempt. One reason could be that it is difficult to understand all the details correctly. P.R. Mueser and D. Graberg give a somewhat redundant description of the game [2]. This problem is further discussed in Exercise 3.2.

Given the commutative property of the intersection, the events A and B play a symmetric role in the definition (3.59). This consideration immediately leads to Bayes' rule.

Theorem (Bayes' rule) If $\mathbb{P}(A) \neq 0$ and $\mathbb{P}(B) \neq 0$, the conditional probabilities $\mathbb{P}(A|B)$ and $\mathbb{P}(B|A)$ are related as follows

$$\mathbb{P}(B|A) = \frac{\mathbb{P}(A|B)\mathbb{P}(B)}{\mathbb{P}(A)}. \tag{3.63}$$

Proof From the definition of conditional probability in (3.59), one has

$$\mathbb{P}(A \cap B) = \mathbb{P}(A|B)\mathbb{P}(B), \tag{3.64}$$

and

$$\mathbb{P}(A \cap B) = \mathbb{P}(B|A)\mathbb{P}(A), \tag{3.65}$$

hence

$$\mathbb{P}(B|A)\mathbb{P}(A) = \mathbb{P}(A|B)\mathbb{P}(B), \tag{3.66}$$

and the thesis follows. □

Given a partition $\mathcal{H} = \{H_i\}_{i=1}^n$ of Ω, one can derive the so-called theorem of *total probability*:

Theorem (Total probability) If $\mathcal{H} = \{H_i\}_{i=1}^n$ is a partition of Ω then for any E in $\mathcal{F}$

$$\mathbb{P}(E) = \sum_{i=1}^{n} \mathbb{P}(E|H_i)\mathbb{P}(H_i). \tag{3.67}$$

Proof The following chain of equalities holds true:

$$E = E \cap \Omega = E \cap (\cup_{i=1}^n H_i) = \cup_{i=1}^n (E \cap H_i), \tag{3.68}$$

then E is written as the union of mutually disjoint sets and

$$\mathbb{P}(E) = \mathbb{P}\left(\cup_{i=1}^n (E \cap H_i)\right) = \sum_{i=1}^{n} \mathbb{P}(E \cap H_i) = \sum_{i=1}^{n} \mathbb{P}(E|H_i)\mathbb{P}(H_i), \tag{3.69}$$

yielding the thesis. □

The result is trivial if E is an event generated by $\mathcal{H}$; for instance consider $E = H_1 \cup H_2$. In this case, $\mathbb{P}(E|H_i) = 1$ if $i = 1,2$, $\mathbb{P}(E|H_i) = 0$ if $i = 3,\ldots,n$, hence $\mathbb{P}(E) = \mathbb{P}(H_1) + \mathbb{P}(H_2)$.

Combining Bayes' rule and total probability, the following interesting result follows:

Corollary (Bayes' theorem) If $\mathcal{H} = \{H_i\}_{i=1}^n$ is a partition of Ω, that is a family of mutually disjoint events, $H_i \cap H_j = \varnothing$ for any $i \neq j$, such that $\Omega = \cup_{i=1}^n H_i$ then for any E in $\mathcal{F}$ and for any j

$$\mathbb{P}(H_j|E) = \frac{\mathbb{P}(E|H_j)\mathbb{P}(H_j)}{\mathbb{P}(E)} = \frac{\mathbb{P}(E|H_j)\mathbb{P}(H_j)}{\sum_{i=1}^{n} \mathbb{P}(E|H_i)\mathbb{P}(H_i)}. \tag{3.70}$$

Proof Bayes' rule states that

$$\mathbb{P}(H_j|E) = \frac{\mathbb{P}(E|H_j)\mathbb{P}(H_j)}{\mathbb{P}(E)}, \tag{3.71}$$

total probability that

$$\mathbb{P}(E) = \sum_{i=1}^{n} \mathbb{P}(E|H_i)\mathbb{P}(H_i). \tag{3.72}$$

The thesis follows by replacement of the second equation into the first. □

Bayes' theorem is a logical consequence of the axioms; it has far-reaching applications, especially in the subjective approach to probability. The H_j are usually interpreted as a set of mutually exclusive and exhaustive hypotheses, which are not observable, and each of which offers a *probabilistic explanation* of the observed event E. The term $\mathbb{P}(H_j)$ is the *prior* or initial probability that H_j is true; the term $\mathbb{P}(E|H_j)$ represents the probability of the observed value given the hypothesis H_j; and finally $\mathbb{P}(H_j|E)$ is the final (*posterior*) probability that the hypothesis H_j is true after having observed the event. In Equation (3.70), the evidence E is fixed, whereas the focus is on any H_j running on all the possibilities. Then the denominator is nothing else than a normalization factor for the final distribution, and one can simply write $\mathbb{P}(H_j|E) \propto \mathbb{P}(E|H_j)\mathbb{P}(H_j)$, where $\sum_{j=1}^n \mathbb{P}(H_j|E) = \sum_{j=1}^n \mathbb{P}(H_j) = 1$. In words, each term of the final distribution is proportional to the corresponding initial one multiplied by $\mathbb{P}(H_j|E)$. Considering two alternative hypotheses H_i and H_j, facing the evidence E, the so-called Bayes factor,

$$\frac{\mathbb{P}(H_i|E)}{\mathbb{P}(H_j|E)} = \frac{\mathbb{P}(E|H_i)}{\mathbb{P}(E|H_j)} \frac{\mathbb{P}(H_i)}{\mathbb{P}(H_j)} \tag{3.73}$$

measures how the initial ratio $\mathbb{P}(H_i)/\mathbb{P}(H_j)$ is modified by the ratio of the hypothetical probabilities of the observed evidence.

The term $\mathbb{P}(E|H_j)$ is accepted under all the conceptions of probability: in the simplest case it represents the (hypothetical) probability of the observed colouring selected from an urn whose composition is described by the hypothesis H_j. On the contrary, in many applications, the prior probability $\mathbb{P}(H_j)$ describes the knowledge on (or the degree of belief in) the hypothesis H_j; however, it has no objective frequentist meaning; therefore, in practice, it is refused by the so-called orthodox statisticians. The refusal of $\mathbb{P}(H_i)$ implies the falling of $\mathbb{P}(H_j|E)$ and the refusal of the whole theorem.

The comparison (3.73) of two alternative hypotheses given some evidence is reduced to $\mathbb{P}(E|H_i)/\mathbb{P}(E|H_j)$, called the *likelihood ratio*, where $\mathbb{P}(E|H_i)$ is called the *likelihood* of the hypothesis H_j after the occurrence of E. Note that for varying H_j, one has that $\sum_{j=1}^{n} \mathbb{P}(E|H_j)$ is different from 1. If E is fixed, the same mathematical function $\mathbb{P}(E|H_i)$ is the likelihood function for the hypotheses $\mathbb{P}(E|\cdot)$, which should not be confused with the hypothetical probability distribution of the observed events $P(\cdot|H_j)$ for H_j given. Orthodox statisticians almost completely refuse to apply Bayes' theorem to statistical inference. On the contrary, the theorem is widely used within the subjective and logical approaches to probability. This is the main reason for the separation between orthodox and Bayesian statistics.

Note that $\mathbb{P}(E|H_j)$ represents the (hypothetical) probability of the observed value in the hypothesis H_j. If $\mathbb{P}(E|H_1) > \mathbb{P}(E|H_2)$, we could say that H_1 is a better probabilistic explanation of E than H_2, being the strength of the explanation expressed in terms of a conditional probability. Laplace's first memoir on the subject is entitled *Probabilitè des causes par les évènements*, and Bayes' theorem appears as the solution of the problem of induction, that is to infer something from the observed 'effect' to the unobservable 'cause'. Actually, conditioning reflects a logical or informational or frequency relationship between events and hypotheses, whereas a causal (deterministic) relationship would give $\mathbb{P}(E|H_1) = 1$ for H_1 sufficient cause, and $\mathbb{P}(E|H_1^c) = 0$ for a necessary cause.

In fact $\mathbb{P}(E|H_j)$ is closer to the intuitive statistical notion of relevance, or correlation. Given two events A and E, a measure of the relevance of E with respect to A is given by the ratio

$$Q_A^E = \frac{\mathbb{P}(A|E)}{\mathbb{P}(A)}; \tag{3.74}$$

if Q_A^E is greater (or smaller) than 1, intuitively the occurrence of E increases (or decreases) the probability of A: it seems a good (a bad) *probabilistic cause* of A.

But notice that one has

$$Q_A^E = \frac{\mathbb{P}(A|E)}{\mathbb{P}(A)} = \frac{\mathbb{P}(A \cap E)}{\mathbb{P}(A)\mathbb{P}(E)} = \frac{\mathbb{P}(E|A)}{\mathbb{P}(E)} = Q_E^A, \tag{3.75}$$

meaning that the relevance is symmetric, thus violating the asymmetry typical of the usual naïve causal relationship. In other words, Laplace's causal interpretation of Bayes' theorem can be misleading. The term *probability of hypotheses* is a more neutral way of interpreting the formula.

3.6.7 Independent events

Definition (Independent events) Two events $A, B \in \mathcal{F}$ are independent if

$$\mathbb{P}(A|B) = \mathbb{P}(A), \tag{3.76}$$

and

$$\mathbb{P}(B|A) = \mathbb{P}(B). \tag{3.77}$$

This definition immediately leads to the following:

Theorem (Probability of the intersection of independent events) Given two independent events A and B, the probability of their intersection is

$$\mathbb{P}(A \cap B) = \mathbb{P}(A)\mathbb{P}(B). \tag{3.78}$$

Proof The proof is an immediate consequence of the definition of conditional probabilities as well as of independent events. Indeed the following chain of equalities holds true:

$$\mathbb{P}(A \cap B) = \mathbb{P}(A|B)\mathbb{P}(B) = \mathbb{P}(A)\mathbb{P}(B).$$

□

If the probability space is built up in order to describe more than a single experiment, one can study if the experiments are connected or not, that is if the result of one of them influences the probability of the other experiments.

Definition (Independence of experiments) Given two partitions $\mathcal{A} = \{A_i\}_{i=1}^n$ and $\mathcal{B} = \{B_i\}_{i=1}^m$ of Ω, the two experiments are independent if

$$\mathbb{P}(A_i \cap B_j) = \mathbb{P}(A_i)\mathbb{P}(B_j), \tag{3.79}$$

for any $A_i \in \mathcal{A}$, $B_j \in \mathcal{B}$. It follows that

$$\mathbb{P}(A_i|B_j) = \mathbb{P}(A_i), \quad (3.80)$$

or, in other words, that the results of $\mathcal{B}$ are irrelevant in the assignment of probability on $\mathcal{A}$. It also follows that if A_i is independent of B_j, then B_j is independent of A_i, i.e $\mathbb{P}(B_j|A_i) = \mathbb{P}(B_j)$: the independence relation is symmetric. The definition (3.79) can be extended to any number of experiments: $\mathbb{P}(A_i \cap B_j \cap \ldots \cap Z_k) = \mathbb{P}(A_i)\mathbb{P}(B_j)\ldots\mathbb{P}(Z_k)$. This definition of independent experiments, due to Kolmogorov, naturally leads to the definition of independent events. The simplest partition is the dichotomous partition: $\mathcal{A}^{(i)} = \{A_i, A_i^c\}$. Hence

Definition (Independence of n events) n events are independent if the n experiments $\mathcal{A}^{(i)}$ are independent. It follows that $\mathbb{P}(A_{j_1}^{(1)} A_{j_2}^{(2)} \ldots A_{j_n}^{(n)}) = \mathbb{P}(A_{j_1}^{(1)})$ $\mathbb{P}(A_{j_2}^{(2)})\ldots\mathbb{P}(A_{j_n}^{(n)})$, where $A_{j_i}^{(i)}$ is either A_i or A_i^c.

3.7 Exercises

3.1 Prove the multiplication rule (3.62) by induction.

3.2 You are in front of three doors. Behind one of these doors there is a prize, a luxury car. You do not know where the car is. The quiz master (who knows perfectly where the prize is) asks you to choose one door. You choose door number 1. Now the quiz master opens door 2: it is empty. Does this information change your probability of winning? The correct answer is no, but this result is often challenged. And what if the quiz master gives you the possibility of changing your choice? Build an event algebra representing the Monty Hall Quiz. Look for a quantitative solution.

3.3 You test positive for a disease whose prevalence is 1/1000; your test has a 5% rate of false positive results. You also know that the rate of false negative results is 0.1%. What is the probability that you are ill?

3.4 If all descriptions belonging to **n** are equiprobable, show that:

1. the events $\left\{A_{\omega_i}^{(i)}\right\}_{i=1,\ldots,n}$ are identically distributed, $\mathbb{P}\left(A_j^{(i)}\right) = n_j/n$ for all objects $i = 1,\ldots,n$;
2. individual events are not independent and negatively correlated;
3. in the case $g = 2$, the predictive probability is

$$\mathbb{P}\left(A_1^{(m+1)}|A_{\omega_1}^{(1)}\ldots A_{\omega_m}^{(m)}\right) = \frac{k-h}{n-m}. \quad (3.81)$$

3.5 Given that the 43 motions of the known-at-that-time celestial bodies belonging to the solar system lay (almost) on the ecliptic plane and rotated in the same direction, Laplace observed that one can bet more than $4 \cdot 10^{12}$ against 1 that the solar system derives from an ancient rotating nebula. How can this bet be justified?

3.8 Summary

In this chapter, Kolmogorov's axioms of probability theory were presented in finitary form. The structure of a Boolean lattice was used instead of the more general σ-field (or σ-algebra) used in measure-theoretic probability. A probability space consists of the sample space Ω, assumed to be a finite set, of a Boolean lattice $\mathcal{F}$ of subsets of Ω (the events) and of a function $\mathbb{P}$ assigning to each event a number in the real interval $[0, 1]$ and obeying the three axioms **A1**, **A2** and **A3**.

Some elementary consequences of the axioms were discussed. The axioms, together with these theorems, allow us to assign probabilities to the union and the intersection of two events as well as to the complement of an event. In natural language, the union is represented by 'or', the intersection by 'and' and the complement by negation ('not').

In order to determine the probability of the intersection, the notion of conditional probability is necessary, leading to the concept of (statistical) independence between events. Two events are statistically independent if and only if the probability of their intersection is given by the product of the probabilities of each event. It is argued that the concept of conditional probability is not trivial at all, often leading to apparent paradoxes.

Contrary to other approaches, Kolmogorov's theory does not include a method to assign probabilities to events. Various probability assignments were explored for the problem of allocating n objects into g categories. Urn models can be used to represent random sampling from populations. Objects are represented by balls and categories are represented by different colours. Drawing objects with replacement (Bernoulli urn) leads to the so-called multinomial distribution where the events denoted by $A_j^{(i)}$ (the ith drawn object belongs to category j) are independent and identically distributed. Drawing objects without replacement (hypergeometric urn) leads to the multivariate hypergeometric distribution where the events $A_j^{(i)}$ are no longer independent. In the multinomial case, the composition of the urn is not changed, whereas in the hypergeometric case it varies with subsequent draws. A third urn model is the so-called Pólya urn. Also in this case the composition of the urn changes with subsequent draws and the events $A_j^{(i)}$ are not independent. Every time that a ball of a certain colour is drawn, it is replaced, but a ball of the same colour is introduced in the urn. In all these three cases, if before the draw there are N balls in the urn, it is assumed that the probability of drawing any given ball is $1/N$. This assumption is equivalent to 'randomness' in sampling.

Further reading

In this chapter, only finite additivity was assumed (axiom **A3**) and, in parallel, only Boolean lattices were used, instead of the more general axiom of countable additivity on σ-algebras. A σ-algebra, $\mathcal{F}$, is a Boolean lattice close with respect to the countable union (and the countable intersection) of sets. Countable additivity means that axiom **A3** must be replaced by axiom

A3″. (Countable additivity) $\mathbb{P}(\cup_{i=1}^{\infty} A_i) = \sum_{i=1}^{\infty} \mathbb{P}(A_i)$, if $\forall i A_i \in \mathcal{F}$ and $\forall i,j, i \neq j$, $A_i \cap A_j = \varnothing$.

Many important results of probability theory cannot be derived if only finite additivity of a probability measure on Boolean lattices is assumed. In order to see what can be achieved when only finite additivity is used, the reader can consult the book by Dubins and Savage.

L.E. Dubins and L.J. Savage, *How to Gamble if You Must: Inequalities for Stochastic Processes*, McGraw-Hill, New York (1965).

The need for defining a probability space in terms of a sample space Ω, a σ-algebra $\mathcal{F}$ and a countable-additive probability measure $\mathbb{P}$ will be clearer in the next chapter, where random variables and stochastic processes will be introduced.

Since the beginning of his celebrated treatise on probability theory, A.N. Kolmogorov has assumed infinite additivity and has described probability theory as a chapter of measure theory, the branch of mathematics concerned with the definition of integrals.

A.N. Kolmogorov, *Foundations of the Theory of Probability* (English translation of *Grundbegriffe der Wahrscheinlichkeitsrechnung*, originally published in German in 1933), Chelsea, New York (1956).

A good knowledge of measure theory is necessary to better understand measure theoretic probability. This is provided by the recent book written by R.L. Schilling.

R.L. Schilling, *Measures, Integrals and Martingales*, Cambridge University Press, Cambridge, UK (2006).

A modern introductory account of measure theoretic probability is the book by D. Pollard.

D. Pollard, *A User's Guide to Measure Theoretic Probability*, Cambridge University Press, Cambridge, UK (2002).

Reading these two books makes study of the standard reference by Billingsley easier.

P. Billingsley, *Probability and Measure*, Wiley, New York (1995).

Laplace was often mentioned in this chapter. His books on probability theory are now in the public domain and can be downloaded using Google Books.

Pierre-Simon, comte de Laplace, *Théorie Analytique des Probabilités*, Paris (1814).

Pierre-Simon, marquis de Laplace, *Essai Philosophique sur les Probabilités*, H. Remy, Bruxelles (1829).

References

[1] L. Wittgenstein, *Tractatus Logico-Philosophicus*, Routledge & P. Kegan, London (1922).
[2] P.R. Mueser and D. Graberg, *The Monty Hall Dilemma Revisited: Understanding the Interaction of Problem Definition and Decision Making*, University of Missouri Working Paper 99-06, `http://129.3.20.41/eps/exp/papers/9906/9906001.html`.

4

Finite random variables and stochastic processes

In this chapter, random variables and stochastic (or random) processes are defined. These concepts allow us to study populations and individuals in a much more natural way. Individual random variables are nothing else than labels attached to individuals, and, if these labels have a quantitative meaning, one can define the usual notions of expected value, variance, standard deviation, covariance, and so on. The results of the previous two chapters can be translated into the new language of random variables and stochastic processes.

After studying this chapter you should be able to work with:

- sequences of exchangeable random variables, including sequences of independent and identically distributed (i.i.d.) random variables, as a particular case;
- the expected value and variance for numerical random variables;
- the covariance and correlation coefficients for multidimensional random variables.

4.1 Finite random variables

4.1.1 Example: five firms in a small town, revisited

In order to illustrate the main theme of this chapter, it is useful to begin with a simple example. To this purpose, reconsider the example of Section 2.1.3, where U represents a set made of five firms, partitioned into three industrial sectors, directly labelled by the integer numbers 1, 2 and 3. Assume that the joint description of all the elements in the population is $\omega_1 = 1$, $\omega_2 = 1$, $\omega_3 = 2$, $\omega_4 = 3$, $\omega_5 = 2$, in short $\boldsymbol{\omega} = (1,1,2,3,2)$. If the labels 1, 2 and 3 have no numerical meaning, they can be replaced by labels such as a, b, c or any other kind of label. But, if the label i is interpreted as the number of plants, or the annual revenue in millions of pounds, one has a case in which all the properties of either natural or real numbers can be exploited. For instance if i denotes the number of plants of each firm, one can sum the five labels to determine that the total number of plants is

9, or one can compute the average number of plants by using the arithmetic mean and get $9/5 = 1.8$, and so on. These concepts will be thoroughly discussed in this chapter.

4.1.2 Measurable maps (or functions) for finite sets

A map, or a set function f, from the finite set E to the finite set E', and denoted by $f : E \to E'$, is a correspondence between E and E' where any $a \in E$ corresponds to one and only one element $u \in E'$; this is written $u = f(a) \in E'$ and u is called the *image* of a. Conversely, the set of elements of $A \subset E$ corresponding to the same element $v \in E'$, called the *inverse image* of the element v, is denoted by $f^{-1}(v) = \{a \in E : f(a) = v\}$. In other words, the map (or function) is nothing other than a labelling (a division into categories) where each member of E has a label taken from E'. Therefore, for any subset of labels $A' \subset E'$ one can also find its inverse image $A \subset E$: the class of all elements whose labels belong to A'; in formulae $A = f^{-1}(A') = \{a \in E : f(a) \in A'\}$.

If some (not necessarily all) subsets of E form a Boolean lattice $\mathcal{F}$, the couple $\langle E, \mathcal{F} \rangle$ is called a *measurable space*. Assume that $A = f^{-1}(A')$ and $B = f^{-1}(B')$ belong to $\mathcal{F}$ (they are events in the probabilistic interpretation of measurable spaces). Then also $A \cup B$ is an event, and the same for $A \cap B$, and so on. Now $A \cup B$ is composed of all elements whose label is at least in A' or in B', that is $A \cup B = f^{-1}(A') \cup f^{-1}(B')$ and this coincides with $f^{-1}(A' \cup B')$. It means that, if A' and B' are measurable (meaning that their inverse images are events), then also $A' \cup B'$, $A' \cap B'$ and A'^c are measurable (their inverse images are also events). In other words, the subsets of E' whose inverse image is an event form a family of sets $\mathcal{F}'$ closed with respect to unions, intersections and complementation. Then $\mathcal{F}'$ is a Boolean lattice and also the couple $\langle E', \mathcal{F}' \rangle$ is a measurable space. More precisely, one says that the map f is $\mathcal{F} \backslash \mathcal{F}'$-measurable.

Note that the sets $f^{-1}(u)$ for all $u \in E'$ are a partition of E into disjoint classes (if a label is not used, one can set $f^{-1}(u) = \varnothing$). This partition of E corresponds to dividing its elements according to the labels they carry. *Belonging to the same class with respect to the label* is an equivalence relation and is discussed in Exercise 4.1.

The concept of mapping was already introduced in Section 2.1, where an α-description is defined as a set function $\boldsymbol{\omega} : U \to X$. All elements that are mapped into the same category belong to the same class, and the inverse image of $\boldsymbol{\omega}$ partitions U into disjoint classes. Returning to the example of five firms, denoting by $\boldsymbol{\omega}^{-1}(x^{(j)})$ the set of all elements mapped into the same value $x^{(j)}$, one has: $\boldsymbol{\omega}^{-1}(1) = \{u_1, u_2\}$, $\boldsymbol{\omega}^{-1}(2) = \{u_3, u_5\}$ and $\boldsymbol{\omega}^{-1}(3) = \{u_4\}$. It is possible immediately to verify that $\boldsymbol{\omega}^{-1}(i) \cap \boldsymbol{\omega}^{-1}(j) = \varnothing$ for $i \neq j$, and that $\cup_{i=1}^{3} \boldsymbol{\omega}^{-1}(i) = U$.

4.1.3 Definition of finite random variable

Given a probability space $\langle \Omega, \mathcal{F}, \mathbb{P} \rangle$ with finite sample space Ω, a *finite random variable*, $X(\omega)$, is a measurable set function whose domain is the sample space Ω and whose range is some finite set of labels $V = \{v_1, \ldots, v_g\}$. If these labels are just used to denote categories, the random variable will be called *categorical*; if, on the other hand, $v_1, \ldots, v_g$ are numbers, the random variable will be called *numerical*. Note that $X(\omega)$ is constant on each element of the decomposition $C_1, \ldots, C_g$ of the sample space corresponding to the inverse images of the labels $v_1, \ldots, v_g$, that is $C_i = \{\omega : X(\omega) = v_i\} \equiv \{\omega : \omega = X^{-1}(v_i)\}$. Therefore, there is a one-to-one correspondence between singletons in V, $\{v_i\}$, and their inverse images C_i in Ω. If $C_i \in \mathcal{F}$, then X is *measurable*, and one can define a probability on subsets of V by means of the formula

$$\mathbb{P}_X(v_i) = \mathbb{P}(X^{-1}(v_i)) = \mathbb{P}(X = v_i) = \mathbb{P}(\{\omega : X(\omega) = v_i\}), \tag{4.1}$$

where the equalities highlight the fact that these formulations are equivalent. Indeed, for any $D \subset V$, the image probability measure

$$\mathbb{P}_X(D) = \mathbb{P}(X^{-1}(D)) = \mathbb{P}(X \in D) = \sum_{v_i \in D} \mathbb{P}_X(v_i) \tag{4.2}$$

satisfies the basic axioms of probability with respect to the set V and the family $\mathcal{P}(V)$ of all subsets of V (including V and $\varnothing$). Therefore, the triplet $\langle V, \mathcal{P}(V), \mathbb{P}_X \rangle$ is a probability space. The reader is invited to check that the axioms of Kolmogorov are satisfied by the choice (4.2), based on the definition (4.1). In other words, as the set of labels (values) is finite, it makes sense to compute the probability of any set belonging to the power set of V. Moreover, $\{v_1\}, \ldots, \{v_g\}$ being a partition of V, and, correspondingly, $C_1, \ldots, C_g$ a partition of Ω, it follows that

$$\sum_{v_i \in V} \mathbb{P}_X(v_i) = \sum_{i=1}^{g} \mathbb{P}(C_i) = 1, \tag{4.3}$$

and $\{\mathbb{P}_X(v_i) = \mathbb{P}(X = v_i),\ v_i \in V\}$ is called the *probability distribution* of the random variable X. If the labels are ordered or, better, numerical values, the function

$$F_X(x) = \mathbb{P}(X \leq x) = \sum_{v_i \leq x} \mathbb{P}_X(v_i) \tag{4.4}$$

is called the *cumulative distribution function* of the random variable X and

$$C_X(x) = 1 - F_X(x) \tag{4.5}$$

is the *complementary cumulative distribution function* of X.

4.1.4 Example: the sum of two dice

Coming back to the discussion of Section 4.1.2, one can see that things are a bit more complicated if the domain of a set function f is a finite probability space Ω. Whereas an α-description is defined as a set function $\boldsymbol{\omega}: U \to X$, a random variable is a set function $f: \Omega \to V$ defined on each possible α-description. Namely, an α-description reflects *one* possible state of affairs for a set of individual objects, whereas a random variable is defined for *all* possible states of affairs. Therefore, while an α-description partitions objects with respect to categories (it is a snapshot of the population), a random variable is a function which, for each snapshot, has a definite value belonging to the set of labels V. If the number of possible states of affairs is finite, $\#\Omega = g^n$, the possible values of f are finite as well, and smaller or equal to g^n. In the example of Section 4.1.1, the total number of plants is a function of the α-description and it equals the sum of all components of $\boldsymbol{\omega}$. The possible values of f are all the integers between 5 (all firms have one plant) and 15 (all firms have three plants). Summarizing, a particular α-description, which can be regarded as a set function, is just a singleton of the sample space Ω representing all the individual descriptions.

In order to further illustrate this point, consider the sum of points shown after a toss of two dice. Here, the set U is made up of two elements representing each die $U = \{u_1, u_2\}$; $X = \{1, \ldots, 6\}$, one label for each individual outcome, whereas the sample space is a cartesian product $\Omega = \{\boldsymbol{\omega} = (\omega_1, \omega_2),\ \omega_i = 1, \ldots, 6\}$, which is the set of all set functions $\boldsymbol{\omega}: U \to X$. Now, focus on the random variable $S(\boldsymbol{\omega}) = \omega_1 + \omega_2$, which is defined for all $\boldsymbol{\omega} \in \Omega$, and whose range is $V = \{2, \ldots, 12\}$. In Fig. 4.1, the value of S is represented in each cell (a description). In order to consider S as a random variable, it is necessary to know the probability of all the inverse images of its values, represented by the cells lying on the counter diagonals of the table. Their number is 11 and they are a decomposition of Ω. As far as the random variable S is concerned, it would be enough to fix directly the 11 values corresponding to the events $C_i = \{\boldsymbol{\omega} : S(\boldsymbol{\omega}) = i\}$, with the constraint $\sum_{i=2}^{12} \mathbb{P}(C_i) = 1$. However, thanks to the finiteness of Ω, and to the fact that the singletons $\boldsymbol{\omega}$ are the thinnest partition of Ω, following what was said in Chapter 3, it suffices to define non-negative numbers $p(\boldsymbol{\omega})$ such that $\sum_{\boldsymbol{\omega} \in \Omega} p(\boldsymbol{\omega}) = 1$, so that the probability function $\mathbb{P}(C) = \sum_{\boldsymbol{\omega} \in C} p(\boldsymbol{\omega})$ for any $C \in \mathcal{P}(\Omega)$ satisfies the basic axioms. This procedure can be used irrespective of the particular random variable defined on Ω. If all individual descriptions are considered equiprobable,

$$\mathbb{P}(\{\boldsymbol{\omega}\}) = \frac{1}{\#\Omega} = \frac{1}{36},$$

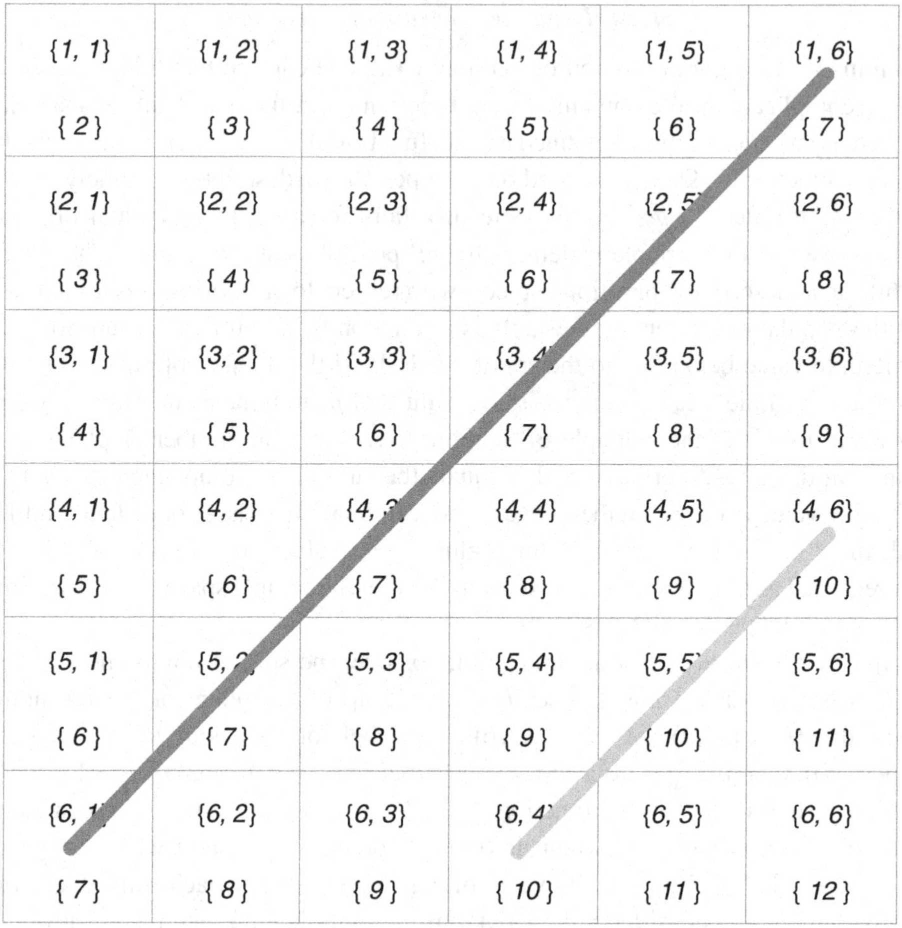

{1, 1} { 2 }	{1, 2} { 3 }	{1, 3} { 4 }	{1, 4} { 5 }	{1, 5} { 6 }	{1, 6} { 7 }
{2, 1} { 3 }	{2, 2} { 4 }	{2, 3} { 5 }	{2, 4} { 6 }	{2, 5} { 7 }	{2, 6} { 8 }
{3, 1} { 4 }	{3, 2} { 5 }	{3, 3} { 6 }	{3, 4} { 7 }	{3, 5} { 8 }	{3, 6} { 9 }
{4, 1} { 5 }	{4, 2} { 6 }	{4, 3} { 7 }	{4, 4} { 8 }	{4, 5} { 9 }	{4, 6} { 10 }
{5, 1} { 6 }	{5, 2} { 7 }	{5, 3} { 8 }	{5, 4} { 9 }	{5, 5} { 10 }	{5, 6} { 11 }
{6, 1} { 7 }	{6, 2} { 8 }	{6, 3} { 9 }	{6, 4} { 10 }	{6, 5} { 11 }	{6, 6} { 12 }

Figure 4.1. The sum of the results of two dice is constant on any counterdiagonal of the square. The results producing the values 7 and 10 are indicated by the tinted bars.

one can calculate, for any value $v_i \in V$, the probability $\mathbb{P}_S(v_i) = \mathbb{P}(S = v_i) = \mathbb{P}(\{\omega : S(\omega) = v_i\}$. This leads to

$$\mathbb{P}_S(i) = \begin{cases} (i-1)/36, \; i = 2, \ldots, 7 \\ (13-i)/36, \; i = 8, \ldots, 12 \end{cases},$$

that is a discrete triangular distribution. As $\{S = 2\}, \{S = 3\}, \ldots, \{S = 12\}$ is a partition of V, and, correspondingly $C_2, C_3, \ldots, C_{12}$ a partition of Ω, it follows that, setting $p_i = \mathbb{P}_S(i)$, one has $\sum_{i=2}^{12} p_i = 1$. It follows also that, given $\mathbb{P}_S(\cdot)$, one can calculate the probability of any possible set of values. For instance the probability that $S > 7$ and S is even is given by $\mathbb{P}(S \in \{8, 10, 12\}) = p_8 + p_{10} + p_{12} = 1/13$.

4.1.5 The mathematical expectation

For a numerical random variable, X, the mathematical expectation is defined as

$$\mathbb{E}(X) = \sum_{\omega\in\Omega} X(\omega)\mathbb{P}(\omega). \tag{4.6}$$

As $\mathbb{P}_X(v_i) = \mathbb{P}(\{\omega : X(\omega) = v_i\} = \sum_{\omega\in C_i} \mathbb{P}(\omega)$, and $X(\omega) = v_i$, for $\omega \in C_i$, the terms in the sum can be grouped

$$\mathbb{E}(X) = \sum_{\omega\in\Omega} X(\omega)\mathbb{P}(\omega) = \sum_{i=1}^{g}\sum_{\omega\in C_i} X(\omega)\mathbb{P}(\omega) = \sum_{i=1}^{g} v_i\mathbb{P}_X(v_i). \tag{4.7}$$

Note that in (4.6) all the contributions $X(\omega)$ coming from any point ω weighted by their probability $\mathbb{P}(\omega)$ are added, whereas in (4.7) all the values v_1, weighted by their probability $\mathbb{P}_X(v_i)$ are added. In general, in order to have a finite expectation, it is necessary to assume that

$$\sum_{\omega\in\Omega} |X(\omega)|\mathbb{P}(\omega) < \infty,$$

but, in the finite case, the expectation (4.6) always exists (it is the sum of a finite number of finite terms). Note that $\mathbb{E}(X)$ is sometimes written as μ_X, or even μ if there is no doubt about the random variable to which it refers. The mathematical expectation is analogous to the concept of centre of mass for a set of points on a line, where v_i is the coordinate of the ith particle and $p_i = \mathbb{P}_X(v_i)$ is its fraction of the total mass. If a value dominates, that is $p_j \simeq 1$, the expectation is close to v_j, while if all values have the same probability, the expectation coincides with the arithmetic mean. Several useful variables can be defined starting from the mathematical expectation.

Now, consider an ordinary function $g : \mathbb{R} \to \mathbb{R}$ of a real variable. As an example, consider the following case: $Y = X^2$. If the values of X are the five integers $\{-2,-1,0,1,2\}$, the values of Y will be the three integers $\{0,1,4\}$. The inverse image of $Y = 0$ is $X = 0$, but the inverse image of $Y = 1$ contains the two events $X = -1$ and $X = 1$, and the inverse image of $Y = 4$ contains $X = -2$ and $X = 2$, so that $\mathbb{P}_Y(0) = \mathbb{P}_X(0)$, $\mathbb{P}_Y(1) = \mathbb{P}_X(-1) + \mathbb{P}_X(1)$, and $\mathbb{P}_Y(4) = \mathbb{P}_X(-2) + \mathbb{P}_X(2)$. In other words, $Y = g(X)$ is a new random variable. The following equality defines Y as a random variable $Y : \Omega \to \mathbb{R}$

$$Y(\omega) = g(X(\omega)) \;\; \omega \in \Omega. \tag{4.8}$$

Given a finite random variable X assuming g_X values, the values of $Y = g(X)$ will be $g_Y \leq g_X$. If V_Y is the space of possible values $v_{Y,j}$ for Y, one can derive the probability distribution $\mathbb{P}_Y(\cdot)$ defined on the measurable space $\langle V_Y, \mathcal{P}(V_Y)\rangle$ from a knowledge of the distribution $\mathbb{P}_X(\cdot)$ and of the inverse image with respect to g for any value $v_{Y,j}$

$$\mathbb{P}_Y(v_{Y,j}) = \sum_{v_{X,i} \in g^{-1}(v_{Y,j})} \mathbb{P}_X(v_{X,i}); \qquad (4.9)$$

then, from a knowledge of the probability of singletons, one can obtain the probability for any set belonging to $\mathcal{P}(V_Y)$, using Equation (4.2). Therefore, the triplet $\langle V_Y, \mathcal{P}(V_Y), \mathbb{P}_Y \rangle$ is a probability space.

From Equations (4.7) and (4.9), one has that

$$\mathbb{E}[g(X)] = \mathbb{E}(Y) = \sum_{v_{Y,j} \in V_Y} v_{Y,j} \mathbb{P}_Y(v_{Y,j}) = \sum_{v_{X,i} \in V_X} g(v_{X,i}) \mathbb{P}_X(v_{X,i}). \qquad (4.10)$$

The result presented in (4.10) is known as the *fundamental theorem of mathematical expectation*. The above discussion shows that there are no major problems in deriving it for the case of finite random variables, whereas in the general case one has to assume that

$$\sum_{v_{X,i} \in V_X} |g(v_{X,i})| \mathbb{P}_X(v_{X,i}) < \infty,$$

meaning that there is absolute convergence of the sum (integral). Returning to the simple example presented above, assume that each of the 5 values of X has the same probability $\mathbb{P}_X(-2) = \mathbb{P}_X(-1) = \mathbb{P}_X(0) = \mathbb{P}_X(1) = \mathbb{P}_X(2) = 1/5$; this means that $\mathbb{P}_Y(0) = 1/5$ and $\mathbb{P}_Y(1) = \mathbb{P}_Y(4) = 2/5$.

4.1.6 The variance

As an important application of the fundamental theorem of mathematical expectation, one can define the variance of a numerical random variable, X, as the mathematical expectation of the squared deviations from $\mathbb{E}(X)$

$$\mathbb{V}(X) = \mathbb{E}[(X - \mathbb{E}(X))^2] = \sum_{\omega \in \Omega} (X(\omega) - \mathbb{E}(X))^2 \mathbb{P}(\omega). \qquad (4.11)$$

It turns out that

$$\mathbb{V}(X) = \mathbb{E}(X^2) - [\mathbb{E}(X)]^2, \qquad (4.12)$$

and this equality is discussed in Exercise 4.3. Note that $\mathbb{V}(X)$ is non-negative, and it is zero only for degenerate (deterministic) distributions. Statisticians usually denote $\mathbb{V}(X)$ by σ_X^2. The concept of variance is similar to the mechanical concept of moment of inertia with respect to the centre of mass. The square root of the variance is the standard deviation, σ_X, a measure of the dispersion for the probability mass around the mathematical expectation.

4.1.7 Covariance and correlation

Given two random variables X and Y defined on the same probability space, their covariance is defined as

$$\mathbb{C}(X,Y) = \mathbb{E}[(X - \mathbb{E}(X))(Y - \mathbb{E}(Y))]. \tag{4.13}$$

When working with more than one random variable, it is useful to introduce the concept of joint distribution. Now, the constituents of the algebra are the events $C_{i,j} = \{\omega : X(\omega) = v_{X,i}, Y(\omega) = v_{Y,j}\}$. The range of the joint random variable is the cartesian product $V_X \times V_Y$. The *joint distribution* of the random variables X and Y is

$$\begin{aligned}\mathbb{P}_{X,Y}(v_{X,i}, v_{Y,j}) &= \mathbb{P}(X^{-1}(v_{X,i}) \cap Y^{-1}(v_{Y,j})) \\ &= \mathbb{P}(X = v_{X,i}, Y = v_{Y,j}) = \mathbb{P}(\{\omega : X(\omega) = v_{X,i}, Y(\omega) = v_{Y,j}\}),\end{aligned} \tag{4.14}$$

and on subsets $A_X \times A_Y$ of $V_X \times V_Y$, as a consequence of (4.2), one immediately gets

$$\mathbb{P}_{X,Y}(A_X \times A_Y) = \sum_{v_{X,i} \in A_X} \sum_{v_{Y,j} \in A_Y} \mathbb{P}_{X,Y}(v_{X,i}, v_{Y,j}). \tag{4.15}$$

Again, the singletons $\{(v_{X,i}, v_{Y,j})\}$ are a partition of $V_X \times V_Y$, therefore

$$\sum_{v_{X,i} \in V_X} \sum_{v_{Y,j} \in V_Y} \mathbb{P}_{X,Y}(v_{X,i}, v_{Y,j}) = 1.$$

From the theorem of total probability, it is true that

$$\mathbb{P}_X(v_{X,i}) = \sum_{v_{Y,j} \in V_Y} \mathbb{P}_{X,Y}(v_{X,i}, v_{Y,j}) \tag{4.16}$$

and

$$\mathbb{P}_Y(v_{Y,j}) = \sum_{v_{X,i} \in V_X} \mathbb{P}_{X,Y}(v_{X,i}, v_{Y,j}) \tag{4.17}$$

are probability distributions, as well. They are called *marginal distributions* of X and Y, respectively.

This construction can be extended to any number n of random variables defined on the same probability space, and it will be useful in defining stochastic processes.

It is possible to use Equation (4.14) in computing (4.13). Using the properties of expectation (see Exercise 4.2), one gets

$$\mathbb{C}(X,Y) = \mathbb{E}[(X - \mathbb{E}(X))(Y - \mathbb{E}(Y))] = \mathbb{E}(XY) - \mathbb{E}(X)\mathbb{E}(Y). \tag{4.18}$$

The mathematical expectation of the product XY can be computed by means of $\mathbb{P}_{X,Y}(\cdot,\cdot)$ as

$$\mathbb{E}(XY) = \sum_{v_{X,i} \in V_X} \sum_{v_{Y,j} \in V_Y} v_{X,i} v_{Y,j} \mathbb{P}_{X,Y}(v_{X,i}, v_{Y,j}).$$

A useful quantity related to covariance is the Bravais–Pearson correlation coefficient $\rho_{X,Y}$. It is defined as

$$\rho_{X,Y} = \frac{\mathbb{C}(X,Y)}{\sigma_X \sigma_Y}. \tag{4.19}$$

It is possible to prove that $-1 \leq \rho_{X,Y} \leq 1$ and this proof is left to the reader as a useful exercise. If $\rho_{X,Y} = -1$, the random variables X and Y are *perfectly anti-correlated*; if $\rho_{X,Y} = 1$, they are *perfectly correlated*. The two random variables are called *uncorrelated* if $\rho_{X,Y} = 0$, meaning that $\mathbb{C}(X,Y) = 0$ and that $\mathbb{E}(XY) = \mathbb{E}(X)\mathbb{E}(Y)$.

It is important to remark that, even if $\rho_{X,Y} \neq 0$, this does not mean that X is a cause of Y or vice versa. In other words the presence of correlation between two random variables does not imply a causal relationship between the variables. There might be a common cause, or the correlation might be there by mere chance. This problem often arises in statistical analyses in the social sciences when the chains of causation among possibly relevant variables are far from clear; this has already been discussed at length by John Stuart Mill in his 1836 methodological essay *On the Definition of Political Economy and the Method of Investigation Proper to It* [1].

Two random variables X and Y defined on the same probability space are *independent* if all the events $X^{-1}(A_X)$ and $Y^{-1}(A_Y)$ are independent. In particular, this means that

$$\begin{aligned} \mathbb{P}_{X,Y}(v_{X,i}, v_{Y,j}) &= \mathbb{P}(X^{-1}(v_{X,i}) \cap Y^{-1}(v_{Y,j})) \\ &= \mathbb{P}(X^{-1}(v_{X,i}))\mathbb{P}(Y^{-1}(v_{Y,j})) = \mathbb{P}_X(v_{X,i})\mathbb{P}_Y(v_{Y,j}). \end{aligned} \tag{4.20}$$

In other words, if X and Y are independent the joint probability distribution has a nice factorization in terms of the product of two marginal distributions. Therefore, for independent random variables it turns out that $\mathbb{E}(XY) = \mathbb{E}(X)\mathbb{E}(Y)$. Independent random variables are *uncorrelated* and their covariance $\mathbb{C}(X,Y)$ vanishes. The converse is not true: not all uncorrelated random variables are independent!

4.1.8 Example: indicator functions

The simplest random variable is the so-called event indicator function. It transforms categorical (qualitative) random variables into numerical random variables. Let A be an event $A \in \mathcal{F}$ and ω an element of Ω. The indicator function $\mathbb{I}_A(\omega)$ is

$$\mathbb{I}_A(\omega) = \begin{cases} 1 & \text{if } \omega \in A \\ 0 & \text{otherwise.} \end{cases} \tag{4.21}$$

The indicator function has the following properties; let A and B be two events, then one has:

1. (Indicator function of intersection)

$$\mathbb{I}_{A \cap B}(\omega) = \mathbb{I}_A(\omega)\mathbb{I}_B(\omega); \tag{4.22}$$

2. (Indicator function of union)

$$\mathbb{I}_{A \cup B}(\omega) = \mathbb{I}_A(\omega) + \mathbb{I}_B(\omega) - \mathbb{I}_A(\omega)\mathbb{I}_B(\omega); \tag{4.23}$$

3. (Indicator function of complement)

$$\mathbb{I}_{A^c}(\omega) = 1 - \mathbb{I}_A(\omega); \tag{4.24}$$

4. (Expectation)

$$\mathbb{E}[\mathbb{I}_A(\omega)] = \mathbb{P}(A); \tag{4.25}$$

5. (Variance)

$$\mathbb{V}[\mathbb{I}_A(\omega)] = \mathbb{P}(A)(1 - \mathbb{P}(A)). \tag{4.26}$$

These properties will be proved in Exercise 4.4.

4.1.9 Example: probability of the union of *n* sets

An immediate consequence of properties (4.22), (4.23) and (4.24) of the indicator function is the theorem giving the probability of the union of two sets. It suffices to take the expectation of (4.23), use (4.24) and (4.22), to get

$$\mathbb{P}(A \cup B) = \mathbb{P}(A) + \mathbb{P}(B) - \mathbb{P}(A \cap B);$$

this proof is much easier and handier than the one presented in Section 3.6.6. The extension to three sets is based on the associative properties of unions and intersections. First of all, one has

$$\begin{aligned}\mathbb{I}_{A\cup B\cup C} &= \mathbb{I}_{(A\cup B)\cup C} = \mathbb{I}_{A\cup B} + \mathbb{I}_C - \mathbb{I}_{A\cup B}\mathbb{I}_C \\ &= \mathbb{I}_A + \mathbb{I}_B - \mathbb{I}_A\mathbb{I}_B + \mathbb{I}_C - (\mathbb{I}_A + \mathbb{I}_B - \mathbb{I}_A\mathbb{I}_B)\mathbb{I}_C \\ &= \mathbb{I}_A + \mathbb{I}_B + \mathbb{I}_C - \mathbb{I}_A\mathbb{I}_B - \mathbb{I}_A\mathbb{I}_C - \mathbb{I}_B\mathbb{I}_C + \mathbb{I}_A\mathbb{I}_B\mathbb{I}_C,\end{aligned}$$

leading to (after taking the expectation)

$$\begin{aligned}\mathbb{P}(A\cup B\cup C) = {} & \mathbb{P}(A) + \mathbb{P}(B) + \mathbb{P}(C) \\ & - \mathbb{P}(A\cap B) - \mathbb{P}(A\cap C) - \mathbb{P}(B\cap C) + \mathbb{P}(A\cap B\cap C). \qquad (4.27)\end{aligned}$$

Equation (4.27) generalizes to

$$\begin{aligned}\mathbb{P}(\cup_{i=1}^n A_i) = {} & \sum_{i=1}^n \mathbb{P}(A_i) - \sum_{i\neq j} \mathbb{P}(A_i\cap A_j) \\ & + \sum_{\text{distinct } i,j,k} \mathbb{P}(A_i\cap A_j\cap A_k) - \ldots \pm \mathbb{P}(\cap_{i=1}^n A_i). \qquad (4.28)\end{aligned}$$

This result is also known as the *inclusion–exclusion* formula.

4.1.10 Example: the matching problem

At a party there are n married couples and men and women are randomly paired for a dance. A matching occurs if a married couple is paired by chance. What is the probability of one matching? And what about two? This is an old problem in probability theory and it is usually referred to as the *Montmort's matching problem* as it was first considered by the French mathematician Pierre-Rémond de Montmort active in the early days of probability theory. His book *Essay d'analyse sur les jeux d'hazard* appeared in 1708 [2].

To fix the ideas, let women be labelled from 1 to n, so that $V = \{1,\ldots,n\}$ is the set of labels. Then men are chosen at random and each man is represented by a random variable X_i where the subscript i refers to the fact that he is married to woman i. Then $X_i = j$ means that the man married to the ith woman will dance with the jth woman. One has a match when $X_i = i$. For instance, if $n = 3$ couples, one has $V = \{1,2,3\}$ and there are $n! = 3! = 6$ possible couplings (constituents). They are

1. $(X_1 = 1, X_2 = 2, X_3 = 3)$ corresponding to 3 matchings;
2. $(X_1 = 1, X_2 = 3, X_3 = 2)$, $(X_1 = 3, X_2 = 2, X_3 = 1)$ and $(X_1 = 2, X_2 = 1, X_3 = 3)$ corresponding to 1 matching;
3. $(X_1 = 2, X_2 = 3, X_3 = 1)$ and $(X_1 = 3, X_2 = 1, X_3 = 2)$ corresponding to no matchings.

If the random variable N_n is introduced, representing the number of matchings over any permutation, the case $n = 3$ is described by the distribution $\mathbb{P}(N_3 = 0) = 2/6 = 1/3$, $\mathbb{P}(N_3 = 1) = 3/6 = 1/2$, $\mathbb{P}(N_3 = 2) = 0$ and $\mathbb{P}(N_3 = 3) = 1/6$, where the following equation is used to determine the matching probability

$$\mathbb{P}(N_n = k) = \frac{b_n(k)}{n!}, \tag{4.29}$$

and where, for ease of notation, the following position is established

$$b_n(k) = \#\{N_n = k\}. \tag{4.30}$$

The total number of couples is $n!$, a rapidly growing function of n. Therefore, the direct enumeration method soon becomes impractical. What can be said in the general case? One can immediately observe that $\mathbb{P}(N_n = n-1)$ must be 0; in fact, if there are $n-1$ matchings the remaining couple must match as well. This fact is sometimes referred to as the *pigeonhole principle*. If attention is fixed on the event $X_1 = 1$, meaning that the first couple is matched and one is not interested in what happens to the other couples, one has that

$$\mathbb{P}(X_1 = 1) = \frac{(n-1)!}{n!} = \frac{1}{n}; \tag{4.31}$$

indeed, if the first husband is paired with his wife, there are still $n-1$ available places that can be filled in $(n-1)!$ ways. Note that this probability is the same for any of the n events $\{X_i = i\}$. Similarly, imposing $X_1 = 1$ and $X_2 = 2$ leads to

$$\mathbb{P}(X_1 = 1, X_2 = 2) = \frac{(n-2)!}{n!} = \frac{1}{n}\frac{1}{n-1}; \tag{4.32}$$

note that $\mathbb{P}(X_1 = 1, X_2 = 2) > \mathbb{P}(X_1 = 1)\mathbb{P}(X_2 = 2)$ meaning that the two events are not independent and positively correlated. Again this probability is the same for any event of the kind $\{X_i = i, X_j = j\}$ with $i \neq j$. The total number of these events is given by the number of ways two objects can be chosen out of n, that is $n!/(2!(n-2)!)$. In general, for $k < n-1$, one can write

$$\mathbb{P}(X_1 = 1, X_2 = 2, \ldots, X_k = k) = \frac{(n-k)!}{n!} = \frac{1}{n}\frac{1}{n-1}\cdots\frac{1}{n-k+1}; \tag{4.33}$$

The number of events of this kind is given by $n!(k!(n-k)!)$ as one has to choose k places out of n. Due to the pigeonhole principle, the event $\{X_1 = 1, \ldots, X_{n-1} = n-1\}$ coincides with $\{X_1 = 1, \ldots, X_n = n\}$ which is only one case over $n!$ and one finds that

$$\mathbb{P}(X_1 = 1, \ldots, X_n = n) = \frac{1}{n!}. \tag{4.34}$$

The event $\{N_n > 0\}$ means that there is at least one matching. It is the union of events of the kind $\{X_i = i\}$. In symbols

$$\{N_n > 0\} = \cup_{i=1}^{n} \{X_i = i\}. \tag{4.35}$$

Using the inclusion–exclusion formula (4.28) and the fact that the probabilities of the intersections with the same number of sets coincide, one gets

$$\begin{aligned}\mathbb{P}(N_n > 0) &= \mathbb{P}\left(\cup_{i=1}^{n} \{X_i = i\}\right) \\ &= n\mathbb{P}(X_1 = 1) - \binom{n}{2}\mathbb{P}(X_1 = 1, X_2 = 2) \pm \ldots \\ &+ (-1)^{n-1}\mathbb{P}(X_1 = 1, \ldots, X_n = n) \\ &= 1 - \frac{1}{2!} \pm \ldots (-1)^{n-1}\frac{1}{n!},\end{aligned} \tag{4.36}$$

and the probability of *derangements* (no matchings) is

$$\mathbb{P}(N_n = 0) = 1 - \mathbb{P}(N_n > 0) = \sum_{i=0}^{n} \frac{(-1)^i}{i!}. \tag{4.37}$$

Note that

$$\lim_{n\to\infty} \mathbb{P}(N_n > 0) = 1 - \mathrm{e}^{-1}, \tag{4.38}$$

and

$$\lim_{n\to\infty} \mathbb{P}(N_n = 0) = \mathrm{e}^{-1}, \tag{4.39}$$

a somewhat paradoxical result meaning that when n diverges, the probability of at least one matching does not vanish but remains positive and quite large (more than 63%). Equation (4.37) can be used to find the probability of the generic event $\{N_n = k\}$. From Equations (4.29), (4.30) and (4.37) one gets

$$b_n(0) = n! \sum_{i=0}^{n} \frac{(-1)^i}{i!}. \tag{4.40}$$

If there are k matchings, there are $b_{n-k}(0)$ possible derangements in the remaining $n-k$ places. The positions of the k matchings can be chosen in $n!/(k!(n-k)!)$ ways. Therefore, the number of ways of obtaining $0 < k \leq n$ matchings is

$$b_n(k) = \frac{n!}{k!(n-k)!} b_{n-k}(0) = \frac{n!}{k!(n-k)!}(n-k)! \sum_{i=0}^{n-k} \frac{(-1)^i}{i!} = \frac{n!}{k!} \sum_{i=0}^{n-k} \frac{(-1)^i}{i!}, \tag{4.41}$$

and, from (4.29), the probability distribution of N_n is

$$\mathbb{P}(N_n = k) = \frac{b_n(k)}{n!} = \frac{1}{k!}\sum_{i=0}^{n-k}\frac{(-1)^i}{i!}. \tag{4.42}$$

Note that Equation (4.42) correctly implies $\mathbb{P}(N_n = n-1) = 0$ in agreement with the previous discussion.

4.1.11 Categorical and numerical random variables

It is now due time to reconsider what was established in the previous chapters in the light of the abstract notion of a random variable. Returning to our sample space $\Omega = \{\boldsymbol{\omega} = (\omega_1, \ldots, \omega_n)\}$, where n elements are classified into g categories, for each $i = 1, \ldots, n$, it is possible to define a function $X_i : \Omega \to V = \{1, \ldots, g\}$ so that $X_i(\boldsymbol{\omega}) = \omega_i$. If the ith element of $\boldsymbol{\omega}$ belongs to the jth category, one has $X_i(\boldsymbol{\omega}) = j$. Note that X_i depends only on the state of the ith individual. It follows that, for each description belonging to the event $A_j^{(i)}$, $X_i = j$ is observed. In this way, the results of Chapter 3, developed there in terms of events, can be translated into the more elegant language of random variables. Usually, $X_i = j$ is considered as an event (without writing $\{X_i = j\}$), and the intersection is represented by a comma, so that $\mathbb{P}(X_1 = i, X_2 = j)$ denotes the probability of $\{X_1 = i\} \cap \{X_2 = j\}$.

The set function X_i decomposes Ω into g sets $\{A_1^{(i)}, \ldots, A_g^{(i)}\}$, defined as $A_j^{(i)} = \{\boldsymbol{\omega} : X_i(\boldsymbol{\omega}) = j\}$, which is the inverse image of the set $\{j\} \subset V = \{1, \ldots, g\}$. Therefore the set function X_i is a random variable if, for every singleton belonging to $\{1, \ldots, g\}$, one knows $\mathbb{P}(A_j^{(i)}) = \mathbb{P}(\{\boldsymbol{\omega} : X_i(\omega) = j\})$, and sets $\mathbb{P}_{X_i}(\{j\}) = \mathbb{P}(X_i = j) = \mathbb{P}(A_j^{(i)})$. For any subset B of V, the usual construction leads to $\mathbb{P}_{X_i}(B) = \sum_{j \in B} \mathbb{P}_{X_i}(\{j\}) = \sum_{j \in B} \mathbb{P}(\{\boldsymbol{\omega} : X_i(\boldsymbol{\omega}) = j\})$.

The singleton $\{\boldsymbol{\omega}\} = A_{\omega_1}^{(1)} \cap A_{\omega_2}^{(2)} \cap \ldots \cap A_{\omega_n}^{(n)}$ can be more easily referred to as $X_1 = \omega_1, X_2 = \omega_2, \ldots, X_n = \omega_n$, taking into account that the set $\{\boldsymbol{\omega} : X_1(\boldsymbol{\omega}) = \omega_1, X_2(\boldsymbol{\omega}) = \omega_2, \ldots, X_n(\boldsymbol{\omega}) = \omega_n\}$ is nothing other than the singleton defined by the intersection $A_{\omega_1}^{(1)} \cap A_{\omega_2}^{(2)} \cap \ldots \cap A_{\omega_n}^{(n)}$. Therefore, the inverse image of the sequence of all individual random variables is the most elementary event one can consider. Given this one-to-one correspondence, one can forget about the Ω-space and work directly on sequences of individual random variables. Then, one writes X_1 instead of $X_1(\boldsymbol{\omega})$, always keeping in mind that the meaning of $\mathbb{P}(X_i = j)$ is $\mathbb{P}(\{\boldsymbol{\omega} : X_i(\boldsymbol{\omega}) = j\})$. In other words, a knowledge of $\mathbb{P}(X_i = j)$ is equivalent to a knowledge of the probability of the state of affairs where $X_i(\boldsymbol{\omega}) = j$.

If labels denote categories, the only advantage of the previous discussion is simplicity. Indeed, for individual descriptions, the values of X_i are nothing other than the components of the elementary descriptions. However, using indicator functions for individual random variables $\mathbb{I}_{X_i=j}(\omega)$, one gets a formal definition for the new random variable *frequency of the ith category*

$$N_i(\omega) = \sum_{j=1}^{n} \mathbb{I}_{X_j=i}(\omega). \tag{4.43}$$

4.1.12 Discrete and continuous random variables

In the following, in the case of large systems, the limiting distributions will be approximated by countable discrete or continuous distributions. In the case of discrete random variables there is no problem in probabilizing all the subsets of a countable set, and many formulae presented above can be straightforwardly extended by replacing n with ∞ and assuming that the corresponding series converge. However, as discussed at the end of Chapter 3, one must replace finite additivity with countable additivity, so that events such as $\cup_{i=1}^{\infty} A_i$ exist and this transforms a Boolean lattice into a σ-algebra. As mentioned above, in the case of discrete random variables nothing else changes, as in terms of some probability distribution $\mathbb{P}(k) \geq 0, \sum_{k=1}^{\infty} \mathbb{P}(k) = 1$ all the subsets of $\mathbb{N}$ have a definite probability satisfying the axioms. In this book, countable random variables will be used as a convenient tool to describe finite, even if large, systems.

There are more difficulties in dealing with continuous random variables. It is possible to show that not all the subsets of $\mathbb{R}$ are measurable and one is obliged to consider classes of measurable subsets such as the so-called Borel σ-algebra, which is the smallest σ-algebra containing all subsets of $\mathbb{R}$ of the kind $(-\infty, a]$. Therefore, in the triplet $\langle \mathbb{R}, \mathcal{F}, \mathbb{P}_X \rangle$, defining a probability space on the image of a real random variable X, $\mathcal{F}$ usually coincides with the Borel σ-algebra $\mathcal{B}$. The cumulative distribution $F_X(x)$ becomes a (piecewise) continuous function and in the cases in which its first derivative exists, one can define the so-called probability density function

$$p_X(x) = \frac{\mathrm{d}F_X}{\mathrm{d}x}, \tag{4.44}$$

and it turns out that

$$F_X(x) = \sum_{u \leq x} p_X(u) = \int_{-\infty}^{u} p_X(u)\,\mathrm{d}u, \tag{4.45}$$

$$\sum_{x \in \mathbb{R}} p_X(x) = \int_{-\infty}^{+\infty} p_X(x)\,\mathrm{d}x = 1, \tag{4.46}$$

where the two notations (sum and integral) will be used in the following as convenient. Incidentally, the expectation for a continuous random variable with density can be written as

$$\mathbb{E}(X) = \sum_{x \in \mathbb{R}} x p_X(x) = \int_{-\infty}^{+\infty} x p_X(x)\,\mathrm{d}x < \infty. \quad (4.47)$$

4.2 Finite stochastic processes

As discussed in Section 4.1.11, an elementary state of affairs is described by $\{X_i : i = 1, \dots, n\}$, a finite sequence of individual random variables whose range (or state space) is discrete as $X_i \in \{1, \dots, g\}$. In general, the labels $\{1, \dots, g\}$ are interpreted as categories and expressions such as $X_1 + X_2$, or $3X_1$ are meaningless.

The sequence $\{X_i : i = 1, \dots, n\}$ is a simple instance of a *stochastic process*. In general a *stochastic process* is a family of random variables $\{X_i \in S, i \in I\}$ defined on the same probability space $\langle \Omega, \mathcal{F}, \mathbb{P} \rangle$, where S is called the *state space*, and I is a set of indices. The state space can be discrete (finite or countable) or continuous, and the set of indices can be discrete or continuous as well. When the value of the index i is fixed, X_i is a random variable, and $\mathbb{P}(X_i \in A)$ means the probability that the ith value of X belongs to some subset of S. When a point of the sample space is fixed, the stochastic process becomes an ordinary function of the index set I.

In the case relevant to the present book, the state space is the range of the individual random variables, i.e. the labels $\{1, \dots, g\}$, and the set of indices are given by the integers from 1 to n: $I = \{1, \dots, n\}$.

The notation $\mathbf{X}^{(n)} = (X_1, X_2, \dots, X_n)$ usually denotes a random vector, interpreted as a sequence of random variables; a sequence of specific values assumed by these random variables is represented by $\mathbf{x}^{(n)} = (x_1, x_2, \dots, x_n)$; the probability of the event $\mathbf{X}^{(n)} = \mathbf{x}^{(n)}$, that is $(X_1 = x_1, \dots, X_n = x_n)$, is $\mathbb{P}(\mathbf{X}^{(n)} = \mathbf{x}^{(n)}) = p_{X_1,\dots,X_n}(x_1, \dots, x_n) = p_{\mathbf{X}^{(n)}}(\mathbf{x}^{(n)})$; the function $p_{X_1,\dots,X_n}(x_1, \dots, x_n)$ is called *n-point finite dimensional distribution*. The probability space of the events $\mathbf{X}^{(n)} = \mathbf{x}^{(n)}$ is finite, and the probability of all the constituents must be fixed, that is $p_{\mathbf{X}^{(n)}}(\mathbf{x}^{(n)}) : p_{\mathbf{X}^{(n)}}(\mathbf{x}^{(n)}) \geq 0, \sum_{\mathbf{x} \in \Omega^{(x)}} p_{\mathbf{X}^{(n)}}(\mathbf{x}^{(n)}) = 1$, where the symbol $\Omega^{(x)}$ is used for the set of all possible elementary outcomes for n joint individual observations. The cardinality of $\Omega^{(x)}$ is given by $\#\Omega^{(x)} = g^n$, as there is a one-to-one mapping between the set Ω of individual descriptions, and $\Omega^{(x)}$. As was discussed in the previous chapter, one can use either the absolute or the predictive method to assign the probabilities $p_{X_1,\dots,X_n}(x_1, \dots, x_n)$.

As the index set I is usually infinite, a full characterization of the stochastic process is possible if all the finite dimensional distributions are known for all $x_i \in S$ and for any integer $n > 0$. This is a consequence of Kolmogorov's extension theorem. It turns out that the knowledge of all the finite dimensional distributions is sufficient for the existence of a stochastic process, even if I is a continuous set of indices.

Coming back to the simple case in which $I = \{1,\dots,n\}$, and S is a finite set of labels $\{1,\dots,g\}$, the full characterization of the n-step processes is given by the knowledge of the m-point finite dimensional distributions with $1 \leq m \leq n$: $p_{X_1}(x_1)$, $p_{X_1,X_2}(x_1,x_2)$, ..., $p_{X_1,\dots,X_m}(x_1,\dots,x_m)$, ..., $p_{X_1,\dots,X_n}(x_1,\dots,x_n)$. These probability functions must satisfy Kolmogorov's *compatibility conditions*. The first condition is a consequence of the symmetry property of intersections in a Boolean algebra,

$$p_{X_1,\dots,X_m}(x_1,\dots,x_m) = p_{X_{i_1},\dots,X_{i_m}}(x_{i_1},\dots,x_{i_m}), \tag{4.48}$$

where $i_1,\dots,i_m$ is any of the $m!$ possible permutations of the indices. The second condition is a consequence of the finite additivity axiom **A3**,

$$p_{X_1,\dots,X_m}(x_1,\dots,x_m) = \sum_{x_{m+1}\in S} p_{X_1,\dots,X_m,X_{m+1}}(x_1,\dots,x_m,x_{m+1}). \tag{4.49}$$

Kolmogorov's compatibility conditions may seem trivial, but they are used as hypotheses in Kolmogorov's extension theorem quoted above. These conditions are automatically satisfied if the finite dimensional distributions are built using the predictive method. As a consequence of the multiplication theorem, one has

$$\begin{aligned} p_{X_1,\dots,X_m}(x_1,\dots,x_m) &= \mathbb{P}(X_1 = x_1,\dots,X_m = x_m) \\ &= \mathbb{P}(X_1 = x_1)\mathbb{P}(X_2 = x_2|X_1 = x_1)\dots \\ &\quad \mathbb{P}(X_m = x_m|X_1 = x_1,\dots,X_{m-1} = x_{m-1}). \end{aligned} \tag{4.50}$$

Hence, a stochastic process is often characterized in terms of the *predictive law* $\mathbb{P}(X_m = x_m|X_1 = x_1,\dots,X_{m-1} = x_{m-1})$ for any $m > 1$ and of the 'initial' distribution $\mathbb{P}(X_1 = x_1)$. The predictive probability has only to fulfil the probability axioms: $\mathbb{P}(X_m = x_m|\mathbf{X}^{(m-1)} = \mathbf{x}^{(m-1)}) \geq 0$, and $\sum_{x_m\in S}\mathbb{P}(X_m = x_m|\mathbf{X}^{(m-1)} = \mathbf{x}^{(m-1)}) = 1$. In other words, one has to fix the probability of the 'next' observation as a function of all the previous results. Predictive laws which are simple functions of x_m and $\mathbf{x}^{(m-1)}$ are particularly interesting.

4.2.1 Sampling stochastic processes

Stochastic processes defined by means of a predictive law have many applications in physics, genetics, economics and the natural sciences, as they naturally represent probabilistic mechanisms evolving in time. The simplest physical model is represented by coin (or die) tossing, or by the game of *roulette*; in all these cases the 'random' device produces sequences of values. The predictive probability takes into account the properties of the device and all the previous results, in order to assess the probability for the next trial. In the framework of stochastic processes, a sample is nothing other than the initial part of a sequence, and does not coincide with the intuitive notion of the part of a whole (the sampled population). On the other side, in statistics sampling is in a strict sense a partial observation of some population. Its physical model is drawing balls from an urn. In this second framework, the only property of the random device is just the set of balls in the urn: its composition with respect to the variates of interest. The probability of an outcome may depend on the previous results, but, first of all, it depends on the composition $\mathbf{n} = (n_1, \ldots, n_g)$ of the urn, where n_i is the number of objects belonging to the ith category. The simplest example is the draw without replacement from an urn of known composition; in this case, the finite dimensional distributions must satisfy the 'final constraint' that $\mathbf{X}^{(n)}$ has the frequency vector $\mathbf{n}$ describing the composition of the urn. This problem has already been discussed in Chapter 2 as a counting problem and in Chapter 3 as a probabilistic urn problem. In the following sections, the main results of Chapter 3 will be re-written in the language of random variables, analyzed in terms of the possible compositions of the population.

4.2.2 n-Exchangeable random processes

The hypergeometric process

The physical process of drawing balls without replacement at random and from an urn of given composition $\mathbf{n} = (n_1, \ldots, n_g)$ can be described by the following formal stochastic process. Note that the state space $\{1, \ldots, g\}$ denotes the g different colours present in the urn. Consider a sequence of individual random variables $X_1, \ldots, X_n$ whose range is the label set $\{1, \ldots, g\}$ and characterized by the following predictive probability:

$$\mathbb{P}_{\mathbf{n}}(X_{m+1} = j | X_1 = x_1, \ldots, X_m = x_m) = \mathbb{P}_{\mathbf{n}}(X_{m+1} = j | \mathbf{m}) = \frac{n_j - m_j}{n - m}. \quad (4.51)$$

Now, based on the rule (4.51), one can find the probability of the fundamental sequence $\mathbf{X}_f^{(m)} = (X_1 = 1, \ldots, X_{m_1} = 1, X_{m_1+1} = 2, \ldots X_{m_1+m_2} = 2, \ldots, X_{m-m_g+1} = g, \ldots, X_m = g)$ with $m \leq n$ balls drawn, of which the first m_1 is of colour 1, the following m_2 of colour 2, and so on, until the last set of m_g balls of colour g, with $\sum_{i=1}^{g} m_i = m$. By repeated applications of (4.51), one gets

$$\begin{aligned} &\mathbb{P}_{\mathbf{n}}(\mathbf{X}_f^{(m)}) \\ &\quad = \frac{n_1}{n} \frac{n_1 - 1}{n - 1} \cdots \frac{n_1 - m_1 - 1}{n - m_1 - 1} \frac{n_2}{n - m_1} \frac{n_2 - 1}{n - m_1 - 2} \cdots \frac{n_2 - m_2 - 1}{n_2 - m_1 - m_2 - 1} \\ &\quad \times \frac{n_g}{n - m + m_g} \frac{n_g - 1}{n - m + m_g - 1} \cdots \frac{n_g - m_g - 1}{n - m - 1}, \end{aligned} \quad (4.52)$$

a result that can be written in a more compact form as

$$\mathbb{P}_{\mathbf{n}}(\mathbf{X}_f^{(m)}) = \frac{(n-m)!}{n!}\prod_{i=1}^{g}\frac{n_i!}{(n_i-m_i)!}. \tag{4.53}$$

The nice point about the probability given by Equation (4.53) is that it is the same for any individual sequence with the sampling vector $\mathbf{m}$, and the number of these sequences is given by the multinomial factor (see the discussion around Equation (2.8) in Chapter 2); therefore one has

$$\mathbb{P}_{\mathbf{n}}(\mathbf{m}) = \frac{m!}{\prod_{i=1}^{g} m_i!}\mathbb{P}_{\mathbf{n}}(\mathbf{X}_f^{(m)}), \tag{4.54}$$

leading to the hypergeometric sampling distribution

$$\mathbb{P}_{\mathbf{n}}(\mathbf{m}) = \frac{m!}{\prod_{i=1}^{g} m_i!}\frac{(n-m)!}{n!}\prod_{i=1}^{g}\frac{n_i!}{(n_i-m_i)!} = \frac{\prod_{i=1}^{g}\binom{n_i}{m_i}}{\binom{n}{m}}. \tag{4.55}$$

The hypergeometric process discussed above is the simplest case of an *n-exchangeable process*. A general definition of this important class of stochastic processes will be given below.

n-Exchangeable processes

Consider again a sequence of individual random variables $X_1,\ldots,X_n$ whose range is the label set $\{1,\ldots,g\}$. As discussed above there are g^n possible sequences $X_1 = x_1,\ldots,X_n = x_n$, whereas the number of possible occupation vectors $\mathbf{n} = (n_1,\ldots,n_g)$ is given by

$$\binom{n+g-1}{n}. \tag{4.56}$$

Moreover, given a particular occupation vector $\mathbf{n}$, the number of corresponding individual sequences $\mathbf{X}^{(n)}$ is

$$\frac{n!}{n_1!\cdots n_g!}. \tag{4.57}$$

Thanks to the construction discussed above in this section, the sequence of random variables defines a stochastic process. If the finite dimensional probabilities are such that, for any integer $m = 2,\ldots,n$, one has

$$\mathbb{P}(X_1 = x_1,\ldots,X_m = x_m) = \mathbb{P}(X_{i_1} = x_1,\ldots,X_{i_m} = x_m), \tag{4.58}$$

for any of the $m!$ possible permutations of the indices, then the stochastic process is called *n-exchangeable*. Note that condition (4.58) differs from condition (4.48). The meaning of (4.58) is that all the individual sequences corresponding to the same occupation vector **m** have the same probability. Therefore, an immediate consequence of the definition (4.58) and of **A3** (finite additivity) is that

$$\mathbb{P}(\mathbf{X}^{(m)}) = \mathbb{P}(\mathbf{m}) \left(\frac{m!}{m_1! \cdots m_g!} \right)^{-1}, \tag{4.59}$$

where a shortcut notation has been used and **m** is the frequency vector of the particular random vector $\mathbf{X}^{(m)} = \mathbf{x}^{(m)}$. Equation (4.59) is compatible with any probability distribution on **m**, $\{\mathbb{P}(\mathbf{m}) : \sum \mathbb{P}(\mathbf{m}) = 1, \sum_{i=1}^{g} m_i = m\}$. It turns out that $(n+g-1)!/(n!(g-1)!) - 1$ parameters are needed to completely define the probability space. They represent the probability of all possible compositions of the population with respect to the g labels (categories). In the urn model, they are all the possible compositions of the urn with respect to g colours. Given that, for any subsequence, the finite dimensional probability $\mathbb{P}(X_{i_1} = x_1, X_{i_2} = x_2, \ldots, X_{i_m} = x_m)$, $i_j \leq n$ can be obtained by marginalization of $\mathbb{P}(X_1 = x_1, X_2 = x_2, \ldots, X_n = x_n)$ using the second compatibility condition (4.49), it follows that all the finite dimensional probabilities are known if $\mathbb{P}(\mathbf{n})$ is known. The resulting predictive law has some interesting properties:

1. the past evidence $\mathbf{X}^{(m)}$ can be replaced by its frequency vector **m**,

$$\mathbb{P}(X_{m+1} = i | \mathbf{X}^{(m)}) = \mathbb{P}(X_{m+1} = i | \mathbf{m}) = p(i, \mathbf{m}); \tag{4.60}$$

2. the event $X_{m+2} = i, X_{m+1} = j$ and the symmetrical one $X_{m+2} = j, X_{m+1} = i$ are equiprobable given any evidence:

$$\mathbb{P}(X_{m+2} = i, X_{m+1} = j | \mathbf{m}) = \mathbb{P}(X_{m+2} = j, X_{m+1} = i | \mathbf{m}). \tag{4.61}$$

Exchangeable processes as mixtures of hypergeometric processes

The hypergeometric process defined above is not extendible to $n+1$ steps. If the model urn contains n balls, the process stops at the nth step. If the individual vector $\mathbf{X}^{(n)}$ defines a sequence whose length is n steps, from Equation (4.53), one has

$$\mathbb{P}_{\mathbf{n}}(\mathbf{X}^{(n)}) = \frac{1}{\dfrac{n!}{\prod_{i=1}^{g} n_i!}}, \tag{4.62}$$

where the notation highlights the fact that all the sequences are equiprobable. Now assume that Equation (4.59) is valid also at the nth step, leading to

$$\mathbb{P}(\mathbf{X}^{(n)}) = \frac{\mathbb{P}(\mathbf{n})}{\dfrac{n!}{\prod_{i=1}^{g} n_i!}}. \tag{4.63}$$

Equation (4.63) represents a *mixture* of hypergeometric processes with *mixing probability* $\mathbb{P}(\mathbf{n})$; indeed, it can be re-written as

$$\mathbb{P}(\mathbf{X}^{(n)}) = \mathbb{P}(\mathbf{X}^{(n)}|\mathbf{n})\mathbb{P}(\mathbf{n}), \tag{4.64}$$

with

$$\mathbb{P}(\mathbf{X}^{(n)}|\mathbf{n}) = \mathbb{P}_{\mathbf{n}}(\mathbf{X}^{(n)}) = \frac{1}{\dfrac{n!}{\prod_{i=1}^{g} n_i!}}, \tag{4.65}$$

corresponding to the hypergeometric process. The mixing probability represents sampling without replacement from an urn of unknown composition with each composition having probability $\mathbb{P}(\mathbf{n})$. Note that a mixture of hypergeometric processes could be extendible to the next step and even to infinity, as will be clear from the examples discussed below. Finally, note that also the converse is true, that is any n-exchangeable stochastic process can be written as a mixture of hypergeometric processes. This is also a consequence of Equation (4.59). This result is known as the finite version of de Finetti's representation theorem for exchangeable processes and it will be further discussed in Section 4.3, where a proof in the tradition of the Italian neo-Bayesian school will be presented.

Examples of exchangeable processes

It is now time to consider some specific examples where $\mathbb{P}(\mathbf{n})$ is explicitly given.

1. (Hypergeometric process again) As already discussed, the hypergeometric process follows from (4.63) setting $\mathbb{P}(\mathbf{n}') = \delta(\mathbf{n}', \mathbf{n})$. The corresponding urn model is sampling without replacement from an urn whose initial composition is given by the vector $\mathbf{n}$. A possible accommodation model amounts to choosing free positions at the instant of accommodation with uniform probability, where $\mathbf{n} = (n_1, \ldots, n_g)$ is the distribution of the initial available positions. Brillouin gave a physical interpretation to such an accommodation model while trying to derive quantum statistics. He used a geometrical analogy and introduced the 'volume' of a cell and set this volume equal to the 'positive volume' of fermions, particles subject to Pauli's exclusion principle. In such an extreme case, where the volume of a cell and the volume of a particle coincide, no more than one particle

can enter a cell. In Brillouin's framework a category (energy level) is made up of several cells. If the ith energy level contains n_i cells and m_i particles, there is still room for no more than $n_i - m_i$ particles. All sequences with occupation vector different from $\mathbf{n}$ are impossible, whereas all the possible individual sequences are equiprobable. The predictive probability (conditioned on $\mathbf{n}$) in this case is:

$$\mathbb{P}_{\mathbf{n}}(X_{m+1} = i|\mathbf{X}^{(m)}) = p_{\mathbf{n}}(i, m_i, m) = \frac{n_i - m_i}{n - m}, \tag{4.66}$$

a reformulation of (3.20).

2. (Multinomial process) If $\mathbb{P}(\mathbf{n})$ is the multinomial distribution, the independent and identically distributed case follows. To make this clear it is sufficient to replace (3.8) in (4.59) leading to

$$\mathbb{P}(\mathbf{X}^{(n)}) = \prod_{i=1}^{g} p_i^{n_i}, \tag{4.67}$$

where n_i is the number of $X_j = i$, that is $n_i = \#\{X_j = i, j = 1, \ldots, n\}$. This coincides with

$$\mathbb{P}(\{\omega\}) = \prod_{i=1}^{g} p_i^{n_i}, \tag{4.68}$$

where p_i has the same meaning. The corresponding urn model is sampling with replacement from a Bernoullian urn, whereas the accommodation model is that of successive accommodations ruled by a probability which is the same for all variables, and which is independent of the values of all other variables. Indeed, in the i.i.d. case, the predictive probability is:

$$\mathbb{P}(X_{m+1} = i|\mathbf{X}^{(m)}) = p_i, \tag{4.69}$$

a reformulation of (3.19). One must choose $g - 1$ probability values, given that $\{p_i \geq 0, \sum_i^g p_i = 1\}$.

3. (Pólya process) The Pólya process follows from (4.59) setting

$$\mathbb{P}(\mathbf{n}) = \text{Polya}(\mathbf{n}; \boldsymbol{\alpha}) = \frac{n!}{\alpha^{[n]}} \prod_{i=1}^{g} \frac{\alpha_i^{[n]}}{n_i!}, \tag{4.70}$$

where $\boldsymbol{\alpha} = (\alpha_1, \ldots, \alpha_n)$ is a vector of parameters, $\alpha = \sum_{i=1}^{g} \alpha_i$ and $\alpha^{[n]}$ is the Pochhammer symbol representing the *rising (or upper) factorial* defined by

$$\alpha^{[n]} = \alpha(\alpha + 1) \cdots (\alpha + n - 1). \tag{4.71}$$

The urn model for the Pólya process is sampling with replacement and unitary prize from an urn whose initial composition is $\boldsymbol{\alpha} = (\alpha_1, \ldots, \alpha_g)$. In this case, the predictive probability is

$$\mathbb{P}_{\boldsymbol{\alpha}}(X_{m+1} = i|\mathbf{X}^{(m)}) = p_{\boldsymbol{\alpha}}(i, m_i, m) = \frac{\alpha_i + m_i}{\alpha + m}. \tag{4.72}$$

The corresponding accommodation model is choosing each category i with probability proportional with $\alpha_i + m_i$, α_i being the initial ith weight, and m_i the number of objects already accommodated in that category. Based on this property, Brillouin created the geometrical interpretation of 'negative volume' for bosons, as their presence increases the 'volume' of the cell they belong to. Observe that (4.72) is meaningful also for a vector $\boldsymbol{\alpha}$ of negative integer parameters, provided $m_i \leq |\alpha_i|$, in which case $|\alpha_i|$ represents the number of 'positions' available to the ith category, or the initial number of balls of colour i. In other words, one can set $n_i = |\alpha_i|$, $n = |\alpha|$ and (4.72) coincides with (3.20), in symbols $p_{\mathbf{n}}(\cdot) = p_{\boldsymbol{\alpha}=-\mathbf{n}}(\cdot)$. Further note that, in the limit $\alpha_j \to \infty$, $\alpha_j/\alpha = p_j$, formula (4.72) becomes (3.19). This means that, within the realm of exchangeable sequences, the predictive form (4.72) encompasses both independent and (both positively and negatively) correlated sequences, corresponding to simple urn models as well as to (both positively and negatively) correlated accommodation models. This family of sequences can be characterized by the invariance of the heterorelevance coefficient, a notion introduced by Keynes in his *Treatise on Probability* (see the references of Chapter 1). The notion of invariance for heterorelevance coefficients is due to Costantini. Given its importance, the Pólya process will be studied in greater detail in Chapter 5.

4.2.3 Example: bingo and conditioning

The game of bingo is based on the sequential draw without replacement of all 90 numbered balls from an urn. We denote by $X_j = i$ the event that the jth drawn number is i, $i, j = 1, \ldots, n$. The game is fair if all 90! sequences which are distinct permutations of $\{1, \ldots, 90\}$ are equiprobable. Therefore, the sequence of random variables $X_1, X_2, \ldots, X_{90}$ defines an exchangeable stochastic process. Do X_1 and X_{90} have the same distribution? From a naïve point of view the two cases look quite different, as X_1 can take 90 equiprobable values, whereas the possibilities for X_{90} are restricted to the only and last ball left in the urn. However, unconditional on the past, the range of X_{90} just coincides with the range of X_1 or of any other variable! It is true that, conditional on the past history, the range of X_{90} is just the last ball left, but if one arrives just before the last draw and does not know the past history, then $\mathbb{P}(X_{90} = i) = \mathbb{P}(X_1 = i) = 1/90$ (see also Exercise 3.4). This example is useful in order to better understand the notion of marginal distribution for a random vector. As a consequence of (4.51), the predictive probability for the process $X_1, X_2, \ldots, X_{90}$ is $\mathbb{P}(X_{m+1} = i|X_1 = x_1, \ldots, X_m = x_m) = (1 - m_i)/(90 - m)$, where $m_i = 0$ if $X_1 \neq i, \ldots, X_m \neq i$, and $m_i = 1$ otherwise.

In the absence of any evidence, one has $\mathbb{P}(X_{m+1} = i) = 1/90$. Note also that, for instance, $\mathbb{P}(X_{25} = 60|X_1 = 2, X_{90} = 33) = 1/88$, that is the evidence does not distinguish the past from the future (as already discussed in Section 3.6.6), and that not only $\mathbb{P}(X_{90} = 60|X_1 = 60) = 0$, but also $\mathbb{P}(X_1 = 60|X_{90} = 60) = 0$. Hence exchangeable processes have a memory independent from the distance between events, and encompassing both the past and the future.

A possible misunderstanding concerns considering conditioning as something involving only information or knowledge. Indeed, arriving *just before the last draw* and thus not knowing the previous evidence $X_1 = x_1, X_2 = x_2, \ldots, X_{89} = x_{89}$ has an objective, frequency counterpart in considering the last outcome for all possible sequences, whatever the initial part may be. If the game is fair, when examining a large number of last balls left in the game of bingo, one expects a uniform frequency distribution on $\{1,\ldots,90\}$. Namely, the expected distribution is $p(x_1, x_2, \ldots, x_{89}, x_{90}) = 1/90!$ for all the possible different sequences; as a consequence of the total probability theorem, $p(x_{90}) = \sum_{x_1} \cdots \sum_{x_{89}} p(x_1, \ldots, x_{89}, x_{90})$ can be seen as the expected value of the conditional probability $p(x_{90}|x_1, \ldots, x_{89})$ with respect to the finite dimensional distribution $p(x_1, \ldots, x_{89})$, representing the probability for all possible draws of the first 89 numbers. The same holds true for any x_k, or for any couple x_k, x_h and so on. For instance $\mathbb{P}(X_k = j) = \sum_{x_i, i \neq k} p(x_1, \ldots, x_k = j, \ldots, x_{90}) = \mathbb{P}(X_1 = j) = 1/90$. This is so because, for any $i \neq j$, $p(x_1 = i, \ldots, x_k = j, \ldots, x_{90}) = p(x_1 = j, \ldots, x_k = i, \ldots, x_{90})$.

4.2.4 Example: an urn with indefinite composition

As discussed above in the example of Section 4.2, in the case of exchangeable random processes, the validity of (4.59) allows a strong simplification, as it is sufficient to give the probability distribution on $\mathbf{n}$ to complete the probability space for the individual random process as well. This procedure has a common application in problems of statistical inference with uncertainty in the composition of a population. Such problems can be usefully described in terms of urn processes.

Consider an urn containing n balls, the only available information is that the balls can be of two colours labelled by $\{1,2\}$ and nothing else. A draw is made and it turns out that $X_1 = x_1 = 1$; the ball is not replaced. What is the probability distribution of the next draw X_2?

In this formulation, the problem is highly indefinite, as nothing is said about the probability structure on this space. However, this is the typical formulation within the so-called *ignorance-based* notion of probability that can be traced back to Laplace, who would apply the indifference principle to all the possible compositions of the urn, denoted by the vector $\mathbf{n} = (n_1, n_2)$, setting $\mathbb{P}(n, 0) = \mathbb{P}(n-1, 1) = \ldots = \mathbb{P}(0, n) = 1/(n+1)$, see Fig. 4.2. Here, this choice is not

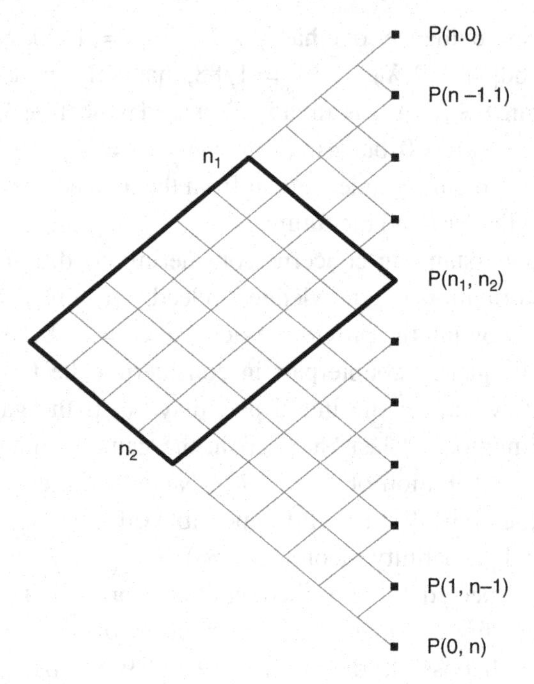

Figure 4.2. Probability distribution on the set of occupation numbers, and the set of sequences leading to a generic value of $(k,n-k)$.

discussed,[1] and the probability of X_2 is computed given the distribution $\mathbb{P}(n_1,n_2)$, with $n_1+n_2=n$, and given that $X_1=1$. Denote by $n_1=k$ the composition of the urn, by $X_2=x_2$ the (generic) second result. The sample space is the union of $n+1$ disjoint events $\cup_{k=0}^{n}\{k\}=\Omega$; in other words, the urn has one of the mutually exclusive $n+1$ compositions with certainty. Moreover, one has that $\{x_2\}=\{x_2\}\cap\Omega=\{x_2\}\cap(\cup_{k=0}^{n}\{k\})=\cup_{k=0}^{n}\{x_2\}\cap\{k\}$: a union of disjoint sets. Therefore, one gets $\mathbb{P}(X_2=x_2)=\sum_{k=0}^{n}\mathbb{P}(\{x_2\}\cap\{k\})$. If X_1 were not given, by total probability, one would have $\mathbb{P}(X_2)=\sum_{k=0}^{n}\mathbb{P}(X_2|k)\mathbb{P}(k)$, and $\mathbb{P}(X_2=1|k)=k/n$. However, the evidence $X_1=1$ is available. Hence in order to obtain $\mathbb{P}(X_2|X_1)=\mathbb{P}(X_2,X_1)/\mathbb{P}(X_1)$, one has to consider that $\mathbb{P}(X_2,X_1)=\sum_{k=0}^{n}\mathbb{P}(X_1,X_2|k)\mathbb{P}(k)$, and that $\mathbb{P}(X_1)=\sum_{k=0}^{n}\mathbb{P}(X_1|k)\mathbb{P}(k)$ leading to

$$\mathbb{P}(X_2|X_1)=\frac{\sum_{k=0}^{n}\mathbb{P}(X_1,X_2|k)\mathbb{P}(k)}{\sum_{k=0}^{n}\mathbb{P}(X_1|k)\mathbb{P}(k)}. \tag{4.73}$$

[1] Why, however, is one indifferent regarding the $n+1$ possible frequencies, but not the 2^n individual description of the balls?

Note that $\mathbb{P}(X_1,X_2|k)$ is the hypothetical probability of the sample, also called conditioned sampling distribution by statisticians, whereas $\mathbb{P}(k)$ denotes the probability of the hypothesis. Note also that the total probability theorem, which can be interpreted as a weighted mean of hypothetical probabilities, is used for both $\mathbb{P}(X_2,X_1)$ and $\mathbb{P}(X_1)$ separately. It would be a mistake to write $\mathbb{P}(X_2|X_1,k) = \mathbb{P}(X_2,X_1|k)/\mathbb{P}(X_1|k)$ and then to find the expected value of this ratio of conditional probabilities. On this point, see the exercise on hypothetical conditional probability below.

In order to determine $\mathbb{P}(X_1|k)$, and $\mathbb{P}(X_1,X_2|k)$, the type of sampling must be specified. Assume that $X_1,\ldots,X_n|k$ is a hypergeometric process, as the drawn balls are not replaced. The probabilities of sequences conditioned on the composition of the urn can be computed. Everybody agrees with the choice $\mathbb{P}(X_1 = 1|k) = k/n$ if there are k balls of colour 1 over n. The joint conditional probability distribution $\mathbb{P}(X_1,X_2|k)$ is given by $\mathbb{P}(X_1,X_2|k) = \mathbb{P}(X_2|X_1,k)\mathbb{P}(X_1|k)$, and $\mathbb{P}(X_2 = 1|X_1 = 1,k) = (k-1)/(n-1)$. This intuitive result is in agreement with the predictive probability (4.66). For the absolute method see Exercise 3.4. Hence $\mathbb{P}(X_1 = 1) = \sum_{k=0}^{n}\mathbb{P}(X_1 = 1|k)\mathbb{P}(k) = (n+1)^{-1}\sum_{k=0}^{n}\mathbb{P}(X_1 = 1|k)$, and $\sum_{k=0}^{n}\mathbb{P}(X_1 = 1|k) = \sum_{k=0}^{n} k/n = (n+1)/2$. In the end, one finds that $\mathbb{P}(X_1 = 1) = 1/2$. As for $\mathbb{P}(X_1 = 1, X_2 = 1)$, one has that $\mathbb{P}(X_1 = 1, X_2 = 1) = (n+1)^{-1}\sum_{k=0}^{n}\mathbb{P}(X_2 = 1, X_1 = 1,k)$, and $\sum_{k=0}^{n}\mathbb{P}(X_2 = 1, X_1 = 1,k) = (n(n-1))^{-1}\sum_{k=0}^{n}(k-1)k = (n+1)/3$. Eventually, all these calculations lead to $\mathbb{P}(X_2 = 1|X_1 = 1) = \mathbb{P}(X_2 = 1, X_1 = 1)/\mathbb{P}(X_1 = 1) = 2/3$.

The approach used to solve this problem can be summarized as follows:

1. set the probability distribution $\mathbb{P}(\mathbf{n})$;
2. consider $X_1,\ldots,X_n|\mathbf{n}$ as a hypergeometric process;
3. compute the probability of any event A as $\sum_{\mathbf{n}}\mathbb{P}(A|\mathbf{n})\mathbb{P}(\mathbf{n})$.

The predictive approach in this case is very simple: let us consider (3.25), the predictive rule used in the example of n energy elements to be divided into 2 molecules in Section 3.4.3. Then, setting

$$\mathbb{P}(X_{m+1} = 1|X_1 = x_1,\ldots,X_m = x_m) = \frac{1+m_1}{2+m}, \tag{4.74}$$

the distribution of $\mathbf{n} = (n_1,n_2)$ is given by $\mathbb{P}(n_1,n_2) = 1/(n+1)$ and is uniform. The probability of $X_1 = 1$ is simply $1/2$, and that of $X_2 = 1$ given that $X_1 = 1$ is $(1+1)/(2+1) = 2/3$.

4.2.5 Example: Laplace's rule of succession

Laplace's succession rule is an example of an application for the predictive probability (4.74). The rule states that 'whenever, in n identical experiments, one has observed k successes (and $n-k$ failures), the probability that the next observation is a success is $(k+1)/(n+2)$'. Such a rule is too general and, not surprisingly, it was already criticized by frequentists in the second half of the nineteenth century. In order to better appreciate Laplace's line of reasoning, it is useful to follow the route he used to derive (4.74).

Laplace's procedure

Assume that the 'true' but unknown probability of success is $0 < \theta \leq 1$. Then the hypothetical probability of a specific sequence with k successes in n trials is $\theta^k(1-\theta)^{n-k}$. Then, introduce the a-priori distribution of $\theta \in (0,1]$: $f(\theta) \geq 0, \sum_\theta f(\theta) = 1$. As a consequence of total probability, $\sum_\theta \theta^k(1-\theta)^{n-k} f(\theta)$ is the probability of a specific sequence with k successes denoted by the random vector $\mathbf{X}^{(n)}(k)$.

Laplace wants to find $\mathbb{P}(X_{n+1} = 1|\mathbf{X}^{(n)}(k))$, and he introduces the random variable $\Theta = \theta$. Now, one has $\mathbb{P}(X_{n+1} = 1|\mathbf{X}^{(n)}(k)) = \sum_\theta \mathbb{P}(X_{n+1} = 1, \Theta = \theta|\mathbf{X}^{(n)}(k)) = \sum_\theta \mathbb{P}(X_{n+1} = 1|\Theta = \theta, \mathbf{X}^{(n)}(k))\mathbb{P}(\Theta = \theta|\mathbf{X}^{(n)}(k))$, where

$$\mathbb{P}(X_{n+1} = 1|\Theta = \theta, \mathbf{X}^{(n)}(k)) = \mathbb{P}(X_{n+1} = 1|\Theta = \theta), \tag{4.75}$$

because given $\Theta = \theta$ the previous results are inessential, and

$$\mathbb{P}(\Theta = \theta|\mathbf{X}^{(n)}(k)) \tag{4.76}$$

is the posterior distribution in the hypothesis $\Theta = \theta$ after k successes and $n-k$ failures. It turns out that

$$\mathbb{P}(X_{n+1} = 1|\Theta = \theta) = \theta, \tag{4.77}$$

as the parameter is equal to the probability of a success, and $\mathbb{P}(\Theta = \theta|\mathbf{X}^{(n)}(k)) = \mathbb{P}(\Theta = \theta|k,n)$ can be found using Bayes' theorem:

$$\mathbb{P}(\Theta = \theta|\mathbf{X}^{(n)}(k)) = A\mathbb{P}(\mathbf{X}^{(n)}(k)|\Theta = \theta)f(\theta) = A\theta^k(1-\theta)^{n-k}f(\theta), \tag{4.78}$$

where $A^{-1} = \sum_\theta \theta^k(1-\theta)^{n-k}f(\theta)$ is a normalization constant. Note that the posterior probability $\mathbb{P}(\theta|\mathbf{X}^{(n)}(k))$ is a function of θ only. Laplace chooses $f(\theta) = c$, a uniform distribution for the random variable Θ, as a consequence of the indifference principle applied to all the possible values of the parameter. This assumption means that for any interval $(a,b) \subset (0,1]$, $\mathbb{P}(\Theta \in (a,b)) = b-a$, and the constant is $c = 1$. However, for the purpose of deriving the rule of succession, it is sufficient to observe that, being constant, $f(\theta)$ cancels from (4.78), and the equation reduces to $\mathbb{P}(\Theta = \theta|\mathbf{X}^{(n)}(k)) = A' \cdot \theta^k(1-\theta)^{n-k}$. Before presenting all the mathematical details, note that the evidence transforms the initial flat prior distribution of Θ in a posterior distribution peaked around $\theta \simeq k/n$, and this effect increases with the size of the sample. The exact calculation shows that

$$\mathbb{P}(X_{n+1} = 1|\mathbf{X}^{(n)}(k)) = A' \int_0^1 \theta \cdot \theta^k (1-\theta)^{n-k} \mathrm{d}\theta = \frac{1+k}{2+n}. \tag{4.79}$$

In other words, the predictive probability of success is equal to the expected value of θ over the posterior distribution. Laplace applied this formula to the outstanding philosophical problem introduced by Hume. What is the degree of confidence for the truth of the sentence *the sun will rise tomorrow*? Having looked for the evidence of all successes since the birth of historical reports, Laplace obtained a number of trials equal to $1,826,213$ and all of them successful. Therefore, he could bet $1,826,214:1$ in favour of the sun rising on the next day. If the evidence is large, $(k+1)/(n+2) \simeq k/n$, and Laplace's estimate cannot be distinguished from the frequency of successes, which frequentists use to assess probability. However, the formula holds also when no evidence is present, a case in which the frequency interpretation of probability is meaningless, and this is the main reason for the criticisms put forward by frequentists.

Jokes against the succession rule

Imagine that somebody is falling from the top floor of the Empire State Building, and passing any floor while falling, he/she observes 'till now, everything is OK'. The probability of success (survival) at the next floor is increasing, if one applies the succession rule. This seems paradoxical, as the falling person will die with probability very close to 1 when hitting the road surface. A more careful analysis of this paradox reveals that:

1. conditioned on the knowledge of the floor number of the building, the process is deterministic;
2. the process is not exchangeable: once the falling person is dead (failure), he/she cannot be resuscitated.

Another common criticism of inductive inference (and of Laplace's rule as a method for quantitatively assessing it) is as follows. You are a turkey living in the USA and you will be killed just before the next Thanksgiving Day. Since your birth, you have observed that some humans come at regular periods in order to feed you. Every day, your confidence in the success (tomorrow they will come and feed me) is increasing, even if Thanksgiving is approaching. Your problem is that you do not know anything about Thanksgiving.

4.2.6 Probability of facts/probability of laws

Now compare Laplace's procedure with the one in Section 4.2.4; the two hypotheses of uniform distributions: $\mathbb{P}(n,0) = \ldots = \mathbb{P}(k,n-k) = \ldots = \mathbb{P}(0,n) = 1/(n+1)$

and $f(\theta) = 1$, for $\theta \in (0,1]$ lead to the same predictive probability. Is this merely due to coincidence? In the first case the probability of the individual descriptions $\mathbb{P}(X_1 = x_1, \ldots, X_m = x_m|k,n)$ follows the hypergeometric distribution. Assume that $n \gg 1$, and introduce the relative frequency $\theta_k = k/n \in \{0, 1/n, 2/n, \ldots, (n-1)/n, 1\}$. Then the distance between two consecutive points $\theta_{k+1} - \theta_k = \Delta\theta_k = \Delta\theta = 1/n$ decreases as n increases. The set $a < k \leq b$, whose probability is $(b-a)/(n+1)$, is mapped into $\theta_a < \theta_k \leq \theta_b$. Then $\mathbb{P}(\theta_a < \theta_k \leq \theta_b) = (b-a)/(n+1) \simeq \theta_b - \theta_a$, that is the continuous uniform distribution in $(0,1]$. In the limit $n \to \infty$, one can write that the random variable $\theta_k = k/n$ converges to the random variable $\theta : \mathbb{P}(c < \theta \leq d) = \int_c^d f(\theta)\mathrm{d}\theta$ with $f(\theta) = 1, \theta \in [0,1]$.[2] Returning to $\mathbb{P}(X_1 = x_1, \ldots, X_m = x_m|k,n)$, for $n \to \infty$, it is possible to prove that the hypergeometric distribution converges to the binomial distribution

$$\mathbb{P}(X_1 = x_1, \ldots, X_m = x_m|k,n) \simeq \left(\frac{k}{n}\right)^h \left(1 - \frac{k}{n}\right)^{m-h} \simeq \theta^h (1-\theta)^{m-h};$$

this result is easy to understand in terms of urn models. When the sample size n is large and both n_1 and n_2 are large, there is no difference in the probabilities resulting from sampling with replacement and sampling without replacement, as, in practice, removing a few balls does not change the composition of the urn. In other words, assuming a uniform density on θ, conditioned on which the sample is i.i.d. (binomial, in this simple case), is equivalent to assuming a uniform probability on the unknown composition $(k, n-k)$, conditioned on which the sample is hypergeometric. The two methods share the same predictive probability. In the history of predictive probabilities, the first route was followed by de Finetti and by subjectivists, whereas the second finitary procedure was used by Johnson, who did not use Bayesian methods, together with Keynes, Carnap and other supporters of logicism, who tried to avoid the introduction of infinities. Eventually, one could say that the finitary predictive characterization of a stochastic process only uses the *probability of facts*, that is the probability of the observables X_i or $\mathbf{n}$, whereas the other method introduces the probability of a hidden variable – the parameter, conditioned on which the variables are i.i.d. – representing the *probability of a law*. This law (the parameter) becomes a fact only in the infinite limit $n \to \infty$, where it represents the value of the relative frequency.

4.3 Appendix 1: a finite version of de Finetti's theorem

It turns out that all n-exchangeable processes can be represented as mixtures (linear superpositions) of hypergeometric processes. Here, this important result

[2] Here one can exchange (with [, as, for a continuous density, single points have zero probability. Further note that this result can be made rigorous by introducing the appropriate definition of convergence for a sequence of random variables.

due to Bruno de Finetti will be further discussed in the case of a dichotomous n-exchangeable process.

4.3.1 Kolmogorov's compatibility conditions for a finite dichotomous process

Let us consider a finite (n-step) dichotomous stochastic process. This is a sequence of n random variables $X_1,\ldots,X_n$ each of which can assume one of two values. For the sake of simplicity, let us assume that, $\forall i$, $X_i = \mathbb{I}_A$ that is $X_i = 1$ if event A occurs at step i and $X_i = 0$ if event A does not take place at step i.

A stochastic process is fully characterized if all its finite dimensional distributions are given. As discussed above, these distributions must satisfy Kolmogorov's compatibility conditions. In the simple case under scrutiny, these conditions become

$$\begin{gathered} p_0 + p_1 = 1, \\ p_{1,1} + p_{1,0} = p_1, \qquad p_{0,1} + p_{0,0} = p_0, \\ p_{1,1,1} + p_{1,1,0} = p_{1,1}, \qquad p_{0,0,1} + p_{0,0,0} = p_{0,0}, \\ \ldots, \end{gathered} \tag{4.80}$$

where, e.g., $p_{1,1,1} = \mathbb{P}(X_1 = 1, X_2 = 1, X_3 = 1)$ and the symmetry compatibility conditions (such as $\mathbb{P}(X_1 = 1, X_2 = 1, X_3 = 0) = \mathbb{P}(X_3 = 0, X_1 = 1, X_2 = 1)$) are implicitly assumed.

4.3.2 Again on exchangeability

An exchangeable finite (n-step) dichotomous stochastic process is a sequence of n dichotomous random variables $X_1,\ldots,X_n$ such that the finite dimensional probabilities do not depend on the particular order of 0s and 1s, but only on the number of 0s and 1s. In other words, for instance, one has $p_{0,0,1} = p_{0,1,0} = p_{1,0,0}$. Notice that, as the sums of the number of 0s and the number of 1s coincide with the number of trials considered, and $S_n = \sum_{i=1}^{n} X_i$ coincides with the number of 1s (usually called the number of successes), one has for the probability of a particular sequence with h successes out of $m \le n$ trials, $X_1,\ldots,X_m$,

$$\mathbb{P}(X_1,\ldots,X_m) = \frac{\mathbb{P}(S_m = h)}{\binom{m}{h}}, \tag{4.81}$$

as $m!/(h!(m-h)!)$ is the number of sequences with h successes out of m trials. This formula is a specialization of (4.59). It is useful to introduce the new symbol $p_h^{(m)} = \mathbb{P}(S_m = h)$. Another useful quantity is $p_h = p_h^{(h)}$, the probability of h successes out of h trials. One has that

$$p_h = \sum_{r=h}^{m} p_r^{(m)} \binom{r}{h} \left[\binom{m}{h} \right]^{-1}; \tag{4.82}$$

in order to prove equation (4.82), one has to count all the possibilities containing h successes in the first h trials. A single sequence with $r \geq h$ successes out of m trials has probability

$$\frac{p_r^{(m)}}{\binom{m}{r}}; \tag{4.83}$$

the number of these sequences is

$$\binom{m-h}{r-h}, \tag{4.84}$$

and $h \leq r \leq m$ so that

$$p_h = \sum_{r=h}^{m} p_r^{(m)} \binom{m-h}{r-h} \left[\binom{m}{r}\right]^{-1}, \tag{4.85}$$

which coincides with (4.82). Note that one conventionally assumes $p_0 = p_0^{(0)} = 1$. One can now write Kolmogorov's compatibility conditions for an exchangeable dichotomous finite random process as follows

$$\frac{p_h^{(m)}}{\binom{m}{h}} = \frac{p_h^{(m+1)}}{\binom{m+1}{h}} + \frac{p_{h+1}^{(m+1)}}{\binom{m+1}{h+1}}, \tag{4.86}$$

which is valid for $m < n$; the meaning of Equations (4.86) is exactly the same as Equations (4.80). For instance if $m = 2$ and $h = 2$, it coincides with $p_{1,1} = p_{1,1,0} + p_{1,1,1}$. These equations can be used to invert Equation (4.82) and find $p_h^{(m)}$ as a function of p_h. If $h = m-1$ and m is replaced by $m-1$, one gets

$$p_{m-1} = \frac{p_{m-1}^{(m)}}{\binom{m}{m-1}} + p_m, \tag{4.87}$$

leading to

$$\frac{p_{m-1}^{(m)}}{\binom{m}{m-1}} = -(p_m - p_{m-1}) = -\Delta p_{m-1}. \tag{4.88}$$

Analogously, from

$$\frac{p_{m-2}^{(m-1)}}{\binom{m-1}{m-2}} = \frac{p_{m-2}^{(m)}}{\binom{m}{m-2}} + \frac{p_{m-1}^{(m)}}{\binom{m}{m-1}}, \tag{4.89}$$

one gets

$$\frac{p_{m-2}^{(m)}}{\binom{m}{m-2}} = -\frac{p_{m-1}^{(m)}}{\binom{m}{m-1}} + \frac{p_{m-2}^{(m-1)}}{\binom{m-1}{m-2}} = -\Delta p_{m-2} + \Delta p_{m-1} = \Delta^2 p_{m-2}, \tag{4.90}$$

and from

$$\frac{p_{m-3}^{(m-1)}}{\binom{m-1}{m-3}} = \frac{p_{m-3}^{(m)}}{\binom{m}{m-3}} + \frac{p_{m-2}^{(m)}}{\binom{m}{m-2}}, \tag{4.91}$$

it turns out that

$$\frac{p_{m-3}^{(m)}}{\binom{m}{m-3}} = -\Delta^2 p_{m-3} + \Delta^2 p_{m-2} = -\Delta^3 p_{m-3}. \tag{4.92}$$

In the end, the following equation can be proved by induction

$$\frac{p_h^{(m)}}{\binom{m}{h}} = (-1)^{m-h} \Delta^{m-h} p_h. \tag{4.93}$$

In other words, a finite exchangeable dichotomous random process can be completely specified either by assigning the probabilities $p_h^{(m)}$ subject to the compatibility conditions (4.86) or by assigning the probabilities p_h such that the signs of the differences $\Delta^{m-h} p_h$ are alternate. Via Equation (4.93), this assignment leads to the knowledge of the probabilities $p_h^{(m)}$ automatically satisfying the compatibility conditions.

Remark (Meaning of the symbol Δ^r) Given a sequence $a_0, a_1, a_2, \ldots$, the first difference is $\Delta a_k = a_{k+1} - a_k$, the second difference is $\Delta^2 a_k = \Delta(\Delta a_k) = a_{k+2} - 2a_{k+1} + a_k$, the rth difference is $\Delta^r a_k = \Delta(\Delta^{r-1} a_k)$. The general expansion formula can be proved by induction

$$\Delta^r a_k = \sum_{j=0}^{r} \binom{r}{j} (-1)^{r-j} a_{k+j}, \tag{4.94}$$

with $\Delta^0 a_k = a_k$.

4.3.3 Representation theorem for finite dichotomous exchangeable processes

Let us consider a process where balls are drawn without replacement from an urn containing n balls of which n_1 are white (drawing a white ball is a success). In this case one has that

$$\widetilde{p}_h^{(n_1,m)} = \frac{\binom{n_1}{h}\binom{n-n_1}{m-h}}{\binom{n}{m}} = \frac{\binom{m}{h}\binom{n-m}{n_1-h}}{\binom{n}{n_1}}, \tag{4.95}$$

where $\widetilde{p}_h^{(n_1,m)}$ is the probability of extracting h white balls in a sequence of n draws with $h \leq n_1$ and $m \leq n$. Equation (4.95) defines a particular finite exchangeable dichotomous process known as the *hypergeometric process* already discussed in Section 4.2.2. Now, we want to prove that all finite dichotomous exchangeable processes can be represented in terms of mixtures of hypergeometric processes.

A mixture of dichotomous stochastic processes is defined as follows. Consider a family of dichotomous random processes $X_1^{(\theta)},\ldots,X_n^{(\theta)},\ldots$ depending on a parameter θ. Each possible value of θ, say $\bar{\theta}$, defines a usual sequence of random variables $X_1^{(\bar{\theta})},\ldots,X_n^{(\bar{\theta})},\ldots$. All the finite dimensional probabilities in Equation (4.80) will depend on θ. Then, for instance, $p_0^{(\theta)}$ denotes the probability of a failure given a specific value of θ, $p_{1,1}^{(\theta)}$ is the probability of a sequence with two successes given θ, and so on. Suppose now that also θ is considered as a random variable and let $F_\theta(x)$ be its cumulative distribution function, then the finite dimensional probabilities of the mixture process will be obtained by means of equations of this type

$$p_0 = \int_{-\infty}^{+\infty} p_0^{(u)} \mathrm{d}F_\theta(u),$$

$$p_1 = 1 - p_0 = 1 - \int_{-\infty}^{+\infty} p_0^{(u)} \mathrm{d}F_\theta(u) = \int_{-\infty}^{+\infty} p_1^{(u)} \mathrm{d}F_\theta(u),$$

$$p_{1,1} = \int_{-\infty}^{+\infty} p_{1,1}^{(u)} \mathrm{d}F_\theta(u),$$

$$\ldots.$$

Note that every process which is a mixture of exchangeable processes is exchangeable. For the case under scrutiny, it suffices to verify that any probability of the kind $p^{(\theta)}_{1,1,0}$ depends only on h, the number of successes, and this is true also of any mixture. A corollary of this result is that

Theorem (Finite representation theorem of de Finetti for dichotomous processes, easy part) Every dichotomous process which is a mixture of n-step hypergeometric processes is an n-step exchangeable process.

Such a process can be written as follows

$$p_h^{(m)} = \sum_{k=0}^{n} c_k \frac{\binom{k}{h}\binom{n-k}{m-h}}{\binom{n}{m}}, \tag{4.96}$$

where c_k are the probabilities for the values of the parameter n_1, that is $c_k = \mathbb{P}(n_1 = k)$ with $\sum_{k=0}^{n} c_k = 1$ and $h \leq m \leq n$. Now, it turns out that also the converse of the previous theorem is true.

Theorem (Finite representation theorem of de Finetti for dichotomous processes, less easy part) Every dichotomous n-step random process is a mixture of hypergeometric processes.

In order to prove this result, notice that every dichotomous n-step random process is characterized by the assignment of the probabilities $p_{n_1}^{(n)}$ for $n_1 = 0, 1, \ldots, n$ with $p_{n_1}^{(n)} \geq 0$ and $\sum_{n_1=0}^{n} p_{n_1}^{(n)} = 1$. Using Kolmogorov's compatibility conditions one can go back and obtain all the probabilities $p_h^{(m)}$ with $h \leq m \leq n$. Therefore, if one proves that the probabilities $p_h^{(m)}$ can be written as mixtures of the analogous probabilities for the hypergeometric process, one gets the result. It is easier to do so for the probabilities p_h whose linear combinations give the probabilities $p_h^{(m)}$. In particular, as a consequence of (4.82), for $m = n$, one has

$$p_h = \sum_{r=h}^{n} p_r^{(n)} \binom{r}{h} \left[\binom{n}{h}\right]^{-1}, \tag{4.97}$$

and

$$\widetilde{p}_h^{(r,n)} = \binom{r}{h} \left[\binom{n}{h}\right]^{-1}, \tag{4.98}$$

is just the probability of h successes in h draws in an n-step hypergeometric process of parameter r (an urn with r white balls). Notice that the weights of the mixture are exactly the $n+1$ assigned probabilities $p_{n_1}^{(n)}$.

4.4 Appendix 2: the beta distribution

In order to evaluate $(A')^{-1} = \sum_\theta \theta^k (1-\theta)^{n-k}$ in Section 4.2.5, it is necessary to evaluate the integral $\int_0^1 \theta^k (1-\theta)^{n-k} \mathrm{d}\theta$. It turns out that

$$(A')^{-1} = \int_0^1 \theta^k (1-\theta)^{n-k} \mathrm{d}\theta = \frac{k!(n-k)!}{(n+1)!} = \frac{1}{n+1} \binom{n}{k}^{-1}; \tag{4.99}$$

the integral can be conveniently expressed in terms of Euler's gamma function and of Euler's beta function. Euler's gamma function is defined by the integral

$$\Gamma(x) = \int_0^\infty \exp(-t) t^{x-1} \mathrm{d}t, \tag{4.100}$$

and given that $\Gamma(x) = (x-1)\Gamma(x-1)$ and $\Gamma(1) = \Gamma(2) = 1$, one gets $\Gamma(n+1) = n!$ for positive integers. Euler's beta can be defined in terms of Euler's gamma as follows

$$B(a,b) = \frac{\Gamma(a)\Gamma(b)}{\Gamma(a+b)}, \tag{4.101}$$

so that

$$(A')^{-1} = \frac{k!(n-k)!}{(n+1)!} = \frac{\Gamma(k+1)\Gamma(n-k+1)}{\Gamma(n+2)} = B(k+1, n-k+1). \tag{4.102}$$

Note that from Equations (4.99) and (4.101) it follows that

$$B(a,b) = \frac{\Gamma(a)\Gamma(b)}{\Gamma(a+b)} = \int_0^1 \theta^{a-1} (1-\theta)^{b-1} \mathrm{d}\theta. \tag{4.103}$$

From this last result, one can define the probability density

$$\mathrm{Beta}(\theta; a, b) = \frac{\Gamma(a+b)}{\Gamma(a)\Gamma(b)} \theta^{a-1} (1-\theta)^{b-1}, a, b > 0, 0 \le \theta \le 1 \tag{4.104}$$

defined on $[0,1]$, as $\mathrm{Beta}(\theta; a, b) \ge 0$, $\int_0^1 \mathrm{Beta}(\theta; a, b) \mathrm{d}\theta = 1$. The first two moments of this distribution are

$$\mathbb{E}[\Theta] = a/(a+b), \tag{4.105}$$

$$\mathbb{V}[\Theta] = ab/[(a+b)^2(a+b+1)], \tag{4.106}$$

and they are further discussed in Exercise 4.6. Note that the prior uniform distribution has probability density function given by $\mathrm{Beta}(\theta; 1, 1)$. The normalized posterior distribution is characterized by the probability density function

$$p(\theta|k,n) = (n+1)\binom{n}{k}\theta^k(1-\theta)^{n-k} = \frac{\Gamma(n+2)}{\Gamma(k+1)\Gamma(n-k+1)}\theta^k(1-\theta)^{n-k}, \tag{4.107}$$

which is nothing other than $\mathrm{Beta}(\theta; k+1, n-k+1)$. Finally, recalling that $\mathbb{P}(X_{n+1} = 1|\mathbf{X}^{(n)}(k)) = \sum_\theta \theta \cdot \mathbb{P}(\Theta = \theta|\mathbf{X}^{(n)}(k)) = \mathbb{E}_{\mathrm{posterior}}(\Theta)$, from (4.105), one immediately gets Laplace's succession rule:

$$P(X_{n+1} = 1|\mathbf{X}^{(n)}(k)) = \frac{k+1}{n+2}. \tag{4.108}$$

4.5 Exercises

4.1 Prove that the inverse image of all labels divides the set Ω into equivalence classes.

4.2 Prove that $\mathbb{E}(a+bX) = a + b\mathbb{E}(X)$.

4.3 Prove that:

1. $\mathbb{V}(X) = \mathbb{E}(X^2) - \mathbb{E}^2(X)$;
2. $\mathbb{V}(X+a) = \mathbb{V}(X)$;
3. $\mathbb{V}(bX) = b^2\mathbb{V}(X)$.

4.4 Prove the properties of the indicator function (see Section 4.1.8).

4.5 Apply the total probability theorem to $\mathbb{P}(X_2 = x_2|X_1 = x_1)$, considering that they are two results of draws from a bivariate urn of indefinite composition.

4.6 Compute the expected value and the variance for the beta distribution.

4.6 Summary

Many results of Chapters 2 and 3 can be conveniently reformulated in the language of random variables and stochastic processes. A random variable is a measurable function mapping a probability space into another probability space. A function is another name for a categorization of objects. Therefore, it is natural to introduce random variables X_i whose range is a label set denoting the g categories into which n objects are divided. Random variables are characterized by their probability distributions. For a numerical random variable, the distribution can often be summarized by two numbers: the expected value or mathematical expectation and the variance. The mathematical expectation denotes a 'typical' value of the distribution, whereas the variance represents typical fluctuations of the random variable around this typical value.

When two or more random variables on the same probability space are considered, it is interesting to study their covariance, indicating how much these variables are correlated. A random vector is characterized by a joint probability distribution. Sums of the joint probability distribution on the possible values of all but one random variable lead to the so-called marginal distribution.

A sequence of random variables is a stochastic process. The index of the sequence often denotes time, but this is not necessarily the case. In many examples of interest, the index set runs on the objects that are studied. Stochastic (or random) processes are characterized by the set of finite dimensional probability distributions (appropriate joint distributions for the variables in the sequence) and these distributions must satisfy Kolmogorov's compatibility conditions.

Among all stochastic processes, exchangeable processes are those for which the probability of a sequence does not depend on the specific outcomes, but only on the number of occurrences of any category. This greatly simplifies the analysis of such processes. It turns out that finite exchangeable processes can be represented as mixtures of the hypergeometric process. The hypergeometric process has a very simple urn model: drawing balls from an urn without replacement. For large sample sizes, this process coincides with sampling with replacement: the usual Bernoullian process. Both processes are instances of another important class of exchangeable stochastic processes: the family of Pólya processes. The analysis of these processes in terms of predictive laws leads to Laplace's rule of succession – a milestone of predictive probabilistic inference.

Further reading

A.N. Kolmogorov and S.V. Fomin discuss the meaning of maps (or functions) as categorizations at the beginning of their book on functional analysis.

A.N. Kolmogorov and S.V. Fomin, *Elements of the Theory of Functions and Functional Analysis*, Dover, New York (1999).

The relevance quotient and its invariance are the subject of an essay by D. Costantini.

D. Costantini, *The Relevance Quotient*, Erkenntniss, **14**, 149–157 (1979).

This issue is discussed in detail at the end of the next chapter.

L. Brillouin used the concepts developed in this chapter to derive quantum statistics. His paper was published in 1927.

L. Brillouin, *Comparaison des différentes statistiques appliquées aux problèmes de quanta*, Annales de Physique, Paris, VII: 315–331 (1927).

D. Costantini and U. Garibaldi noticed the analogy between the tools used by Brillouin and the tools used in probabilistic predictive inference.

D. Costantini and U. Garibaldi, *Una formulazione probabilistica del Principio di esclusione di Pauli nel contesto delle inferenze predittive*, Statistica, anno LI, n.1, p. 21 (1991).

An account in English of these results is available:

D. Costantini and U. Garibaldi, *The Ehrenfest Fleas: From Model to Theory*, Synthese, **139** (1), 107–142 (2004).

There are many good books on random variables and stochastic processes. A classical reference is the book by J.L. Doob.

J.L. Doob, *Stochastic Processes* (Wiley Classics Library), Wiley, New York (1990).

The introductory textbooks by S. Karlin and H.M. Taylor

S. Karlin and H.M. Taylor, *A First Course in Stochastic Processes*, Academic Press, London (1975).

and by S.M. Ross

S.M. Ross, *Stochastic Processes*, Wiley, New York (1983).

are also useful classical references. Cambridge University Press has just published a book on stochastic processes for physicists including some elements of modern probability theory.

K. Jacobs, *Stochastic Processes for Physicists*, Cambridge University Press, Cambridge, UK (2009).

The proof in Section 4.3 is adapted from

L. Daboni and A. Wedlin, *Statistica. Un'introduzione all'impostazione neo-bayesiana*, UTET, Turin (1982).

a book on neo-Bayesian statistics.

References

[1] J.S. Mill, *On the Definition of Political Economy and the Method of Investigation Proper to It*, essay 5 in J.S. Mill, *Essays on Some Unsettled Questions of Political Economy*, John W. Parker, West Strand, London (1844).

[2] P.-R. de Montmort, *Essay d'analyse sur les jeux d'hazard*, J. Quillau, Paris (1708). Reprinted by the American Mathematical Society in 2006.

5

The Pólya process

The multivariate generalized Pólya distribution plays a key role in the finitary analysis of many interesting problems. It gives the distribution of the statistical descriptions for the so-called Pólya stochastic process. As discussed in Chapter 4, such a process has a simple urn model interpretation. Therefore, this chapter is devoted to a detailed analysis of the generalized multivariate Pólya distribution and of the related stochastic process.

After studying this chapter you should be able to:

- define the Pólya process;
- derive the multivariate generalized Pólya distribution as the sampling distribution of the Pólya process;
- compute the expected value and the variance of the multivariate generalized Pólya distribution;
- discuss some remarkable limits of the multivariate generalized Pólya distribution;
- marginalize the multivariate generalized Pólya distribution;
- characterize the Pólya process in terms of the invariance of the heterorelevance coefficient.

5.1 Definition of the Pólya process

The sequence $X_1, X_2, \ldots, X_n$, with $X_i \in \{1, \ldots, g\}$, is a generalized Pólya process if the predictive probability of the process is given by (4.72), rewritten below for the sake of clarity

$$\mathbb{P}(X_{m+1} = j | X_1 = x_1, \ldots, X_m = x_m) = \frac{\alpha_j + m_j}{\alpha + m}, \tag{5.1}$$

where $m_j = \#\{X_i = j, i = 1, \ldots, m\}$ is the number of occurrences of the jth category in the evidence $(X_1 = x_1, \ldots, X_m = x_m)$, and m is the number of observations or trials. As for the parameters α and α_j, in the usual Pólya process, α_j is a positive integer, representing the number of balls of type (colour) j in the auxiliary Pólya urn; α is the total number of balls in the urn and it is given by $\alpha = \sum_{j=1}^{g} \alpha_j$. If α_j is positive but real, it can be interpreted as the initial weight of the jth category, it being understood that the prize (the added weight) of the drawn category is always just 1. The meaning of α_j becomes clearer if one defines $p_j = \alpha_j/\alpha$, then Equation (5.1) can be written

$$\mathbb{P}(X_{m+1} = j|X_1 = x_1,\ldots,X_m = x_m) = \frac{\alpha p_j + m_j}{\alpha + m} = \frac{\alpha}{\alpha + m} p_j + \frac{m}{\alpha + m}\frac{m_j}{m}, \quad (5.2)$$

and p_j has a natural interpretation as prior (or initial) probability for the jth category: $p_j = \alpha_j/\alpha = \mathbb{P}(X_1 = j)$. Note that in this way, the limit $\alpha \to \infty$ has an immediate interpretation as (5.2) becomes

$$\mathbb{P}(X_{m+1} = j|X_1 = x_1,\ldots,X_m = x_m) = p_j, \quad (5.3)$$

leading to the i.i.d. multinomial distribution. In other words, the distribution of the generalized Pólya process converges to the distribution of the multinomial process when the parameter α diverges. When α_j is negative, if $|\alpha_j|$ is integer and $m_j \leq |\alpha_j| = |\alpha| p_j$, Equation (5.1) still represents a probability and it can be written as

$$\mathbb{P}(X_{m+1} = j|X_1 = x_1,\ldots,X_m = x_m) = \frac{|\alpha| p_j - m_j}{|\alpha| - m}, \quad (5.4)$$

but here, the number of variables (observations or trials) is limited by $|\alpha|$. As discussed in Section 4.2.2, Equation (5.4) leads to the hypergeometric distribution. Therefore, in the following, the general case (5.1) will be studied, with the total weight α belonging to the set $\alpha \in (0,\infty) \cup \{-n, -n-1, -n-2, \ldots, -\infty\}$, which is the union of the generalized Pólya and the hypergeometric domain, with the multinomial case appearing as both $\alpha \to -\infty$ and $\alpha \to +\infty$.

5.1.1 The finite dimensional distributions

As discussed in Section 4.2, Equation (4.50) shows that a knowledge of predictive probabilities leads to knowledge of the finite dimensional distributions $\mathbb{P}(X_1 = x_1,\ldots,X_n = x_n)$. In the specific case under scrutiny, it suffices to consider the *fundamental* sequence $\mathbf{X}_f^{(m)} = (1,\ldots,1,\ldots,g,\ldots,g)$, consisting of m_1 labels 1 followed by m_2 labels 2, and so on, ending with m_g occurrences of label g. As a direct consequence of Equations (4.50) and (5.1), the probability of this sequence is given by

$$\frac{\alpha_1}{\alpha}\frac{\alpha_1+1}{\alpha+1}\cdots\frac{\alpha_1+m_1-1}{\alpha+m_1-1}\frac{\alpha_2}{\alpha+m_1}\frac{\alpha_2+1}{\alpha+m_1+1}\cdots\frac{\alpha_2+m_2-1}{\alpha+m_1+m_2-1}\cdots$$
$$\cdots\frac{\alpha_g}{\alpha+m_1+m_2+\ldots+m_{g-1}}\frac{\alpha_g+1}{\alpha+m_1+m_2+\ldots+m_{g-1}+1}\cdots$$
$$\cdots\frac{\alpha_g+m_g-1}{\alpha+m_1+m_2+\ldots+m_g-1} = \frac{\alpha_1^{[m_1]}\cdots\alpha_g^{[m_g]}}{\alpha^{[m]}}. \quad (5.5)$$

Recalling that $m_1 + m_2 + \ldots + m_g = m$, the previous formula can be re-written in a more compact way, in terms of rising factorials. Note that this result does not depend on the order of appearance of the categories. Indeed, the probability of a different sequence with the same values of $\mathbf{m} = (m_1, \ldots, m_g)$ would have the same denominator and identical (but permuted) terms in the numerator. This means that the Pólya process is exchangeable and that the finite dimensional distributions depend only on the frequency vector $\mathbf{m}$. The finite dimensional distribution is

$$\mathbb{P}(\mathbf{X}^{(m)}) = \frac{1}{\alpha^{[m]}} \prod_{i=1}^{g} \alpha_i^{[m_i]}. \tag{5.6}$$

So the predictive probability (5.1) yields the law (5.6); it turns out that the converse is also true, as from the definition of conditional probability

$$\mathbb{P}(X_{m+1} = j | \mathbf{X}^{(m)}) = \frac{\mathbb{P}(\mathbf{X}^{(m)}, X_{m+1} = j)}{\mathbb{P}(\mathbf{X}^{(m)})} = \frac{\alpha_j + m_j}{\alpha + m}, \tag{5.7}$$

where $\mathbf{X}^{(m)}$ is an evidence with frequency vector $\mathbf{m} = (m_1, \ldots, m_g)$, both the numerator and the denominator follow (5.6), and the calculation is left to the reader as a useful exercise on Pochhammer symbols.

The *sampling distribution* is the probability of the frequency vector $\mathbf{m}$. In order to evaluate it as a function of the size m, it is sufficient to multiply (5.6) by the number of distinct equiprobable sequences leading to $\mathbf{m}$

$$\mathbb{P}(\mathbf{m}) = \text{Polya}(\mathbf{m}; \boldsymbol{\alpha}) = \frac{m!}{\prod_{i=1}^{g} m_i!} \frac{\prod_{i=1}^{g} \alpha_i^{[m_i]}}{\alpha^{[m]}} = \frac{m!}{\alpha^{[m]}} \prod_{i=1}^{g} \frac{\alpha_i^{[m_i]}}{m_i!}; \tag{5.8}$$

this equation defines the Pólya distribution which has already been informally discussed in Section 4.2.2.

5.1.2 Example: the dichotomous generalized Pólya distribution and its limits

The dichotomous (or bivariate) case, $g = 2$, is sufficient to illustrate all the essential properties of the Pólya process. Let $X_i = 0, 1$, and define $(m_1 = h, m_0 = m - h)$ as the frequency vector of $X_1, \ldots, X_m$. Then $h = \#\{X_i = 1\}$ is the number of observed successes and $S_m = h$ is a random variable whose values h vary according to $h \in \{0, 1, \ldots, m\}$. Moreover, one can write $S_m = \sum_{i=1}^{m} X_i$. In this simple case the multivariate distribution (5.8) simplifies to

$$\mathbb{P}(S_m = h) = \frac{m!}{m_0! m_1!} \frac{\alpha_0^{[m_0]} \alpha_1^{[m_1]}}{\alpha^{[m]}} = \binom{m}{h} \frac{\alpha_1^{[h]} \alpha_0^{[m-h]}}{\alpha^{[m]}}. \tag{5.9}$$

The sampling distribution of the dichotomous Pólya process is also known as the *beta-binomial* distribution. The following remarks are noteworthy.

1. (Uniform distribution) One has $1^{[n]} = n!$ and $2^{[n]} = (n+1)!$; if $\alpha_0 = \alpha_1 = 1$, then $\alpha = \alpha_0 + \alpha_1 = 2$, and it turns out that

$$\mathbb{P}(S_m = h) = \frac{m!}{h!(m-h)!}\frac{h!(m-h)!}{(m+1)!} = \frac{1}{m+1}; \tag{5.10}$$

   this is the uniform distribution on all frequency vectors of m elements and $g = 2$ categories.
2. (Binomial distribution) For $x \gg n$, one has $x^{[n]} = x(x+1)\dots(x+n-1) \simeq x^n$; moreover, one can set $\alpha_0 = \alpha p_0$, and $\alpha_1 = \alpha p_1$. Therefore, one obtains that

$$\lim_{\alpha\to\infty} \mathbb{P}(S_m = h) = \lim_{\alpha\to\infty} \binom{m}{h}\frac{\alpha_1^h \alpha_0^{m-h}}{\alpha^m} = \binom{m}{h} p_1^h (1-p_1)^{m-h}; \tag{5.11}$$

   this is the binomial distribution of parameters m and p_1. In other words, if the initial urn has a very large number of balls, the Pólya prize is inessential to calculate the predictive probability, they are all equal for each category and the Pólya process cannot be distinguished from repeated Bernoullian trials, as already discussed above.
3. (Hypergeometric distribution) Given that $x^{[n]} = x(x+1)\dots(x+n-1)$, if x is negative, and $n \leq |x|$, one gets $x^{[n]} = (-1)^n |x|(|x|-1)\dots(|x|-n+1)$, meaning that $x^{[n]} = (-1)^n\, |x|_{[n]}$, where $x_{[n]} = x(x-1)\dots(x-n+1)$ is the so-called *falling (or lower) factorial*, and setting $\alpha_1 = -N_1 = -Np_1$, $\alpha_0 = -N_0 = -Np_0$, with $p_1 + p_0 = 1$, one gets:

$$\mathbb{P}(S_m = h) = \frac{m!}{h!(m-h)!} \times \frac{(-1)^h N_1 \cdots (N_1 - h + 1)(-1)^{m-h} N_0 \cdots (N_1 - n + h + 1)}{(-1)^m N \cdots (N - m + 1)}, \tag{5.12}$$

   leading to the hypergeometric distribution

$$\mathbb{P}(S_m = h) = \frac{\binom{N_1}{h}\binom{N_0}{m-h}}{\binom{N}{m}} = \binom{m}{h}\frac{\binom{N-m}{N_1-h}}{\binom{N}{N_1}}, \tag{5.13}$$

which is the usual sampling distribution from a hypergeometric urn with N_0, N_1 balls and $N = N_1 + N_0$.

4. (Limit of the hypergeometric distribution) Also for the falling factorial $x \gg n$ yields $x_{[n]} \simeq x^n$. Therefore, one obtains

$$\lim_{\alpha \to -\infty} \mathbb{P}(S_m = h) = \binom{m}{h} p_1^h (1 - p_1)^{m-h}; \tag{5.14}$$

in other words, sampling without replacement from a very large urn cannot be distinguished from sampling with replacement.

5.2 The first moments of the Pólya distribution

Consider the evidence vector $\mathbf{X}^{(m)} = (X_1 = x_1, \ldots, X_m = x_m)$. In the dichotomous case, $h = \#\{X_i = 1\}$ can also be written as $S_m = h = \sum_{i=1}^m X_i$. In order to generalize this definition to the general case of g categories, it is natural to introduce the indicator function $\mathbb{I}_{X_i = j}(\omega) = I_i^{(j)}$, and define $S_m^{(j)} = \sum_{i=1}^m I_i^{(j)}$. Therefore, the random variable $S_m^{(j)}$ gives the number of successes for the jth category out of m observations or trials and $S_m^{(j)} = m_j$. The expected value and the variance for $S_m^{(j)}$ can be computed with the help of two simple theorems, which will be further discussed in Exercise 5.1 and Exercise 5.2.

Theorem (Expected value for the sum of m random variables) One has

$$\mathbb{E}\left(\sum_{i=1}^m X_i\right) = \sum_{i=1}^m \mathbb{E}(X_i). \tag{5.15}$$

This theorem will be proven in Exercise 5.1.

Theorem (Variance of the sum of m random variables)

$$\mathbb{V}\left(\sum_{i=1}^m X_i\right) = \sum_{i=1}^m \mathbb{V}(X_i) + 2\sum_{i=1}^m \sum_{j=i+1}^m \mathbb{C}(X_i, X_j). \tag{5.16}$$

The proof of this theorem will be presented in Exercise 5.2.

One has to determine $\mathbb{E}(I_i^{(k)})$ and $\mathbb{E}(I_i^{(k)} I_j^{(k)})$. As for the expected value, one has that $\mathbb{E}(I_i^{(k)}) = 1 \cdot \mathbb{P}(I_i^{(k)} = 1) + 0 \cdot \mathbb{P}(I_i^{(k)} = 0) = \mathbb{P}(I_i^{(k)} = 1)$ coinciding with the marginal probability of success, that is of observing category k at the ith step. From Equation (5.7), in the absence of any evidence, one has $\mathbb{P}(I_i^{(k)} = 1) = \mathbb{P}(X_i = k) = \alpha_k / \alpha = p_k$. Therefore, the random variables $I_i^{(k)}$ are equidistributed (they are exchangeable), and $\mathbb{E}(S_m^{(k)}) = \sum_{i=1}^m \mathbb{E}(I_i^{(k)}) = m\mathbb{E}(I_1^{(k)})$, yielding

$$\mathbb{E}(S_m^{(k)}) = mp_k. \tag{5.17}$$

Considering the variance $\mathbb{V}(S_m^{(k)})$, as a consequence of Equation (5.16), the covariance matrix of $I_1^{(k)},\ldots,I_m^{(k)}$ is needed. Due to the equidistribution of $I_1^{(k)},\ldots,I_m^{(k)}$, the moment $\mathbb{E}[(I_i^{(k)})^2]$ is the same for all i, and because of exchangeability $\mathbb{E}(I_i^{(k)}I_j^{(k)})$ is the same for all i,j, with $i \neq j$. Note that $(I_i^{(k)})^2 = I_i^{(k)}$ and this means that $\mathbb{E}[(I_i^{(k)})^2] = p_k$; it follows that

$$\mathbb{V}(I_i^{(k)}) = \mathbb{E}[(I_i^{(k)})^2] - \mathbb{E}^2(I_i^{(k)}) = p_k - p_k^2 = p_k(1-p_k). \tag{5.18}$$

In Exercise 5.3, it is shown that

$$\mathbb{E}(I_i^{(k)}I_j^{(k)}) = \mathbb{P}(X_i = k, X_j = k); \tag{5.19}$$

now, from exchangeability, from (5.19) and from (5.7), one gets

$$\begin{aligned}\mathbb{E}(I_i^{(k)}I_j^{(k)}) &= \mathbb{P}(X_i = k, X_j = k) = \mathbb{P}(X_1 = k, X_2 = k)\\ &= \mathbb{E}(I_1^{(k)}I_2^{(k)}) = \mathbb{P}(X_1 = k)\mathbb{P}(X_2 = k|X_1 = k) = p_k\frac{\alpha_k + 1}{\alpha + 1}.\end{aligned} \tag{5.20}$$

Therefore, the covariance matrix is given by

$$\begin{aligned}\mathbb{C}(I_i^{(k)}, I_j^{(k)}) &= \mathbb{C}(I_1^{(k)}, I_2^{(k)})\\ &= \mathbb{E}(I_1^{(k)}I_2^{(k)}) - \mathbb{E}(I_1^{(k)})\mathbb{E}(I_2^{(k)}) = \frac{p_k(1-p_k)}{\alpha + 1}.\end{aligned} \tag{5.21}$$

From this equation and from Equation (5.18), it follows that the correlation coefficient is given by

$$\rho(I_1^{(k)}, I_2^{(k)}) = \frac{1}{\alpha + 1}; \tag{5.22}$$

therefore, if $\alpha \to 0$, the correlation coefficient converges to 1, whereas if $\alpha \to \pm\infty$, the correlation coefficient vanishes. The variance of the sum $S_m^{(k)}$ follows from Equations (5.16), (5.18) and (5.21)

$$\mathbb{V}(S_m^{(k)}) = m\mathbb{V}(I_1^{(k)}) + m(m-1)\mathbb{C}(I_1^{(k)}, I_2^{(k)}) = mp_k(1-p_k)\frac{\alpha + m}{\alpha + 1}. \tag{5.23}$$

5.2.1 Remarks on the first moments; the coefficient of variation and lack of self-averaging

In the previous derivation, the initial probability p_k and the ratio α_k/α were used indifferently for computational convenience. Regarding the analysis of the moments, it is useful to consider p_k and α as independent parameters. One can see that the expected value of successes only depends on p_k, the probability of the category, and not on the total weight α; on the contrary, the variance strongly depends on α; Moreover, one has that:

1. if $\alpha \to \infty$, $\mathbb{V}_\infty(S_m^{(k)}) = mp_k(1-p_k)$, recovering the variance of the binomial distribution, as expected;
2. if $0 < \alpha < \infty$, a factor $\dfrac{\alpha+m}{\alpha+1}$ multiplies $\mathbb{V}_\infty(S_m^{(k)})$ and this factor is $O(m)$ for large m;
3. if $\alpha < 0$, with $|\alpha| \geq m$, a factor $(|\alpha|-m)/(|\alpha|-1)$ multiplies $\mathbb{V}_\infty(S_m^{(k)})$; this factor is a decreasing function of m, and it vanishes for $m = |\alpha|$. This is expected in sampling without replacement from an urn containing $|\alpha|$ balls.

In the binomial case ($|\alpha| \to \infty$), the relative frequency of successes, given by $S_m^{(k)}/m$, has expected value $\mathbb{E}_\infty(S_m^{(k)}/m) = \mathbb{E}_\infty(S_m^{(k)})/m = p_k$, whereas its variance is $\mathbb{V}_\infty(S_m^{(k)}/m) = \mathbb{V}_\infty(S_m^{(k)})/m^2 = p_k(1-p_k)/m$. The *coefficient of variation* of a positive random variable is defined as

$$CV_X = \frac{\sigma_X}{\mathbb{E}(X)}, \tag{5.24}$$

where $\sigma_X = \sqrt{\mathbb{V}(X)}$ is the standard deviation of X, and CV_X is a dimensionless number, expressing the relative deviation with respect to the expected value. For the random variable $S_m^{(k)}/m$, in the case $|\alpha| \to \infty$, one has

$$CV_{S_m^{(k)}/m} = \frac{\sigma_{S_m^{(k)}/m}}{\mathbb{E}(S_m^{(k)}/m)} \approx \frac{1}{\sqrt{m}}\sqrt{\frac{\alpha-\alpha_k}{\alpha_k}}, \tag{5.25}$$

yielding a vanishing relative deviation as m increases. Therefore, the sequence is *self-averaging*, meaning that expected relative fluctuations vanish as m increases, so that the behaviour of the sequence becomes closer and closer to the deterministic case. However, in the general case $0 < \alpha < \infty$, one gets

$$CV_{S_m^{(k)}/m} = \frac{1}{\sqrt{m}}\sqrt{\frac{\alpha-\alpha_k}{\alpha_k}\frac{\alpha+m}{\alpha+1}}. \tag{5.26}$$

Therefore, one finds that the coefficient of variation does not vanish in the limit $m \to \infty$:

$$\lim_{m\to\infty} CV_{S_m^{(k)}/m} = \sqrt{\frac{\alpha-\alpha_k}{\alpha_k}\frac{1}{\alpha+1}}. \tag{5.27}$$

In other words, the behaviour is fully probabilistic also in the limit $m \to \infty$. Note that the case $\alpha < 0$ is not interesting, as after $m = |\alpha|$ steps the process ends and $CV_{S_m^{(k)}/m} = 0$ for $m = |\alpha|$.

5.2.2 Again on Ehrenfest and Boltzmann

Returning to the examples of Chapter 3, if n molecules independently and symmetrically partition into $g = 2$ categories (two symmetrical parts of a box) labelled as $0, 1$, the predictive probability becomes

$$\mathbb{P}(X_{m+1} = j | X_1 = x_1, \ldots, X_m = x_m) = p_j = \frac{1}{2}, \tag{5.28}$$

where $j = 0, 1$. This predictive probability leads to the limit of the generalized Pólya distribution for $\alpha_j \to \infty$, $\alpha \to \infty$ and $p_j = \alpha_j/\alpha = 1/2$. Let $S_m = k_m$ describe the number of elements in category 1. One has

$$\mathbb{E}(S_n) = np_1 = n/2, \tag{5.29}$$

and

$$\mathbb{V}(S_n) = np_1(1 - p_1) = n/4, \tag{5.30}$$

and Equation (5.26) eventually leads to

$$CV_{S_n/n} = \frac{\sigma_{S_n/n}}{\mathbb{E}(S_n/n)} = \frac{1}{\sqrt{n}}. \tag{5.31}$$

For a gas in normal conditions occupying a volume of 1 cm^3, the number of molecules is $n \approx 10^{20}$; therefore $CV_{S_n/n} \approx 10^{-10}$, or one over 10 USA billions. Without entering the exact calculation, one could bet some billions against 1 that the fraction of molecules in side 1 is 0.5 ± 10^{-10}. In this simple case the expected value summarizes the whole phenomenon, and probability leads to the same conclusion as a deterministic reasoning, claiming that particles are 50 : 50-distributed.

On the contrary, in the case of n energy elements to be divided into $g = 2$ molecules, in Section 3.4.3, it was assumed that they follow a uniform generalized Pólya distribution with $\alpha_1 = \alpha_2 = 1$ (indeed, this is how the result of Section 3.4.3 can be re-interpreted now). In this case one has again that $\mathbb{E}(S_n) = n/2$, but now

$$\mathbb{V}(S_n) = \frac{n}{4}\frac{2+n}{2+1} = \frac{n(n+2)}{12}. \tag{5.32}$$

The variance of the relative frequency is

$$\mathbb{V}(S_n/n) = \frac{n(n+2)}{12n^2} \approx 1/12, \tag{5.33}$$

and

$$\sigma_{S_n/n} \approx \frac{1}{2}\sqrt{\frac{1}{3}}. \tag{5.34}$$

There is no self-averaging, as $CV_{S_n/n} \approx \sqrt{1/3}$. In this case, the distribution of S_n is uniform on the $n+1$ values $0,1,\dots,n$, and the distribution of S_n/n is uniform on $0,1/n,\dots,1$. It is always true that the expected value of S_n/n is $1/2$, but this does not summarize the phenomenon at all. If one were to bet on a value, one should put the same odds $1:n$ on each of the $n+1$ numbers $0,1,\dots,n$.

5.3 Label mixing and marginal distributions

5.3.1 Label mixing

The g-variate generalized Pólya distribution has a nice property which is useful for *label mixing* as defined by R. von Mises. Starting from (5.1), how can $\mathbb{P}(X_{m+1} \in A|X_1 = x_1,\dots,X_m = x_m)$ be obtained, where the set A is a set of categories $A = \{j_1,\dots,j_r\}$? The answer to this question is not difficult: it turns out that

$$\mathbb{P}(X_{m+1} \in A|\mathbf{X}^{(m)}) = \sum_{i=1}^{r} \mathbb{P}(X_{m+1} = j_i|\mathbf{X}^{(m)}), \tag{5.35}$$

where, as usual, $\mathbf{X}^{(m)} = (X_1 = x_1,\dots,X_m = x_m)$ summarizes the evidence. In the Pólya case, $\mathbb{P}(X_{m+1} = j|\mathbf{X}^{(m)})$ is a linear function of both the weights and the occupation numbers; therefore one gets

$$\mathbb{P}(X_{m+1} \in A|\mathbf{X}^{(m)}) = \frac{\sum_{j\in A}\alpha_j + \sum_{j\in A} m_j}{\alpha + m} = \frac{\alpha_A + m_A}{\alpha + m}, \tag{5.36}$$

where $\alpha_A = \sum_{j\in A}\alpha_j$ and $m_A = \sum_{j\in A} m_j$. As a direct consequence of (5.36), the marginal distributions of the g-variate generalized Pólya distribution are given by the dichotomous Pólya distribution of weights α_i and $\alpha - \alpha_i$, where i is the category with respect to which the marginalization is performed. In other words, one finds that

$$\sum_{m_j, j\neq i} \text{Polya}(\mathbf{m};\boldsymbol{\alpha}) = \text{Polya}(m_i, m - m_i; \alpha_i, \alpha - \alpha_i). \tag{5.37}$$

5.3.2 The geometric distribution as a marginal distribution of the uniform multivariate Pólya distribution

Consider a g-variate generalized Pólya distribution with parameters $\alpha_i = 1$, $i = 1,\dots,g$ and $\alpha = g$. Its predictive probability is

$$\mathbb{P}(X_{m+1} = j|X_1 = x_1,\dots,X_m = x_m) = \frac{1+m_j}{g+m}, \tag{5.38}$$

and the sampling distribution after n steps is

$$\mathbb{P}(\mathbf{n}) = \frac{n!}{\alpha^{[n]}} \prod_{i=1}^{g} \frac{\alpha_i^{[n_i]}}{n_i!} = \frac{n!}{g^{[n]}} = \frac{n!(g-1)!}{(n+g-1)!} = \binom{n+g-1}{n}^{-1}, \tag{5.39}$$

where the fact that $1^{[x]} = x!$ was used. In the usual Boltzmann's example, one is in the presence of n energy elements accommodating on g molecules so that any configuration of molecules has the same probability. When the focus is on a single molecule, say the one numbered 1, all the other $g-1$ molecules can be considered as a *thermostat*. What is the probability that $n_1 = k$, meaning that the first molecule receives k energy elements (and the thermostat receives $n-k$ energy elements)? This is given by the marginalization of the uniform multivariate Pólya distribution, leading to

$$\begin{aligned}\mathbb{P}(k) &= \binom{n}{k} \frac{1^{[k]}(g-1)^{[n-k]}}{g^{[n]}} \\ &= \frac{n!}{g^{[n]}} \frac{(g-1)^{[n-k]}}{(n-k)!} = \frac{\binom{n-k+(g-1)-1}{n}}{\binom{n+g-1}{n}}.\end{aligned} \tag{5.40}$$

In other words, all molecule descriptions being equiprobable, the number of descriptions assigning k energy elements to the chosen molecule coincides with the number of descriptions of $n-k$ elements on the $g-1$ molecules of the thermostat. Equation (5.40) has an interesting limit for $n, g \gg 1$ with $\chi = n/g$ constant. First, one has that

$$\begin{aligned}\frac{(g-1)^{[n-k]}}{g^{[n]}} &= \frac{(g-1)g(g+1)\cdots(g+n-k-2)}{g(g+1)\cdots(g+n-1)} \\ &= \frac{(g-1)}{(g+n-k-1)\cdots(g+n-1)} \simeq \frac{(g-1)}{(g+n)^{k+1}},\end{aligned} \tag{5.41}$$

whereas, in the same limit,

$$\frac{n!}{(n-k)!} \simeq n^k; \tag{5.42}$$

therefore, combining the two previous equations leads to

$$\mathbb{P}(k) \simeq \frac{gn^k}{(g+n)^{k+1}} = \frac{g}{g+n}\left(\frac{n}{g+n}\right)^k. \tag{5.43}$$

The meaning of $\chi = n/g$ is the average number of elements for each molecule; in the limit, one is led to the geometric distribution of parameter χ

$$\mathbb{P}(k) \simeq \text{Geo}(k;\chi) = \frac{1}{1+\chi}\left(\frac{\chi}{1+\chi}\right)^k, \tag{5.44}$$

where k can now vary from 0 to ∞. This is no longer a finite distribution, but still a discrete distribution. This sort of limit, where $n, g \to \infty$ and $n/g = \chi$ stays constant is often called *thermodynamic limit* by physicists.

The geometric distribution is often introduced in terms of the distribution for the (discrete) time of occurrence for the first success in a binomial process of parameter $p = 1/(1+\chi)$, in which case one has

$$\mathbb{P}(X = h) = p(1-p)^{h-1},\ h = 1,2,\ldots \tag{5.45}$$

The alternative definition (5.44) describes the distribution of the waiting time of (the number of failures before) the first success; in the latter case

$$\mathbb{P}(Y = k) = p(1-p)^k,\ k = 0,1,2,\ldots \tag{5.46}$$

The two definitions are connected by the transformation of random variables $Y = X - 1$.

5.3.3 The negative binomial distribution as a marginal distribution of the general multivariate Pólya distribution

In the general case, the chosen category has weight α_1 and the thermostat's weight is $\alpha - \alpha_1$; therefore the term $\alpha_1^{[k]}/k!$ does not simplify, $g-1$ is replaced by $\alpha - \alpha_1$, and the thermodynamic limit becomes $n, \alpha \gg 1$ with $\chi = n/\alpha$. Considering that $x^{[m]} = \Gamma(x+m)/\Gamma(x)$, one has

$$\begin{aligned}\frac{(\alpha-\alpha_1)^{[n-k]}}{\alpha^{[n]}} &= \frac{\Gamma(\alpha-\alpha_1+n-k)}{\Gamma(\alpha-\alpha_1)}\frac{\Gamma(\alpha)}{\Gamma(\alpha+n)} \\ &= \frac{\Gamma(\alpha+n-\alpha_1-k)}{\Gamma(\alpha+n)}\frac{\Gamma(\alpha)}{\Gamma(\alpha-\alpha_1)}.\end{aligned} \tag{5.47}$$

In the limit $x \gg m$, $\Gamma(x-m)/\Gamma(x) \simeq x^{-m}$. Therefore one finds

$$\frac{(\alpha-\alpha_1)^{[n-k]}}{\alpha^{[n]}} = \frac{\alpha^{\alpha_1}}{(\alpha+n)^{\alpha_1+k}}; \tag{5.48}$$

and, multiplying by $n!/(n-k)! \simeq n^k$ one finally has

$$\mathbb{P}(k) \simeq \text{NegBin}(k;\alpha_1,\chi)$$

$$= \frac{\alpha_1^{[k]}}{k!}\left(\frac{1}{1+\chi}\right)^{\alpha_1}\left(\frac{\chi}{1+\chi}\right)^{k}, k = 0,1,2,\ldots; \tag{5.49}$$

this distribution is called the *negative binomial distribution*; the geometric distribution (5.44) is a particular case in which $\alpha_1 = 1$ and $\alpha = g$ of (5.49). If α_1 is an integer, the usual interpretation of the negative binomial random variable is the description of the (discrete) waiting time of (i.e. the number of failures before) the first α_1th success in a binomial process with parameter $p = 1/(1+\chi)$. The moments of the negative binomial distribution can be obtained from the corresponding moments of the Polya$(n_1, n-n_1;\alpha_1,\alpha-\alpha_1)$ in the limit $n,\alpha \gg 1$, with $\chi = n/\alpha$ yielding

$$\mathbb{E}(n_1 = k) = n\frac{\alpha_1}{\alpha} \to \alpha_1\chi, \tag{5.50}$$

$$\mathbb{V}(n_1 = k) = n\frac{\alpha_1}{\alpha}\frac{\alpha-\alpha_1}{\alpha}\frac{\alpha+n}{\alpha+1} \to \alpha_1\chi(1+\chi). \tag{5.51}$$

Note that if α_1 is an integer, k can be interpreted as the sum of α_1 independent and identically distributed geometric variables (and the expected value and the variance simply sum).

If one considers two categories against the thermostat, one finds that the two categories are described by two independent negative binomial distributions. In other words, originally, the two categories were negatively correlated due to the constraint $\sum_{i=1}^{g} n_i = n$. If the limit $n \to \infty$ is considered, the thermostat removes this entanglement, and the two occupation numbers become independent. Indeed, starting from

$$\mathbb{P}(n_1,n_2,n-n_1-n_2;\boldsymbol{\alpha}) = \frac{\alpha_1^{[n_1]}}{n_1!}\frac{\alpha_2^{[n_2]}}{n_2!}\frac{(\alpha-\alpha_1-\alpha_2)^{[n-n_1-n_2]}}{(n-n_1-n_2)!}\frac{n!}{\alpha^{[n]}}, \tag{5.52}$$

in the limit $n,\alpha \gg 1$, and $\chi = n/\alpha$, the first two terms do not change, while the last term can be dealt with as before, replacing k by n_1+n_2, and α_1 by $\alpha_1+\alpha_2$. Therefore, one gets

$$\mathbb{P}(n_1,n_2;\alpha_1,\alpha_2,\chi) = \frac{\alpha_1^{[n_1]}}{n_1!}\frac{\alpha_2^{[n_2]}}{n_2!}\left(\frac{1}{1+\chi}\right)^{\alpha_1+\alpha_2}\left(\frac{\chi}{1+\chi}\right)^{n_1+n_2}$$

$$= \text{NegBin}(n_1;\alpha_1,\chi)\text{NegBin}(n_2;\alpha_2,\chi). \tag{5.53}$$

5.3.4 Example: coins and agents

Consider a system of:

1. $n = 10$ coins, distributed over $g = 2$ agents;
2. $n = 100$ coins, distributed over $g = 20$ agents;
3. $n = 1000$ coins, distributed over $g = 200$ agents.

Let the predictive law be

$$\mathbb{P}(X_{m+1} = j | X_1 = x_1, \ldots, X_m = x_m) = \frac{1 + m_j}{g + m}. \tag{5.54}$$

For all the three cases described above, one has $n/g = 5$. How does the marginal distribution for the first agent change? In the first case $\mathbb{P}_1(n_1 = k)$ is the uniform dichotomous Polya$(n_1, 10 - n_1; 1, 1)$; therefore, one has $\mathbb{P}_1(n_1 = k) = 1/11, k = 0, 1, \ldots, 10$, $\mathbb{E}_1(n_1) = 5 = n/g$, and from Equation (5.23), one finds $\mathbb{V}_1(n_1) = 10$. In the second case $\mathbb{P}_2(n_1 = k)$ is the marginal of the uniform 20-variate Pólya, that is Polya$(n_1, 100 - n_1; 1, 19)$; therefore $\mathbb{E}_2(n_1) = 5$, but now $\mathbb{V}_2(n_1) = 27.1$. Here the situation dramatically changes: the domain of the distribution enlarges from $0, \ldots, 10$ to $0, \ldots, 100$, but the expected value is always 5. The distribution is no longer uniform; it is rapidly decreasing. There is a probability equal to 0.13 that $n_1 > 10$. In the third case $\mathbb{P}_3(n_1 = k) =$ Polya$(n_1, 1000 - n_1; 1, 199)$, again $\mathbb{E}_3(n_1) = 5$, but now $\mathbb{V}_3(n_1) = 29.7$ and $\mathbb{P}(n_1 > 100) = 5 \cdot 10^{-9}$. If NegBin$(k; 1, 5)$ is used instead of Polya$(n_1, 1000 - n_1; 1, 199)$ one finds $\mathbb{E}_{\text{NegBin}}(n_1) = 5$, and $\mathbb{V}_{\text{NegBin}}(n_1) = \alpha_1 \chi (1 + \chi) = 30$. Note that, usually, the negative binomial distribution is parameterized by α_1 and $p = 1/(1 + \chi)$; in the above case this means $\alpha_1 = 1, p = 1/6$.

5.3.5 The Poisson distribution

In the case $\alpha \to \infty$ and $\alpha_i/\alpha = p_i$ for $i = 1, \ldots, g$, the g-variate generalized Pólya distribution converges to the multinomial distribution, and its marginal is the binomial distribution

$$\mathbb{P}(k) = \text{Bin}(k; n, p_1) = \binom{n}{k} p_1^k (1 - p_1)^{n-k}, \tag{5.55}$$

with $p_1 = \alpha_1/\alpha$ (and focus, as usual, on the first category). In the symmetric case one has $p_i = 1/g = p$, from which one readily gets

$$\mathbb{P}(k) = \text{Bin}(k; n, 1/g) = \binom{n}{k} \left(\frac{1}{g}\right)^k \left(1 - \frac{1}{g}\right)^{n-k}. \tag{5.56}$$

In the thermodynamic limit $n, g \gg 1$, with $\lambda = n/g$, one also has $p \to 0$, and $n \to \infty$ while $np = \lambda$ remains finite. Writing $p = \lambda/n$, one gets the so-called Poisson distribution

$$\mathrm{Bin}(k;n,p) = \binom{n}{k} p^k (1-p)^{n-k} \simeq \frac{n^k}{k!}\left(\frac{\lambda}{n}\right)^k \left(1-\frac{\lambda}{n}\right)^{n-k}$$

$$\to \frac{\lambda^k}{k!} e^{-\lambda} = \mathrm{Poi}(k;\lambda). \tag{5.57}$$

The moments of this distribution can be easily recovered from the moments of the binomial:

$$\mathbb{E}_{\mathrm{Poi}}(K) = \lim_{n\to\infty, p\to 0} np = \lambda, \tag{5.58}$$

$$\mathbb{V}_{\mathrm{Poi}}(K) = \lim_{n\to\infty, p\to 0} np(1-p) = \lambda. \tag{5.59}$$

These results can also be directly obtained from the definition in (5.57), but by means of rather cumbersome calculations.

In the above calculations, first the limit $\alpha \to \infty$ is performed, and then the thermodynamic limit. The converse is also possible, also leading to the Poisson distribution from the negative binomial distribution. From (5.49) one can see what happens when $\alpha_1 \to \infty$, $\chi \to 0$ and the product $\alpha_1\chi = \lambda$ stays constant. The various pieces of (5.49) behave as follows: $\alpha_1^{[k]} \simeq \alpha_1^k$, $\chi = n/\alpha \ll 1$, $(1/(1+\chi))^{\alpha_1} \simeq (1-\chi)^{\alpha_1} = (1-\lambda/\alpha_1)^{\alpha_1}$, $(\chi/(1+\chi))^k \simeq \chi^k$, then one finally gets

$$\mathrm{NegBin}(k;\alpha_1,\chi) = \frac{\alpha_1^{[k]}}{k!}\left(\frac{1}{1+\chi}\right)^{\alpha_1}\left(\frac{\chi}{1+\chi}\right)^k$$

$$\simeq \frac{\alpha_1^k}{k!}\left(1-\frac{\lambda}{\alpha_1}\right)^{\alpha_1}\chi^k \simeq \frac{(\alpha_1\chi)^k}{k!}e^{-\alpha_1\chi}$$

$$= \frac{\lambda^k}{k!}e^{-\lambda} = \mathrm{Poi}(k;\lambda). \tag{5.60}$$

5.3.6 Binomial distribution as the thermodynamic limit of the g-variate hypergeometric distribution

The generalized g-variate Pólya distribution with negative integer weights coincides with the g-variate hypergeometric distribution; in this case, α_1 and $\alpha - \alpha_1$ are negative integers. Therefore $\alpha_1^{[k]} = (-1)^k(|\alpha_1|)(|\alpha_1|-1)\cdots(|\alpha_1|-k+1)$.

Hence, the domain of $n_1 = k$ is limited, $0 \le k \le |\alpha_1|$, with negative $\chi = n/\alpha$, and (5.49) becomes

$$(-1)^k \binom{|\alpha_1|}{k} \left(\frac{1}{1-|\chi|}\right)^{\alpha_1} \left(\frac{-|\chi|}{1-|\chi|}\right)^k$$
$$= \binom{|\alpha_1|}{k} \left(\frac{1}{1-|\chi|}\right)^{-|\alpha_1|} \left(\frac{|\chi|}{1-|\chi|}\right)^k. \quad (5.61)$$

Note that $|\chi| \le 1$, as $n \le |\alpha|$. For the sake of clarity, one can set $|\alpha_1| = m$, $|\chi| = p$, and get

$$\binom{|\alpha_1|}{k} \left(\frac{1}{1-|\chi|}\right)^{-|\alpha_1|} \left(\frac{|\chi|}{1-|\chi|}\right)^k = \binom{m}{k} (1-p)^m \frac{p^k}{(1-p)^k}$$
$$= \binom{m}{k} p^k (1-p)^{m-k} = \mathrm{Bin}(k;m,p). \quad (5.62)$$

5.3.7 Example: random distribution of students into classrooms

Assume that in assigning students to classrooms in a high school, teachers initially have $n = 100$ students to allocate, and they have to distribute them into $g = 5$ classrooms; teachers choose an accommodation probability equal to

$$\mathbb{P}(X_{m+1} = j | X_1 = x_1, \ldots, X_m = x_m) = \frac{25 - m_j}{25 \cdot 5 - m}. \quad (5.63)$$

This means that no classroom can contain more than 25 students, and each student, once called, has a uniform probability of ending up in one of the available places. The resulting generalized Pólya is a hypergeometric distribution with $n = 100$, $g = 5$, $\alpha_i = -25$, and $\alpha = -125$. The expected number of students per classroom is $n|\alpha_i|/\alpha = 20$, the variance is given by (5.23) leading to

$$\mathbb{V}(n_1) = 100 \cdot \frac{1}{5} \cdot \frac{4}{5} \cdot \frac{125 - 100}{125 - 1} = 3.2, \quad (5.64)$$

with a standard deviation equal to $\sigma = \sqrt{3.2} = 1.8$. Note that if the allocations were independent, the variance would have been $100 \cdot (1/5) \cdot 4/5 = 16$, and the standard deviation $\sigma = 4$. From (5.13), the probability that a class (the first one, say, as they are all equidistributed) receives k students is given by

$$\mathbb{P}(k) = \frac{\binom{25}{k}\binom{100}{100-k}}{\binom{125}{100}}. \quad (5.65)$$

Now imagine that some school closes in the neighbourhood, so that teachers have to further allocate 100 or 1000 students, and they are obliged to look for 5 or 50 additional classrooms. Using the second form of (5.13), one can write

$$\mathbb{P}(k) = \binom{|\alpha_1|}{k} \frac{\binom{|\alpha| - n}{|\alpha_1| - k}}{\binom{|\alpha|}{|\alpha_1|}} \simeq \binom{|\alpha_1|}{k} p^k (1-p)^{n-k}, \tag{5.66}$$

where $p = |\chi| = n/|\alpha| = 0.8$. Hence the probability of allocating k students to each classroom is approximated by the probability of making $|\alpha_1| = 25$ independent and identically distributed trials – the observations of the 25 possible places – each of them with probability $p = 0.8$ of being occupied. Note that this binomial process is not the student accommodation process (which is hypergeometric), but the place occupation process, which in the thermodynamic limit becomes Bernoullian, and p is the expected occupation number of each place in a classroom.

5.3.8 Example: flying bombs on London

The previous example was somewhat artificial. On the contrary, the example of this section uses data from the real world and it is very important as it anticipates the methods of analysis which will be used in following sections. However, the example is rather sad as it uses data on flying bombs (V1) hitting London and launched by the Germans from the European continent during World War II. The data are taken from Parzen's book (see the suggestions for further reading at the end of Chapter 1), page 260. Below, the number of flying-bomb hits recorded is given for each of 576 small zones with approximately the same area in South London. The variable z_k denotes the number of regions hit by k bombs:

k	0	1	2	3	4	≥ 5
z_k	229	211	93	35	7	1

Note that Parzen does not give the total number of bombs. Assuming, for the sake of simplicity, that $z_5 = 1$ and $z_i = 0$ for $i \geq 6$, one gets $n = \sum_{k=0}^{5} k z_k = 535$, whereas one knows that $g = \sum_{k=0}^{5} z_k = 576$. It is possible to show that these data are compatible with the following hypotheses:

1. the flying bombs were launched at random; more precisely: they are i.i.d. with symmetric (uniform) probability with regards to categories (zones);
2. the marginal distribution of hits on any area is the Poisson distribution;
3. the expected number of zones with k hits is given by $\mathbb{E}(Z_k) = g\mathrm{Poi}(k;\lambda)$, where $\lambda = n/g$.

Note that $X_m = j$ means that the mth flying bomb hits the jth zone, and the hypothesis is made that

$$\mathbb{P}(X_{m+1} = j | X_1 = x_1, \ldots, X_m = x_m) = \frac{1}{g}. \tag{5.67}$$

Then the sampling distribution on $\mathbf{n} = (n_1, \ldots, n_g)$ is the symmetric multinomial distribution

$$\mathbb{P}(\mathbf{n}) = \frac{n!}{n_1! \cdots n_g!} g^{-n}. \tag{5.68}$$

Indeed, $\mathbb{P}(\mathbf{n})$ is exchangeable with respect to the g categories meaning that

$$\mathbb{P}(n_1 = k_1, \ldots, n_g = k_g) = \mathbb{P}(n_{i_1} = k_1, \ldots, n_{i_g} = k_g), \tag{5.69}$$

where $i_1, \ldots, i_g$ is one of the $g!$ permutations of $1, \ldots, g$. Thus

$$\mathbb{P}(\mathbf{n}) = \left(\frac{g!}{z_0! \cdots z_n!} \right)^{-1} \mathbb{P}(\mathbf{z}), \tag{5.70}$$

where $\mathbf{z} = (z_0, z_1, \ldots, z_n)$ is the frequency vector for $\mathbf{n}$, or the partition vector for $\mathbf{X}^{(n)}$. Equation (5.70) is analogous to Equation (4.59) for individual sequences and occupation vectors. Note that for all vectors belonging to the same partition $\mathbf{z}$, one has

$$\prod_{i=1}^{g} n_i! = \prod_{k=0}^{n} k!^{z_k}, \tag{5.71}$$

as z_k is the number of terms equal to $k!$ appearing in $\prod_{i=1}^{g} n_i!$. Therefore, one gets the following important equation

$$\mathbb{P}(\mathbf{z}) = \frac{g!}{\prod_{i=0}^{n} z_i!} \mathbb{P}(\mathbf{n}) = \frac{g!}{\prod_{i=0}^{n} z_i!} \frac{n!}{\prod_{k=1}^{g} n_k!} g^{-n} = \frac{g! n!}{\prod_{i=0}^{n} z_i! (i!)^{z_i}} g^{-n}, \tag{5.72}$$

where $\mathbf{n}$ is any vector corresponding to $\mathbf{z}$. The domain of $\mathbb{P}(\mathbf{z})$ is $A = \{\mathbf{z} : \sum_{k=0}^{535} z_k = 576, \sum_{k=0}^{535} k z_k = 535\}$; $\#A$ is large, but a closed analytical formula is not available. Moreover, $\mathbb{P}(\mathbf{z})$ is an $n+1$-dimensional distribution, not easily compared with empirical observations of $\mathbf{z}$. There are two routes to derive numbers that can be compared with the available data:

1. calculation of $\mathbb{E}(Z_k)$, an exact procedure if the marginal distribution is known for $\mathbb{P}(\mathbf{n})$;
2. an approximate search for the most probable vector, an approximate method which can be traced back to Boltzmann.

The random variable Z_k counts the number of zones hit by k bombs. Denoting by $I^{(k)}_{n_j} = \mathbb{I}_{n_j=k}$ the indicator function of the event $\{n_j = k\}$, the random variable Z_k can be written as follows

$$Z_k = I^{(k)}_{n_1} + I^{(k)}_{n_2} + \ldots + I^{(k)}_{n_g}. \tag{5.73}$$

Therefore, one has that

$$\mathbb{E}(Z_k) = \sum_{j=1}^{g} \mathbb{E}(I^{(k)}_{n_j}) = \sum_{j=1}^{g} \mathbb{P}(n_j = k), \tag{5.74}$$

where $\mathbb{P}(n_j = k)$ is the marginal distribution for the jth category. As a consequence of the symmetry of the multinomial distribution (5.68) the marginal distributions coincide for all the variables n_i, so that Equation (5.74) simplifies to

$$\mathbb{E}(Z_k) = g\mathbb{P}(n_1 = k). \tag{5.75}$$

Marginalizing the multinomial distribution (5.68), it turns out that

$$\mathbb{P}(n_1 = k) = \binom{n}{k}\left(\frac{1}{g}\right)^k \left(1 - \frac{1}{g}\right)^{n-k} \simeq \frac{\lambda^k}{k!}e^{-\lambda}, \tag{5.76}$$

with $\lambda = n/g = 0.93$. Eventually, from (5.75), one arrives at

$$\mathbb{E}(Z_k) = g\binom{n}{k}\left(\frac{1}{g}\right)^k \left(1 - \frac{1}{g}\right)^{n-k} \simeq g\frac{\lambda^k}{k!}e^{-\lambda}. \tag{5.77}$$

Note that the only approximation used is the one yielding the Poisson limit; all the other steps in this method are exact.

The second approximate method makes use of Lagrange multipliers in order to find the most probable vector. Indeed, one has to maximize $\mathbb{P}(\mathbf{z})$ given by Equation (5.72) and subject to the following constraints

$$\sum_{k=0}^{n} z_k = g, \tag{5.78}$$

$$\sum_{k=0}^{n} k z_k = n. \tag{5.79}$$

Of the $n+1$ variables $z_0, \ldots, z_n$, only $n-1$ of them are free, due to the two conditions (5.78) and (5.79). Two free parameters a and b are introduced and one searches the stationary values (the maximum in our case) of the function

$$L(\mathbf{z}) = \ln(\mathbb{P}(\mathbf{z})) + a\left(\sum_{k=0}^{n} z_k - g\right) + b\left(\sum_{k=0}^{n} k z_k - n\right), \tag{5.80}$$

with respect to all $z_0, \ldots, z_n$. The solution of this problem will be a vector $\mathbf{z}^*$ depending on a and b, these parameters to be chosen so that the solution obeys (5.78) and (5.79). The use of $\ln(\mathbb{P}(\mathbf{z}))$ is convenient as the logarithm is a monotonically increasing function and the probabilities are strictly positive in this case, therefore a constrained maximum of the logarithm coincides with a constrained maximum of $\mathbb{P}(\mathbf{z})$; moreover, one can use Stirling's approximation for the logarithm of the factorial, $\ln x! \simeq x\ln x - x$, leading to

$$\ln \mathbb{P}(\mathbf{z}) = \ln g! + \ln n! - n\ln g - \sum_{k=0}^{n} \ln(z_k! k!^{z_k}). \tag{5.81}$$

Further assuming that $z_i \gg 1$, one gets the following chain of approximate equalities

$$\begin{aligned}\sum_{k=0}^{n} \ln(z_k! k!^{z_k}) &= \sum_{k=0}^{n} \ln(z_k!) + \sum_{k=0}^{n} z_k \ln k! \\ &\simeq \sum_{k=0}^{n} z_k \ln z_k - \sum_{k=0}^{n} z_k + \sum_{k=0}^{n} z_k \ln k!.\end{aligned} \tag{5.82}$$

The $n+1$ condition

$$\frac{\partial L(\mathbf{z_k})}{\partial z_k} = 0 \tag{5.83}$$

leads to the set of equations

$$-\ln z_k - \ln k! + a + bk = 0, \tag{5.84}$$

eventually yielding the most probable vector with components

$$z_k^* = A\frac{B^k}{k!}, \tag{5.85}$$

with $A = \mathrm{e}^{-a}$, and $B = \mathrm{e}^{-b}$. Now from

$$\sum_{k=0}^{n} A\frac{B^k}{k!} = g, \tag{5.86}$$

and from

$$\sum_{k=0}^{n} kA\frac{B^k}{k!} = n, \tag{5.87}$$

for $n \gg 1$, one finds $B = n/g = \lambda$, and $A = ge^{-\lambda}$. The conclusion of all these calculations is that, under all the approximations discussed above, the most probable occupancy vector has components given by

$$z_k^* = g\frac{\lambda^k}{k!}\mathrm{e}^{-\lambda}. \tag{5.88}$$

Therefore, one has that $z_k^* = \mathbb{E}(Z_k)$. The third row below contains the theoretical values for $\mathbb{E}(Z_k) = g\mathrm{e}^{-\lambda}\lambda^k/k!$, with $g = 576$, $n = 535$ and $\lambda = n/g = 0.93$.

k	0	1	2	3	4	≥ 5
z_k	229	211	93	35	7	1
$\mathbb{E}(Z_k)$	227.5	211.3	98.15	30.39	7.06	1.54

Without going into the details of goodness-of-fit tests, it is possible to see that the theoretical values are quite close to the data. A short-sighted statistician could claim that, given the observed z_k, the probability that a particular zone was hit by k bombs is nothing other than z_k/g, all the zones being equivalent, and z_k being the number of the g zones hit by k bombs. This is fine for past events, but it says nothing about the next hit. Formally, one can write

$$\mathbb{P}(n_i = k|I_{n_1}^{(k)} + I_{n_2}^{(k)} + \ldots + I_{n_g}^{(k)} = z_k) = z_k/g, \tag{5.89}$$

and this is true for all the hits contained in the evidence. A gambler would not need anything more for a bet on the observed 535 hits. On the contrary, a long-sighted statistician, after seeing that the frequency distribution z_k is close to Poisson, would rather estimate the best value of the parameter λ, say $\widehat{\lambda}(\mathbf{z})$, and he/she could conclude that for any region

$$\mathbb{P}(k) \simeq \frac{\widehat{\lambda}^k}{k!}\mathrm{e}^{-\widehat{\lambda}}, \tag{5.90}$$

with $\widehat{\lambda} = n/g$. This is not exactly true for the past, where z_k was observed and not $\mathbb{E}(Z_k)$, but the introduction of a probability distribution gives instruments to predict the future. Moreover, the model proposed above is even deeper, as it describes single hits in the way they occur. Without any doubt, it is the simplest model to understand and the one richest in information. For instance, hits were random and independent; therefore, there was no advantage in avoiding places that had not yet been hit. In other words, the Germans were not able to aim at specific targets with the flying bombs described by Parzen, and the only effect of their action was random destruction and intimidation of the British population without reaching any further specific military goal.

5.4 A consequence of the constraint $\sum n_i = n$

Consider the occupation number random variables $Y_1, \ldots, Y_g$, such that $Y_i = n_i$. The constraint $\sum_{i=1}^{g} n_i = n$ implies that $\mathbb{V}(\sum_{i=1}^{g} Y_i) = 0$. Therefore, one gets

$$0 = \mathbb{V}\left(\sum_{i=1}^{g} Y_i\right) = \sum_{i=1}^{g} \mathbb{V}(Y_i) + 2\sum_{i=1}^{g} \sum_{j=i+1}^{g} \mathbb{C}(Y_i, Y_j). \tag{5.91}$$

If the random variables Y_i are equidistributed, it turns out that

$$g\mathbb{V}(Y_i) + g(g-1)\mathbb{C}(Y_i, Y_j) = 0, \tag{5.92}$$

leading to

$$\mathbb{C}(Y_i, Y_j) = -\frac{1}{g-1}\mathbb{V}(Y_i). \tag{5.93}$$

Given that $\mathbb{V}(Y_i) = \mathbb{V}(Y_j)$, the correlation coefficient (4.19) becomes

$$\rho_{YiYj} = -\frac{1}{g-1}. \tag{5.94}$$

This result is quite general; in particular, for a dichotomous Pólya distribution one finds that $\rho_{Y_iY_j} = -1$; moreover, the correlation coefficient vanishes for $g \to \infty$.

5.5 The continuum limits of the multivariate Pólya distribution

5.5.1 The gamma limit from the negative binomial

Consider the bivariate generalized Pólya sampling distribution given by Equation (5.9), re-written below for the convenience of the reader

$$\mathbb{P}(S_n = k) = \frac{a^{[k]}}{k!}\frac{b^{[n-k]}}{(n-k)!}\frac{n!}{(a+b)^{[n]}}, \; k = 0, 1, \ldots, n. \tag{5.95}$$

The expected value of S_n is

$$\mathbb{E}(S_n) = n\frac{a}{a+b}; \tag{5.96}$$

if one changes the system so that $n \to n' \gg 1$, $b \to b' \gg 1$, with $n'/b' = \chi$ finite and equal to the original $n/(a+b)$, then

$$\mathbb{E}(S_{n'}) = n'\frac{a}{a+b'} \simeq a\chi = \mathbb{E}_\chi(S_{n'}), \tag{5.97}$$

where $\mathbb{E}_\chi(S_{n'})$ denotes the expected value of the same category in the new limiting situation. This important limit of the finite Pólya distribution is the so-called thermodynamic limit already discussed in Sections 5.3.2 and 5.3.3; it can be traced back to Boltzmann, who considered one molecule against the remaining $g-1$ molecules, when $n, g \to \infty$, where n is the number of energy elements (see also Section 2.3.1). This limit corresponds to increasing the size of the system, while keeping constant the expected value of the subsystem under consideration. One expands the system keeping the density constant. If the subsystem is the category whose element occupation number is described by Equation (5.95), the discrete thermodynamic limit is given by the generalized negative-binomial distribution, which was discussed in Section 5.3.3. This limit can be performed for any value of α, and it gives either a negative-binomial distribution ($\alpha > 0$) or a Poisson distribution as the limit of the binomial distribution ($\alpha \to \infty$) or a binomial distribution ($\alpha < 0$). Re-writing the asymptotic form for $\mathbb{P}(S_n = k)$, derived in Section 5.3.3, where the new parameters are a and $\chi = n'/b'$, representing the 'density', or the number of elements *per* unit weight, one has

$$\mathbb{P}(k) = \frac{a^{[k]}}{k!}\left(\frac{1}{1+\chi}\right)^a \left(\frac{\chi}{1+\chi}\right)^k, \; k = 0, 1, \ldots, \infty. \tag{5.98}$$

A different modification of the system is adding more elements to the original population, keeping the number of categories constant. This is possible only if a and b are positive, as the sampling process is extendible to infinity, and nothing prevents increasing the number of elements at will. This second type of transformation increases the 'density' of the system, and, with some rescaling, leads to the 'continuous limit'. Assume we start from Equation (5.98); one cannot increase n, which has already gone to infinity, but it is possible to increase the density $\chi \gg 1$, and, consequently, also most of the values k are $\gg 1$; in the limit $k \gg 1$, $\chi \gg 1$, given that

$$\frac{a^{[k]}}{k!} \simeq \frac{k^{a-1}}{\Gamma(a)}, \text{ for } k \gg a \tag{5.99}$$

$$\left(\frac{1}{1/\chi + 1}\right)^k \simeq (1 - 1/\chi)^k \simeq \exp(-k/\chi), \text{ for } \chi \gg 1, \tag{5.100}$$

one gets the asymptotic form

$$\mathbb{P}(k) \simeq \frac{k^{a-1}\exp(-k/\chi)}{\Gamma(a)\chi^a} = \text{Gamma}(k; a, \chi), \tag{5.101}$$

that is the gamma distribution. To be precise, (5.101) is just an asymptotic approximation: it does not work for small values of k, and it is not the usual gamma distribution, as it is only defined on integers. This is consistent with the discrete model, where k is the number of elements in the chosen category (the number of successes), which stays discrete also if the total number of elements n has grown to infinity. To obtain a true gamma random variable, with domain $y \in [0,\infty)$, one must introduce the value ε of the element, so that a number of elements k corresponds to some extensive quantity $y = k\varepsilon$ pertaining to the category. The gamma limit is what Boltzmann was looking for in his seminal papers, where the elements were a trick to discretize a continuous quantity: energy. In his case, the continuous limit was essential to recover classical continuous distributions. In economics this is not needed in general. An exception could be the case wherein, if elements represent tokens of wealth, one wishes to pass from hundreds of euros to euros, and then to eurocents, and so on. Introducing the quantum ε, wealth is given by $y = k\varepsilon$. Therefore $\mathbb{P}(Y = k\varepsilon)$ is equal to $\mathbb{P}(k = y/\varepsilon)$, and in the limit $\varepsilon \to 0$, for any y finite the corresponding k is very large. Then Equation (5.101) holds true in all the domain. At the same time the random variable Y has a continuous limit, as the distance between two consecutive points is $\varepsilon \to 0$. Then applying (5.101) to $k = y/\epsilon$, one finds

$$\mathbb{P}(k) \simeq \frac{\left(\frac{y}{\varepsilon}\right)^{a-1} \exp\left(-\frac{y/\varepsilon}{\chi}\right)}{\Gamma(a)\chi^a} = \frac{y^{a-1} \exp\left(-\frac{y}{\chi\varepsilon}\right)}{\Gamma(a)\varepsilon^{a-1}\chi^a} = \varepsilon \frac{y^{a-1} \exp\left(-\frac{y}{w}\right)}{\Gamma(a)w^a}, \tag{5.102}$$

where $w = \chi\varepsilon$ is the new rescaled parameter; from $\mathbb{P}(y = k\varepsilon)\varepsilon = \mathbb{P}(k = y/\varepsilon)$, it follows that $Y \sim \text{Gamma}(a, w)$, for $y \in [0,\infty)$. In other words, in the continuous limit, the random variable Y follows the gamma distribution with probability density

$$p(y) = \text{Gamma}(y; a, w) = \frac{y^{a-1} \exp\left(-\frac{y}{w}\right)}{\Gamma(a)w^a}. \tag{5.103}$$

This route to the gamma distribution is well summarized by considering the various limits for the moments of the corresponding distributions. At each stage the relevant parameters are evidenced. Consider the number of elements S_n in the chosen category (number of successes) out of n elements:

1. (Pólya) If $S_n \sim \text{Polya}(n, a, b)$ (Equation 5.95), then

$$\mathbb{E}(S_n) = n\frac{a}{a+b}, \tag{5.104}$$

and

$$\mathbb{V}(S_n) = n\frac{a}{a+b}\frac{b}{a+b}\frac{a+b+n}{a+b+1}. \tag{5.105}$$

2. (Negative Binomial) The thermodynamic limit is obtained for $b \to \infty$, $n \to \infty$ and $n/b = \chi$; in this limit the number of successes S follows the negative-binomial distribution ($S \sim \mathrm{NegativeBinomial}(a, \chi)$, Equation (5.98)) and one has

$$\mathbb{E}(S) = a\chi, \tag{5.106}$$

and

$$\mathbb{V}(S) = a\chi(1+\chi). \tag{5.107}$$

3. (Gamma) The continuous limit is obtained as follows: introducing $\varepsilon \to 0$, then if $\chi \to \infty$, $\chi\epsilon = w$, the *new* variable $S\epsilon$ converges to a gamma-distributed random variable $Y \sim \mathrm{Gamma}(a, w)$, with $y \in [0, \infty)$, and one further has that

$$\mathbb{E}(Y) = aw, \tag{5.108}$$

and

$$\mathbb{V}(Y) = aw^2. \tag{5.109}$$

The limits of the moments can be directly derived from the values of the original distribution; for instance, in the gamma case, one has

$$\mathbb{E}(Y) = \lim_{\varepsilon \to 0, \chi \to \infty} \varepsilon \mathbb{E}(S), \tag{5.110}$$

and

$$\mathbb{V}(Y) = \lim_{\epsilon \to 0, \chi \to \infty} \varepsilon^2 \mathbb{V}(S). \tag{5.111}$$

5.5.2 The gamma limit from the beta

An alternative route to the gamma limit is to take the continuum limit first, and then the thermodynamic one. The continuum limit of the dichotomous Pólya distribution is obtained by setting $n \to \infty$, $k \to \infty$ with constant $u = k/n$; one finds that u converges to a continuous variable in the domain $[0, 1]$, whose probability density function is known. Introducing the left-continuous cumulative distribution function for S_n, that is $F_n(t) = \mathbb{P}(S_n < t)$, and the same cumulative distribution function for $U = S_n/n$, that is $G_n(t) = \mathbb{P}(U < t)$, one finds that $F_n(t)$ is a step function whose jumps occur at the integers $0, 1, \ldots, n$, whereas, for $G_n(t)$, the same jumps are at the points $0, 1/n, \ldots, n/n$, whose distance is $\Delta u = 1/n$. In formulae, $F_n(t) = G_n(t/n)$; in words, the probability of a frequency of successes less than k coincides with the probability of a relative success frequency less than k/n. Then, one has that

$$\Delta F_n(k) = F_n(k+1) - F_n(k) = \Delta G_n(k/n) = G_n((k+1)/n) - G_n(k/n). \tag{5.112}$$

While the first term is exactly $\mathbb{P}(S_n = k)$, the second can be written as:

$$\Delta G_n(k/n) = G_n(u + \Delta u) - G_n(u) = \frac{\Delta G_n(u)}{\Delta u}\Delta u = \frac{\Delta G_n(u)}{n\Delta u}, \tag{5.113}$$

so that

$$n\mathbb{P}(S_n = nu) = \frac{\Delta G_n(u)}{\Delta u}. \tag{5.114}$$

In the limit $n \to \infty$, one finds that $\Delta u = 1/n \to 0$, and $\lim_{n\to\infty}[\Delta G_n(u)/\Delta u] = \mathrm{d}G(u)/\mathrm{d}u = p(u)$, with $u \in [0,1]$. Then $G_n(u)$ becomes continuous and derivable in $[0,1]$, and its derivative is the probability density function of the (continuous) relative frequency.

Now, we consider the asymptotic behaviour $n\mathbb{P}(S_n = k)$, in the limit $n \to \infty$, and $k \to \infty$ with constant $u = k/n$. From Equation (5.99) one has:

$$\begin{aligned}\mathbb{P}(S_n = k) &= \frac{a^{[k]}}{k!}\frac{b^{[n-k]}}{(n-k)!}\frac{n!}{(a+b)^{[n]}} \approx \frac{k^{a-1}}{\Gamma(a)}\frac{(n-k)^{b-1}}{\Gamma(b)}\frac{\Gamma(a+b)}{n^{a+b-1}} \\ &= B(a,b)^{-1}\frac{1}{n}\left(\frac{k}{n}\right)^{a-1}\left(1-\frac{k}{n}\right)^{b-1},\end{aligned} \tag{5.115}$$

and setting $k = nu$, it turns out that $n\mathbb{P}(S_n = nu) \to p(u)$, where

$$p(u) = B(a,b)^{-1}u^{a-1}(1-u)^{b-1}, \tag{5.116}$$

that is $U \sim \mathrm{Beta}(a,b)$. The moments of $\mathrm{Beta}(a,b)$ are easily recovered for those of the corresponding Pólya distribution (see also Exercise 4.6). Recalling Equations (5.104) and (5.105), one finds

$$\mathbb{E}(U) = \lim_{n\to\infty}\frac{\mathbb{E}(S_n)}{n} = \frac{a}{a+b}, \tag{5.117}$$

$$\mathbb{V}(U) = \lim_{n\to\infty}\frac{\mathbb{V}(S_n)}{n^2} = \frac{a}{a+b}\frac{b}{a+b}\frac{1}{a+b+1}. \tag{5.118}$$

Extending the same procedure to the multivariate Pólya distribution, one can prove that, setting $u_i = n_i/n$, with $\sum_{i=1}^{g} u_i = 1$, then

$$n^{g-1}\mathrm{Polya}(\mathbf{n};\boldsymbol{\alpha}) \to \frac{\Gamma(\alpha)}{\prod_{i=1}^{g}\Gamma(\alpha_i)}\prod_{i=1}^{g}u_i^{\alpha_i-1}du_1\cdots du_{g-1}, \tag{5.119}$$

a multidimensional probability density function defined on the simplex $0 \le u_i \le 1$ and $\sum_{i=1}^{g} u_i = 1$ for all the $i \in \{1,\ldots g\}$. The probability density (5.119) defines the *Dirichlet distribution*. Let $\mathbf{U} \sim \mathrm{Dir}(\boldsymbol{\alpha})$ denote that the random vector $\mathbf{U}$ is distributed according to the Dirichlet distribution of parameter vector $\boldsymbol{\alpha}$. As a consequence of the Pólya label mixing property (Equation (5.36)), one immediately derives the *aggregation* property of the Dirichlet distribution. Note that $\mathbf{U} = U_1, \ldots, U_g$ is a sequence of random variables with values on the simplex $\sum_{i=1}^{g} u_i$ with $0 \leq u_i \leq 1, \forall i \in \{1, \ldots g\}$ whose joint distribution is $\mathrm{Dir}(\alpha_1, \ldots, \alpha_i, \ldots, \alpha_{i+k}, \ldots, \alpha_g)$. If one introduces $U_A = \sum_{j=i}^{i+k} U_j$, then the new sequence $(U_1, \ldots, U_A, \ldots, U_g) \sim \mathrm{Dir}(\alpha_1, \ldots, \alpha_A, \ldots, \alpha_g)$, where $\alpha_A = \sum_{j=i}^{i+k} \alpha_j$. The aggregation property immediately leads to the marginal distribution of the Dirichlet distribution, whose probability density function is nothing other than the beta distribution. If $\mathbf{U} \sim \mathrm{Dir}(\alpha_1, \ldots, \alpha_g)$ then

$$U_i \sim \mathrm{Beta}(\alpha_i, \alpha - \alpha_i). \tag{5.120}$$

Here, a continuous variable was obtained introducing the relative frequency, which is adimensional. Formally, the gamma limit of the beta distribution is straightforward. If, in Equation (5.116), u is replaced by y/y_0, and given that $p(u) = g(y)y_0$, one gets:

$$g(y) = B(a,b)^{-1} \frac{1}{y_0} \left(\frac{y}{y_0} \right)^{a-1} \left(1 - \frac{y}{y_0} \right)^{b-1}, \tag{5.121}$$

for $y \in [0, y_0]$; this is the probability density function of the so-called *beta–Stacy distribution* $Y \sim \mathrm{betaStacy}(a, b, y_0)$. The beta–Stacy distribution, often used in modern Bayesian statistics, is nothing other than the beta distribution rescaled on the interval $y \in [0, y_0]$. Now the gamma limit is obtained by letting $y_0 \to \infty$, $b \to \infty$, and $y_0/b = \chi$ remaining constant. A little thought is sufficient to see that, if no 'quantum' is introduced, both y and y_0 are integer numbers (they are nothing other than the old k and n), so that one ends with the same impasse as in Equation (5.101). In order to obtain a Gamma variable, with domain $y \in [0, \infty)$, one must restart from Equation (5.115)

$$\mathbb{P}(S_n = k) \simeq \frac{\Gamma(a+b)}{\Gamma(a)\Gamma(b)} \frac{1}{n} \left(\frac{k}{n} \right)^{a-1} \left(1 - \frac{k}{n} \right)^{b-1}, \tag{5.122}$$

which works for $k \gg 1$ and $n \gg 1$. One introduces the value of the element again, so that total number of elements n corresponds to some extensive quantity $n\varepsilon = y_0$ and $k\varepsilon = y$ is the part pertaining to the category under scrutiny. Now if $n \to \infty$, $\varepsilon \to 0$, with $n\epsilon = y_0$, then $\mathbb{P}(y = k\varepsilon)$, with $y \in [0, y_0]$ converges to a continuous distribution, and one finds the following probability density function:

$$g(y)\varepsilon \simeq B(a,b)^{-1} \frac{1}{n} \left(\frac{k}{n} \right)^{a-1} \left(1 - \frac{k}{n} \right)^{b-1} = B(a,b)^{-1} \frac{\varepsilon}{y_0} \left(\frac{y}{y_0} \right)^{a-1} \left(1 - \frac{y}{y_0} \right)^{b-1}. \tag{5.123}$$

Now, from Equations (5.99) and (5.100), and from the fact that for $b \gg a$ one has

$$\frac{\Gamma(b+a)}{\Gamma(b)b^a} = \frac{b^{[a]}}{b^a} \to 1, \tag{5.124}$$

considering the limit $y_0 \to \infty, b \to \infty$, with $y_0/b = w$ constant, one eventually finds

$$g(y) = \text{Gamma}(y; a, w) = \frac{y^{a-1}\exp\left(-\frac{y}{w}\right)}{\Gamma(a)w^a}, \tag{5.125}$$

which coincides with $p(y)$ in Equation (5.103).
This alternative route to gamma is summarized by again considering the various limits and the moments of the corresponding distributions. At each stage the relevant parameters are indicated.

1. (Pólya) If $S_n \sim \text{Polya}(n, a, b)$ (Equation 5.95), then $\mathbb{E}(S_n)$ is given by Equation (5.104) and $\mathbb{V}(S_n)$ by Equation (5.105).
2. (Beta) The continuous limit is obtained as follows: if $n \to \infty$, $k \to \infty$ and $k/n = u$; then $U \sim \text{beta}(a, \chi)$ and

$$\mathbb{E}(U) = \frac{a}{a+b}, \tag{5.126}$$

$$\mathbb{V}(U) = \frac{a}{a+b}\frac{b}{a+b}\frac{1}{a+b+1}. \tag{5.127}$$

3. (Gamma) Finally, the thermodynamic limit is obtained as follows: introducing $y = y_0 u$, if $b \to \infty$, $y_0 \to \infty$, with $y_0/b = w$, the *new* variable $y_0 U$ converges to $Y \sim \text{Gamma}(a,w)$, with $y \in [0, \infty)$, and

$$\mathbb{E}(Y) = aw, \tag{5.128}$$

$$\mathbb{V}(Y) = aw^2. \tag{5.129}$$

As before, the limits of the moments are directly derived by the previous values; in the gamma case again, it is easy to check that

$$\mathbb{E}(Y) = \lim_{b\to\infty, y_0\to\infty} y_0 \mathbb{E}(U), \tag{5.130}$$

and

$$\mathbb{V}(Y) = \lim_{b\to\infty, y_0\to\infty} y_0^2 \mathbb{V}(U). \tag{5.131}$$

The scale factor of the gamma distribution is w, corresponding to the scale factor χ of the negative-binomial distribution. Their intuitive meaning is that, if a category has initial weight a, and the expected value of marginal variable is μ, then $\chi = \mu/a$

if the variable is the occupation number, or $w = \mu/a$ if the variable is the continuous magnitude. In the example of Section 5.3.4, one has $a = 1$, $b = g - 1$ and $n/g = 5$, that is the scale parameter of the negative-binomial approximation is equal to the average number of coins. When coins represent euros, NegativeBinomial(1,5) is just Geometric(5) in the domain $\{0,1,\ldots,\infty\}$. If an authority retires euros, and replaces them with eurocents, whose distribution follows the same predictive probability, now there are $n' = 100n$ coins in circulation, but the expected wealth is the same. The new variable is closer to the continuous limit, and the numerical value of w is always equal to 5 euros = 500 cents. For $\epsilon \to 0$, the wealth is Exponential(5)-distributed in the domain $[0,\infty)$.

5.5.3 Further remarks on the limits

The most important limit of the finite Pólya distribution is the thermodynamic limit, which corresponds to increasing the *size* (the number of categories) of the system, keeping the density constant, which is the expected value of the subsystem under consideration. If the subsystem is a category whose occupation number is described by Equation (5.95), the discrete thermodynamic limit is given by the generalized negative-binomial distribution, which has been discussed in Section 5.3.3. This limit can be performed for any value of α, and it gives a usual negative-binomial distribution ($\alpha > 0$), a Poisson distribution ($\alpha \to \infty$), or a binomial distribution ($\alpha < 0$). If correlations are non-negative ($\alpha > 0$), a continuous limit can be performed, whose meaning is increasing the *density of elements*. The continuum limit is a useful simplification of the high density case (large numbers of elements in few categories), or can be used to refine the degree of precision of finite magnitudes, reducing at will the 'quanta' assumed at the very beginning. This focus on the introduction of continuous variables in an essentially discrete landscape can be traced back to Boltzmann's deep conviction that the continuum is nothing other than a (sometimes) useful device to describe a world which is always finite and discrete. Among contemporary probability scholars, a similar deep conviction was shared by E.T. Jaynes, who always underlined the importance of the limiting procedure leading to continuous quantities. His opinion can be summarized as follows: if one wishes to use continuous quantities, one must show the exact limiting procedure from the finite assumptions which are the necessary starting point.

5.6 The fundamental representation theorem for the Pólya process

The predictive form

$$\mathbb{P}(X_{m+1} = j|\mathbf{X}^{(m)}) = \frac{\alpha_j + m_j}{\alpha + m}, \tag{5.132}$$

whose historical root is in Laplace, has been justified from many points of view in the past, without any appeal to urns and manipulation of balls. Given that, in a celebrated paper, Karl Pearson himself claimed that the fundamental problem in practical statistics is assigning probability to an event based on the empirical evidence and the previous information, one can see that Equation (5.132) is a possible answer to the fundamental problem. It is usually justified (see Zabell, Further reading) following Johnson, a logician from Cambridge who was Keynes' teacher, under the postulate of *sufficientness*, which imposes that the predictive form must depend only on the number of successes and the size of the sample, and not on what happens in a failure. Johnson's proof does not say anything in the case of two categories, where the assumptions are trivial, and so it adds nothing to the basic probability axioms. Followers of this approach are then always obliged to state that the number of categories must be greater than two. Rudolph Carnap went around the obstacle, introducing an additional 'linearity principle' for the case of only two categories. A more satisfactory derivation, which works in any case, is due to Costantini, later improved to encompass negative correlations. In the following proof of the fundamental representation theorem in Costantini's version (see Further reading, Chapter 4, Costantini (1979)), the symbol λ will be used instead of α, following Costantini, who used it just in memory of Carnap's 'λ-principle'. The proof is based on the notion of relevance quotient, similar to the 'coefficient of influence' introduced by Keynes. Here the proof is given in order to show also how it is possible to axiomatize a predictive probability, without introducing urn models which can be used later just to interpret the probabilistic conditions.

Definition (Heterorelevance coefficient) The coefficient of heterorelevance of category j with respect to i is

$$Q_i^j(\mathbf{X}^{(m)}) = \frac{\mathbb{P}(X_{m+2} = i|\mathbf{X}^{(m)}, X_{m+1} = j)}{\mathbb{P}(X_{m+1} = i|\mathbf{X}^{(m)})}, \tag{5.133}$$

where $\mathbb{P}(X_{m+1} = i|\mathbf{X}^{(m)})$ gives the probability of observing category i at step $m+1$ based on the observed evidence $\mathbf{X}^{(m)}$. Note that $Q_i^j(\mathbf{X}^{(m)})$ represent how a failure $j \neq i$ at the $(m+1)$th step influences the probability of observing the ith category at the $(m+2)$th step. It is a measure for re-assessing the predictive probability for a category after a failure. Note that independence would imply $Q_i^j(\mathbf{X}^{(m)}) = 1$. In terms of $Q_i^j(\mathbf{X}^{(m)})$, we can introduce the following conditions:

C1. *Exchangeability* (or symmetry) of $Q_i^j(\mathbf{X}^{(m)})$ with respect to the exchange of i and j, that is

$$Q_i^j(\mathbf{X}^{(m)}) = Q_j^i(\mathbf{X}^{(m)}). \tag{5.134}$$

Indeed, **C1** and (5.133) imply

$$\mathbb{P}(X_{m+1} = i, X_{m+2} = j|\mathbf{X}^{(m)}) = \mathbb{P}(X_{m+1} = j, X_{m+2} = i|\mathbf{X}^{(m)}), \tag{5.135}$$

that is the recursive definition of exchangeability.

C2. *Invariance* of $Q_i^j(\mathbf{X}^{(m)})$, that is

$$Q_i^j(\mathbf{X}^{(m)}) = Q_m. \tag{5.136}$$

This means that, provided that $i \neq j$, $Q_i^j(\mathbf{X}^{(m)})$ does not depend on either i and j, or on the evidence $\mathbf{X}^{(m)}$; it is a function of m alone. If **C2** holds true, then one can write

$$\mathbb{P}(X_{m+1} = i|\mathbf{X}^{(m)}) = p(i|m_i, m), \tag{5.137}$$

which represents Johnson's sufficientness principle.

Theorem (Fundamental representation theorem for the Pólya process) If **C2** holds true, that is $Q_i^j(\mathbf{X}^{(m)}) = Q_m$, then one has

$$\mathbb{P}(X_{m+1} = j|\mathbf{X}^{(m)}) = p(j|m_j, m) = \frac{\lambda p_j + m_j}{\lambda + m}, \tag{5.138}$$

where

$$\lambda = \frac{Q_0}{1 - Q_0}, \tag{5.139}$$

and

$$p_j = \mathbb{P}(X_1 = j). \tag{5.140}$$

Proof Whatever the number g of categories, denote with a a given category, and with b its complement. Let $p(a|m,s)$ represent the predictive probability that the $(m+1)$th observation is in a after m successes and s trials. Equation (5.133) simplifies to

$$p(a|m, s+1) = Q_s p(a|m, s). \tag{5.141}$$

For any evidence, denote with $p(a|\cdot)$ the predictive probability of the category, and with $p(b|\cdot) = 1 - p(a|\cdot)$ its complement. Note that predictive probabilities starting from 'pure' evidences such as $(0, s)$ (all failures), or (s, s) (all successes), are easily written in terms of the initial probability $p_a = p(a|0,0)$ and of $Q_0, \ldots, Q_{s-1}$. Let us denote by $Q_{(r,t)}$ the following product

$$Q_{(r,t)} = Q_r Q_{r+1} \cdot \ldots \cdot Q_t = \prod_{i=r}^{t} Q_i. \tag{5.142}$$

If the evidence is a sequence of only failures for the category under scrutiny, then one has

$$p(a|0,s) = Q_0 Q_1 \cdot \ldots \cdot Q_{s-1} p(a|0,0) = Q_{(0,s-1)} p_a; \tag{5.143}$$

if the evidence is a sequence of all successes, then for the complementary category b they have been all failures, and one can write

$$p(b|0,s) = Q_{(0,s-1)} p_b. \tag{5.144}$$

But, due to the fact that $p(b|\cdot) = 1 - p(a|\cdot)$, one is led to

$$p(a|s,s) = 1 - p(b|0,s) = 1 - Q_{(0,s-1)}(1 - p_a) = 1 - Q_{(0,s-1)} + Q_{(0,s-1)} p_a, \tag{5.145}$$

where $p_a = 1 - p_b$ has been used. Summarizing, one has

$$p(a|0,s) = Q_{(0,s-1)} p_a, \tag{5.146}$$

$$p(a|s,s) = 1 - Q_{(0,s-1)} + Q_{(0,s-1)} p_a. \tag{5.147}$$

As for the generic predictive probability $p(a|m,s)$, one is able to transport the initial probability (p_a, p_b) along the borders of the allowed region of individual descriptions, considering first the sequence

$$S_1 = (a_1, \ldots, a_m, b_{m+1}, \ldots, b_s), \tag{5.148}$$

joining $(0,0)$ to (m,s) through (m,m), and then the reversed sequence

$$S_2 = (b_1, \ldots, b_{s-m}, a_{s-m+1}, \ldots, a_s), \tag{5.149}$$

connecting $(0,0)$ to (m,s) through $(0, s-m)$.

As for S_1, the initial m successes give

$$p(a|m,m) = 1 - Q_{(0,m-1)} + Q_{(0,m-1)} p_a. \tag{5.150}$$

The next $n - m$ failures give

$$p(a|m,s) = Q_{(m,s-1)} p(a|m,m), \tag{5.151}$$

leading to

$$p(a|m,s) = Q_{(m,s-1)}(1 - Q_{(0,m-1)} + Q_{(0,m-1)} p_a), \tag{5.152}$$

and from

$$Q_{(0,m-1)} Q_{(m,s-1)} = Q_{(0,s-1)}, \tag{5.153}$$

one can finally write

$$p(a|m,s) = Q_{(m,s-1)} - Q_{(0,s-1)} + Q_{(0,s-1)} p_a. \tag{5.154}$$

By the same reasoning, for S_2, one gets

$$p(b|s-m,s) = Q_{(s-m,s-1)} - Q_{(0,s-1)} + Q_{(0,s-1)} p_b. \tag{5.155}$$

Considering the predictive distribution at (m,s), we have

$$p(a|m,s) + p(b|s-m,s) = 1, \tag{5.156}$$

so that the sum of Equations (5.154) and (5.155) leads to a recurrence relation on $Q_{(m,s-1)}$:

$$Q_{(m,s-1)} - Q_{(0,s-1)} + Q_{(0,s-1)} p_a + Q_{(s-m,s-1)} - Q_{(0,s-1)} + Q_{(0,s-1)} p_b = 1, \tag{5.157}$$

from which, one finds

$$Q_{(m,s-1)} + Q_{(s-m,s-1)} - Q_{(0,s-1)} = 1. \tag{5.158}$$

The general solution of Equation (5.158) is

$$Q_{(m,s)} = \frac{\lambda + m}{\lambda + s + 1}, \tag{5.159}$$

and, eventually, one finds that

$$Q_s = \frac{\lambda + s}{\lambda + s + 1}. \tag{5.160}$$

Replacing (5.159) in (5.154), we get the thesis

$$p(a|m,s) = \frac{\lambda + m}{\lambda + s} - \frac{\lambda}{\lambda + s} + \frac{\lambda p_a}{\lambda + s} = \frac{\lambda p_a + m}{\lambda + s}, \tag{5.161}$$

for $m = 1, \ldots, s-1$. □

5.6.1 Remark on the recurrence formula

Considering that

$$Q_{(a,b)} = Q_a Q_{a+1} \ldots Q_b = \frac{Q_{(0,b)}}{Q_{(0,a-1)}}, \tag{5.162}$$

the bidimensional equation (5.158) can be transformed into the unidimensional relation

$$\frac{Q_{(0,s-1)}}{Q_{(0,m-1)}} + \frac{Q_{(0,s-1)}}{Q_{(0,s-m-1)}} - Q_{(0,s-1)} = 1, \tag{5.163}$$

and setting $Q^{-1}_{(0,a)} = R(a)$, one gets

$$R(m-1) + R(s-m-1) = 1 + R(s-1). \tag{5.164}$$

For instance, take the symmetric case $m = s - m$, that is $s = 2m$. Then $R(2m-1) = 2R(m-1) - 1$, which is satisfied only if $R(x)$ is a linear function of x. Posing $R(x) = a + bx$, where

$$a = Q_0^{-1} = \frac{\lambda+1}{\lambda}, \tag{5.165}$$

one gets $a + b(2m-1) = 2a + 2b(m-1) - 1$, and

$$b = a - 1 = \frac{1}{\lambda}. \tag{5.166}$$

Therefore, one has

$$R(s) = \frac{\lambda+1}{\lambda} + \frac{s}{\lambda} = \frac{\lambda+s+1}{\lambda}, \tag{5.167}$$

and

$$Q_{(0,s)} = \frac{\lambda}{\lambda+s+1}. \tag{5.168}$$

One finds that

$$R(s) = \frac{\lambda+s+1}{\lambda}, \tag{5.169}$$

satisfies (5.164) for any $1 \leq m \leq s-1$, as is easily verified.

5.6.2 Remarks on the fundamental theorem

The main applications of Equation (5.1) were in the field of predictive inference. Therefore, the case $\alpha > 0$ was mainly studied, which is the usual Pólya case. If $\alpha > 0$, then, using $\alpha_j = \alpha p_j$, the ratio $(\alpha p_j + m_j)/(\lambda + m)$ is always positive, and the n-step process can be extended to infinity; moreover, any frequency vector (m_j, m) can be reached from $(0,0)$ with positive probability (except in the trivial case $p_j = 0$, that is not considered here). Now one has that $\alpha = Q_0/(1 - Q_0)$, and

$Q_0 = \alpha/(\alpha+1)$, and $Q_s = (\alpha+s)/(\alpha+s+1)$; therefore a positive α implies $Q_s < 1$. This is not surprising, because if the observations are positively correlated, after a failure the predictive probability decreases. If $\alpha \to \infty$, then one gets $Q_0 \to 1$, and a failure does not influence the predictive probability in the next trial. In the case $\alpha \to 0$, it turns out that $Q_0 \to 0$; this means that if a category does not appear in the first trial, in the second trial it becomes almost impossible to see it. Then, in the limit $\alpha \to 0$, one expects to see 'pure states' almost exclusively. If $\alpha = 0$ the first category determines all the future process, which is no longer probabilistic after the first trial. Only occupation vectors of the form $(0,\ldots,0,s,0,\ldots,0)$ are possible. For this reason, the value $\alpha = 0$ was refused by Carnap. Remarkably, $\alpha = 0$ leads to the frequentist view of probability, where the predictive rule is

$$\mathbb{P}(X_{m+1} = j|\mathbf{X}^{(m)}) = \frac{m_j}{m}. \tag{5.170}$$

This difficulty is only overcome in the following limit $m \gg \alpha < \infty$, when

$$\frac{\lambda p_j + m_j}{\lambda + m} \simeq \frac{m_j}{m}. \tag{5.171}$$

That the predictive form $(\alpha p_j + m_j)/(\alpha + m)$ is richer of consequences than its frequentist counterpart is evident when writing

$$\frac{\alpha p_j + m_j}{\alpha + m} = \frac{\alpha p_j + m\dfrac{m_j}{m}}{\alpha + m}, \tag{5.172}$$

where α and m are understood as weights applied to p_j (the prior probability) and m_j/m (the empirical frequency distribution). Therefore, the predictive probability is a weighted average of the initial probability (with weight α) and of the frequency distribution (with weight m). One can imagine that the predictive assignment results in two stages: the first step is choosing either the initial probability t with probability $\alpha/(\alpha+m)$ or the empirical distribution e with probability $m/(\alpha+m)$; the second step is selecting from the chosen distribution, with probability p_j or m_j/m, respectively. In formulae

$$\mathbb{P}(X_{m+1} = j|\cdot) = \mathbb{P}(X_{m+1} = j|t,\cdot)\mathbb{P}(t|\cdot) + \mathbb{P}(X_{m+1} = j|e,\cdot)\mathbb{P}(e|\cdot), \tag{5.173}$$

where

$$\mathbb{P}(t|\cdot) = \frac{\alpha}{\alpha+m}, \mathbb{P}(e|\cdot) = \frac{m}{\alpha+m}, \tag{5.174}$$

and

$$\mathbb{P}(X_{m+1} = j|t,\cdot) = p_j, \mathbb{P}(X_{m+1} = j|e,\cdot) = \frac{m_j}{m}. \tag{5.175}$$

Then a large value of α means a small influence of the empirical distribution, whereas a small value of α means a large influence of the already accommodated units on the choice of the new object leading to a sort of herding behaviour.

Some caution is needed in the case $\alpha < 0$. The general form

$$\frac{|\alpha|p_j - m_j}{|\alpha| - m}, \tag{5.176}$$

evidences that $|\alpha|$ is the maximum number of possible observations and that $|\alpha|p_j$ must be integer, in order to satisfy probability axioms. Equation (5.146) means that the prior probability p_j multiplied by the coefficient $Q_{(0,s-1)}$ gives the probability of observing the jth category at the next step after having observed failures only. In the case under scrutiny, one has

$$Q_s = \frac{|\alpha| - s}{|\alpha| - s - 1} > 1; \tag{5.177}$$

therefore, as a consequence of Equation (5.146), one arrives at

$$Q_{(0,s)} = \frac{|\alpha|}{|\alpha| - s - 1} > 1. \tag{5.178}$$

Thus $p(j|0,s) = Q(0,s-1)p_j$ increases with s. There must be an $s^* : p(j|0,s^*) = 1$, and $p(B|0,s^*) = 0$, where B represents the set of all categories not coinciding with j. It is intuitive that $s^* = |\alpha|p_B = |\alpha|(1-p_j)$, which must be integer. Now, if one considers $p(B|0,s)$ one reaches the same conclusion leading to the fact that $|\alpha|p_j$ is integer. The domain of definition of $p(j|m_j,s)$ is such that the possible frequency vectors are those crossed by the sequences connecting $(0,0)$ to $(|\alpha|p_j, |\alpha|(1-p_j))$.

5.7 Exercises

5.1 Prove that

$$\mathbb{E}(X_1 + X_2) = \mathbb{E}(X_1) + \mathbb{E}(X_2). \tag{5.179}$$

5.2 Prove that

$$\mathbb{V}(X_1 + X_2) = \mathbb{V}(X_1) + \mathbb{V}(X_2) + 2\mathbb{C}(X_1, X_2). \tag{5.180}$$

5.3 Prove that

$$\mathbb{E}(I_i^{(k)} I_j^{(k)}) = \mathbb{P}(X_i = k, X_j = k). \tag{5.181}$$

5.4 Consider a dichotomous Pólya process with parameters α_0 and α_1 and define the random variable $Y(m) = +1$ if $X(m) = 1$ and $Y(m) = -1$ if $X(m) = 0$. In other words, one has

$$Y = 2X - 1. \tag{5.182}$$

Then define the random variable $S(m)$ (Pólya random walk) as

$$S(m) = \sum_{i=1}^{m} Y(i). \quad (5.183)$$

Write a program that simulates the Pólya random walk and discuss some interesting cases.

5.5 In the previous exercise, a qualitative discussion was presented for the binomial limit of the dichotomous Pólya distribution. But how close is the symmetric dichotomous Pólya to the binomial distribution when $\alpha_0 = \alpha_1 = 100$? Try to answer this question by means of a Monte Carlo simulation.

5.6 A classical problem in probability theory is often stated as follows. Imagine there are n persons in a room, with $n < 365$. What is the probability that at least two of them have a common birthday? This problem and indeed many other problems related to the birthday distribution can be solved by means of the method discussed in Section 5.3.8. How?

5.8 Summary

In this chapter, the Pólya process and the corresponding generalized multivariate Pólya distribution were analyzed in detail. The Pólya process is an exchangeable process whose predictive probability is a mixture of a priori probability and of the observed frequency for the category under scrutiny.

The generalized multivariate Pólya sampling distribution encompasses many well-known distributions including the multinomial distribution and the multivariate hypergeometric distribution. In the thermodynamic limit, the generalized Pólya becomes the generalized negative binomial, encompassing the Poisson distribution ($\alpha \to \infty$). For positive α, the continuous limit of the multivariate Pólya distribution is the Dirichlet distribution and its marginal distribution is the beta distribution already discussed in Section 4.4. The thermodynamic limit of the beta is the gamma distribution.

Due to exchangeability on the frequency vectors, from the multivariate generalized Pólya distribution one can derive the probability of partition vectors. This probability can be marginalized and the mathematical expectation of any component of the partition vector can be obtained. Exemplified in Section 5.3.8, this procedure will often be used in the following chapters.

Based on Lagrange multipliers, an alternative method allows us to determine the most probable partition vector. Also this procedure will be used further in this book.

The invariance of the heterorelevance coefficient defined in Section 5.6 is a sufficient condition for obtaining the predictive probability characterizing the Pólya process. This provides a complete characterization for this class of exchangeable stochastic processes.

Further reading

Pearson states the fundamental problem of practical statistics as follows:

An 'event' has occurred p times out of $p+q=n$ trials, where we have no a priori knowledge of the frequency of the event in the total population of occurrences. What is the probability of its occurring r times in a further $r+s=m$ trials?

The reader can consult the following paper:

K. Pearson, *The Fundamental Problem of Practical Statistics*, Biometrika, **13** (1), 1–16 (1920).

Johnson's original paper on the sufficientness postulate:

W.E. Johnson, *Probability: the Deductive and Inductive Problem*, Mind, **49**, 409–423 (1932).

is discussed in a paper by Zabell:

S.L. Zabell, *W.E. Johnson's "Sufficientness" Postulate*, The Annals of Statistics, **10** (4), 1090–1099 (1982).

Johnson's paper was published posthumously. The term *sufficientness postulate* was introduced by I.J. Good, who first used the expression *sufficiency postulate* and later used sufficientness to avoid confusion with the usual statistical meaning of sufficiency:

I.J. Good, *The Estimation of Probabilities: An Essay on Modern Bayesian Methods*, MIT Press, Research Monograph No. 30 (1965).

Zabell completed the proof sketched in Johnson's paper. Note that Johnson's sufficientness postulate assumes that

$$\mathbb{P}(X_{m+1}=j|X_1=x_1,\ldots,X_m=x_m)=f_j(m_j); \tag{5.184}$$

therefore, originally, it was not a condition on the heterorelevance coefficient. Note that the postulate is vacuous for $g=2$ and a further assumption of linearity is necessary in the dichotomous case in order to derive the fundamental theorem. Costantini's paper on the heterorelevance coefficient is listed in further reading in Chapter 4.

According to Zabell, Carnap and his collaborators independently rediscovered various results due to Johnson. The dichotomous case and the linearity assumption are discussed in a book by Carnap and Stegmüller:

R. Carnap and W. Stegmüller, *Induktive Logik und Wahrscheinlichkeit*, Springer, Vienna (1959).

W. Brian Arthur used the Pólya distribution in economics. This application is discussed in the following book:

W. Brian Arthur, *Increasing Returns and Path Dependence in the Economy*, The University of Michigan Press, Ann Arbor MI (1994).

A website is available with applets and exercises on the Pólya urn scheme

`http://www.math.uah.edu/stat.`

6

Time evolution and finite Markov chains

Up to now, index set I for a stochastic process can indifferently denote either successive draws from an urn or a time step in a time evolution. In this chapter, a probabilistic dynamics for n objects divided into g categories will be defined. The simplest case is *Markovian* dynamics in discrete time, and so this chapter is devoted to the theory of finite Markov chains.

After studying this chapter you should be able to:

- define Markov chains;
- discuss some properties of Markov chains such as irreducibility, periodicity, stationarity and reversibility;
- determine the invariant distribution for Markov chains either by means of the Markov equation or by using Monte Carlo simulations;
- understand the meaning of statistical equilibrium for aperiodic irreducible Markov chains;
- use the detailed balance principle to compute the invariant distribution, if possible;
- exemplify the above properties using the Ehrenfest urn as a prototypical Markov chain.

6.1 From kinematics to Markovian probabilistic dynamics

6.1.1 Kinematics

Consider a population composed of n elements and g categories which partition the variate under study. In economics, the n elements may represent economic agents and the g categories can be seen as strategies, but other interpretations are possible. For instance, one can speak of n individuals working in g factories or n firms active in g economic sectors, and so on. In physics, the n elements may be particles and the g categories energy levels or, following Boltzmann, one can consider n energy elements to be divided into g particles, etc. In any case, the state of the system is described by the individual description $\mathbf{X}^{(n)} = \mathbf{x}^{(n)}$, a vector shortcut for the equalities $X_1 = x_1, \ldots, X_n = x_n$, where x_i denotes the specific category to which the ith element belongs. The following frequency description corresponds to the individual description: $\mathbf{Y}^{(n)} = \mathbf{n} = (n_1, \ldots, n_i, \ldots, n_g)$, with $n_i \geq 0$ and $\sum_{i=1}^{g} n_i = n$. Now, imagine that the individual description is recorded in a snapshot, with the date impressed. The date is important, as one would like to introduce the possibility of a time evolution for the population. Assume that, during the interval between two snapshots, the ith element passes from the jth to the kth category. This move can be described as follows: the element abandons the population for a while, and, immediately after that, it joins the population again, re-accommodating into the kth category. See Fig. 6.1 for a visual description of the move. In symbols, the system starts from $\mathbf{X}_0^{(n)} = \mathbf{X}^{(n)}(t = 0) = (x_1, \ldots, x_i = j, \ldots, x_n)$, which becomes $\mathbf{X}_1^{(n)} = \mathbf{X}^{(n)}(t = 1) = (x_1, \ldots, x_i = k, \ldots, x_n)$, whereas the intermediate state is $\mathbf{X}^{(n-1)} = (x_1, \ldots, x_{i-1}, \circ, x_{i+1}, \ldots, x_n)$. The symbol $\circ$ is used to remind the reader that the ith element is temporarily out of the population, in order to reconsider its position with regard to categories. Regarding the occupation number vector, the first part of the move corresponds to the transition from

$$\mathbf{Y}_0^{(n)} = \mathbf{n} = (n_1, \ldots, n_j, \ldots, n_k, \ldots, n_g), \tag{6.1}$$

to

$$\mathbf{Y}^{(n-1)} = \mathbf{n}_j = (n_1, \ldots, n_j - 1, \ldots, n_k, \ldots, n_g), \tag{6.2}$$

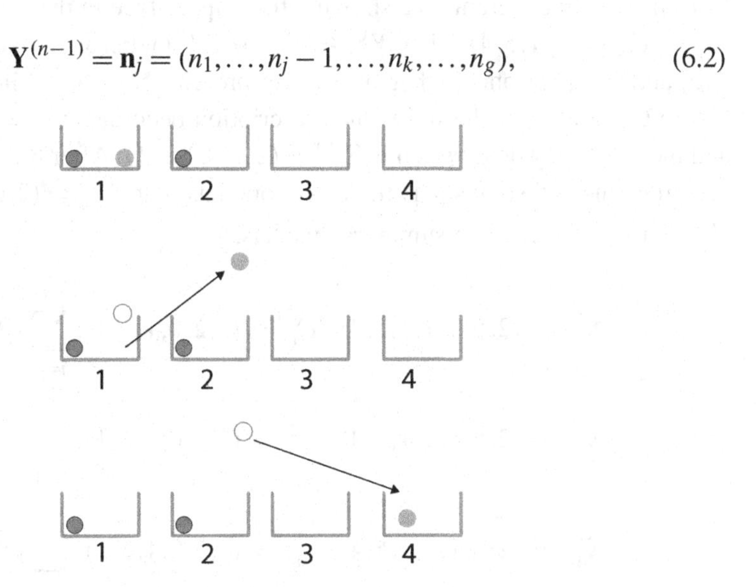

Figure 6.1. Schematic representation of a destruction–creation process. At the beginning (top figure), the light grey ball belongs to category 1 and the occupation vector is $\mathbf{n} = (2, 1, 0, 0)$. Then it is selected and removed (middle figure) and, after the 'destruction', the occupation vector becomes $\mathbf{n}_1 = (1, 1, 0, 0)$. Eventually, it is moved to category 4 (creation) leading to the final occupation vector $\mathbf{n}_1^4 = (1, 1, 0, 1)$.

that is the destruction of an element in the jth category. The second part is the transition from $\mathbf{n}_j = (n_1,\ldots,n_j-1,\ldots,n_k,\ldots,n_g)$ to

$$\mathbf{Y}_1^{(n)} = \mathbf{n}_j^k = (n_1,\ldots,n_j-1,\ldots,n_k+1,\ldots,n_g), \tag{6.3}$$

that is a creation of an element in the kth category. The kinematics is modelled to describe the category changes of the n elements. At each time step, one can imagine that an element is selected, and temporarily abandons the system; after this 'destruction', the same agent re-accommodates into the site system ('creation'). If the final category is different from the initial one a category change takes place. Such a move can be straightforwardly generalized assuming that m agents are selected in a sequence, and then they re-accommodate in a sequence within the time interval from t to $t+1$. Then the size of the population shrinks as long as the selection occurs, and it gradually returns to the original size after all the accommodations. Due to this shrinking–enlarging of the population it is useful to define more precisely the random variable called *occupation number vector* as $\mathbf{Y}^{(n)} = (y_1^{(n)},\ldots,y_g^{(n)}) = (N_{1,n},\ldots,N_{g,n})$, with $\sum_{i=1}^{g} y_i^{(n)} = n$.

6.1.2 Example: seven dice on a table

Consider seven different dice showing their upper face in the following sequence: $\mathbf{X}_0^{(7)} = (2,6,2,1,1,5,4)$. Then $\mathbf{Y}_0^{(7)} = (2,2,0,1,1,1)$, because 1 and 2 appear twice, 4, 5, and 6 appear once, whereas 3 is not present. Now, the third die is removed, before being re-cast, the individual description becomes $\mathbf{X}^{(6)} = (2,6,\circ,1,1,5,4)$, and the frequency description is $\mathbf{Y}^{(6)} = (2,1,0,1,1,1)$. After the third die has been cast, imagine it displays a 1; therefore, one has that $\mathbf{X}_1^{(7)} = (2,6,1,1,1,5,4)$, and $\mathbf{Y}_1^{(7)} = (3,1,0,1,1,1)$. In summary, one has:

$$\mathbf{X}_0^{(7)} = (2,6,2,1,1,5,4);\ \mathbf{Y}_0^{(7)} = (2,2,0,1,1,1);\ \sum_{i=1}^{6} y_i^{(7)} = 7$$

$$\mathbf{X}^{(6)} = (2,6,\circ,1,1,5,4);\ \mathbf{Y}^{(6)} = (2,1,0,1,1,1);\ \sum_{i=1}^{6} y_i^{(6)} = 6$$

$$\mathbf{X}_1^{(7)} = (2,6,1,1,1,5,4);\ \mathbf{Y}_1^{(7)} = (3,1,0,1,1,1).\ \sum_{i=1}^{6} y_i^{(7)} = 7. \tag{6.4}$$

Note that $\mathbf{X}_1^{(7)}$ differs from $\mathbf{X}_0^{(7)}$ in that the third element has changed from category 2 to category 1. As far as the occupation number description is concerned: an element was removed (destroyed) from the second category and an element was added (created) to the first category. If the same operation is repeated, that is a die is picked, and then it is re-cast, and the new situation is recorded (by means of an individual or a statistical vector), after at most 7 trials all possible situations can be reached. Of these situations the most extreme case is $\mathbf{X}^{(7)} = (3,3,3,3,3,3,3)$, corresponding to $\mathbf{Y}^{(7)} = (0,0,7,0,0,0)$. In fact starting from $\mathbf{X}_0^{(7)} = (2,6,2,1,1,5,4)$, $\mathbf{Y}^{(7)}(0) = (2,2,0,1,1,1)$ one has to pick all 7 dice in a row, and always get the result 3. (Note that a possibility is considered, even if this possibility would appear very improbable.)

6.1.3 Individual and/or statistical descriptions

The individual description is a sequence as long as the size of the population n, whereas the statistical description is usually much shorter; in other words, one often has $g \ll n$. For instance, if the number of dice in the previous example were 10^{10} (an order of magnitude typical of economic populations), their statistical description would continue to be a 6-dimensional vector. Even if, in principle, nothing prevents following the time evolution of the individual description (see the example of the Ehrenfest urn model below), focusing on a statistical description is not only practical, but is also a consequence of the probabilistic dynamics. Indeed, if the probabilistic time evolution is not distinguishing the individual names, a statistical description is enough to study all the relevant properties of the evolving population. Further, consider that economic concepts such as the income distribution, the GDP, and so on are statistical in character. Hence, it is natural to define as the *state of the system at time step t*, its statistical distribution $\mathbf{Y}_t^{(n)}$. The superscript can be omitted, simply writing $\mathbf{Y}_t$ when no confusion is possible. The state space is then the set of g-ples summing up to n, denoted by $S_g^n = Y^U = \{(n_1,\ldots,n_g), n_i \geq 0, \sum n_i = n\}$. The discrete-time evolution is represented by the sequence $\mathbf{Y}_0, \mathbf{Y}_1, \ldots, \mathbf{Y}_s, \ldots$, where $\mathbf{Y}_s$ is an element of $S_g^n = Y^U$. A sequence is a possible history of the population, and it can be conceived as a realization of a stochastic process, if the dynamics is probabilistic. If each change from $\mathbf{Y}_s$ to $\mathbf{Y}_{s+1}$ is probabilistic, and the transition probability only depends on the state $\mathbf{Y}_s$ and not on the full history $\mathbf{Y}_{s-1}, \mathbf{Y}_{s-2}, \ldots, \mathbf{Y}_0$, the probabilistic dynamics is said to be a *Markov chain*, whose *transition probability* is a matrix of elements $\mathbb{P}(\mathbf{Y}_{t+1} = \mathbf{n}' | \mathbf{Y}_t = \mathbf{n}) = w(\mathbf{n}, \mathbf{n}')$, $\mathbf{n}', \mathbf{n} \in S_g^n$. If these matrix elements do not depend on time t, the Markov chain is called *homogeneous*. This chapter is devoted to the theory of homogeneous Markov chains.

6.1.4 A Markov chain for the seven dice

Referring to the previous example of dice casting, assume that at each time step a die is chosen at random (all dice being on a par); then it is cast, and the probability of each result is $1/6$ for any die. Let $\mathbf{Y}_0^{(7)} = (2,2,0,1,1,1) = \mathbf{n}$. The probability that a die is selected belonging to the ith category is then n_i/n. The probability of getting the intermediate state $\mathbf{Y}^{(6)} = (2,1,0,1,1,1) = \mathbf{n}_2 = (n_1, n_2 - 1, \ldots, n_6)$ is $2/7$. Casting this die (it was the third one), the probability of obtaining a 1 is $1/6$. Then, one finds $\mathbb{P}(\mathbf{Y}_1^{(7)} = (3,1,0,1,1,1)|\ \mathbf{Y}_0^{(7)} = (2,2,0,1,1,1)) = (2/7)(1/6) = 1/21$. By means of the same procedure, one can determine the transition probability $\mathbb{P}(\mathbf{Y}_{t+1} = \mathbf{n}'|\mathbf{Y}_t = \mathbf{n})$, where $\mathbf{n}' = \mathbf{n}_i^j = (n_1, \ldots, n_i - 1, \ldots, n_j + 1, \ldots, n_6)$; this is given by

$$\mathbb{P}(\mathbf{Y}_{t+1} = \mathbf{n}_i^j|\mathbf{Y}_t = \mathbf{n}) = \frac{n_i}{n}\frac{1}{6}, \; j \neq i. \tag{6.5}$$

If $j \neq i$, a true change takes place, and this can occur in only one way (selecting a die displaying an i which results in a die displaying a j after being cast). But the die could end in the same category as it was before being selected, in this case, $\mathbf{n}' = \mathbf{n}$, and this can occur in 6 alternative ways, each of them through a different destruction $\mathbf{n}_i$, $i = 1, \ldots, 6$. More formally, it is useful to write the two parts of Equation (6.5) separately. If $j \neq i$, then $\mathbb{P}(\mathbf{Y}_{t+1} = \mathbf{n}_i^j|\mathbf{Y}_t = \mathbf{n}) = \mathbb{P}(\mathbf{n}_i|\mathbf{n})\mathbb{P}(\mathbf{n}_i^j|\mathbf{n}_i)$, where $\mathbb{P}(\mathbf{n}_i|\mathbf{n}) = n_i/n$ is the destruction term, transforming a vector $\mathbf{Y}^{(n)}$ of size n into a vector $\mathbf{Y}^{(n-1)}$, whereas the creation term $\mathbb{P}(\mathbf{n}_i^j|\mathbf{n}_i)$ transforms a vector $\mathbf{Y}^{(n-1)}$ of size $n-1$ into a vector $\mathbf{Y}^{(n)}$. Then the probability of no change can be written as follows:

$$P(\mathbf{Y}_{t+1} = \mathbf{n}|\mathbf{Y}_t = \mathbf{n}) = \sum_{i=1}^{6} P(\mathbf{n}_i|\mathbf{n})P(\mathbf{n}_i^i|\mathbf{n}_i) = \sum_{i=1}^{6} \frac{n_i}{n}\frac{1}{6} = \frac{1}{6}. \tag{6.6}$$

Given the rules of the game, all possible sequences of states are those sequences of consecutive occupation numbers differing by no more than two components, that is

$$|\mathbf{n}' - \mathbf{n}| = \sum_{i=1}^{g} |n_i' - n_i| \in \{0, 2\}, \tag{6.7}$$

where 2 denotes a true change and 0 no change. Then, given the initial state any possible sequence $\mathbf{Y}_1, \ldots, \mathbf{Y}_t|\mathbf{Y}_0$ has probability

$$\mathbb{P}(\mathbf{Y}_1 = \mathbf{y}_1|\mathbf{Y}_0 = \mathbf{y}_0)\mathbb{P}(\mathbf{Y}_2 = \mathbf{y}_2|\mathbf{Y}_1 = \mathbf{y}_1) \cdots \mathbb{P}(\mathbf{Y}_t = \mathbf{y}_t|\mathbf{Y}_{t-1} = \mathbf{y}_{t-1}), \tag{6.8}$$

where each term comes from (6.5) if $|\mathbf{y}_t - \mathbf{y}_{t-1}| = 2$, and from (6.6) if $|\mathbf{y}_t - \mathbf{y}_{t-1}| = 0$. Therefore, in this game, the probability of every event can be determined.

Interesting quantities are objects such as $\mathbb{P}(\mathbf{Y}_t = \mathbf{n}'|\mathbf{Y}_0 = \mathbf{n})$ when t is very large, whose meaning is the probability of the states that can be observed after many moves. Formally, one can calculate $\mathbb{P}(\mathbf{Y}_t = \mathbf{n}'|\mathbf{Y}_0 = \mathbf{n})$ adding the probability of all sequences $\mathbf{Y}_1,\ldots,\mathbf{Y}_{t-1}$ connecting $\mathbf{Y}_0 = \mathbf{n}$ to $\mathbf{Y}_t = \mathbf{n}'$. But a little thought is enough to see that, after a number of trials large compared with the number of dice, all dice will be selected at least once, so that their initial state will be cancelled. As they have been recast once at least, an observer that has turned his/her head and has missed all values of $\mathbf{Y}_1,\ldots,\mathbf{Y}_{t-1}$, when turning to the table again, is going to observe seven dice which were cast, the number of castings being irrelevant. It is quite reasonable to expect that the probability of the states will be governed by a $\mathbb{P}(\mathbf{Y}_t = \mathbf{n}'|\mathbf{Y}_0 = \mathbf{n})$ which does not depend on $\mathbf{Y}_0$, and that $\lim_{t\to\infty}\mathbb{P}(\mathbf{Y}_t = \mathbf{n}'|\mathbf{Y}_0 = \mathbf{n}) = \pi(\mathbf{n}')$ is the usual multinomial distribution which describes casting of seven fair dice. In the following, it will be shown that the chain indeed reaches its equilibrium distribution $\pi(\mathbf{n}')$.

Another interesting statistical property is the fraction of time in which the system visits any given state as time goes by. For instance, considering the state $\mathbf{y} = (6,0,0,0,0,0)$ where all the dice display a 1, its visiting fraction is the random variable counting the number of times in which all the dice show 1 divided by the total number of trials. It can be shown that the fraction of visits to any state converges (in probabilistic terms) for $t \to \infty$, and the limiting value is just $\pi(\mathbf{n})$, its probability of being observed at a large but fixed time. This property (also called *ergodicity*), appearing rather commonsensical for probabilistic systems such as the one under scrutiny, is very important in the foundations of classical statistical mechanics based on Newton's deterministic dynamics.

6.2 The Ehrenfest urn model

In the words of Kac, the Ehrenfest urn model is 'probably one of the most instructive models in the whole of Physics' (see Further reading at the end of the chapter). It was designed to support a probabilistic vindication of Boltzmann's H-theorem, and to give a qualitative account of notions like reversibility, periodicity and tendency to equilibrium (concepts that will be further explained below). The model, often referred to as the 'Ehrenfest dog-flea model', is mentioned in many textbooks on probability, stochastic processes and statistical physics. It is probably the simplest *random walk with reflecting barriers* where increments are dependent on the state of the system and it is an example of a *birth–death process* which is a homogeneous Markov chain. For a birth–death Markov chain only the following transition probabilities are non-vanishing: $w(i,i+1)$ (birth), $w(i,i-1)$ (death) and $w(i,i)$. Considering fleas and dogs, in an informal way, one has $g = 2$ dogs and n fleas, of which i are on the first dog and $(n-i)$ on the second. At each discrete time step an integer number between 1 and n is chosen at random, and the corresponding flea is forced to hop towards the other dog. This mechanism can be formalized as follows. A discrete stochastic process $\{Y_t\}_{t\geq 0} = Y_0, Y_1,\ldots,Y_t,\ldots$ is introduced whose values are in $Y^U = (0,\dots,i,\dots,n)$, representing the occupation number of, or the number of balls in, the first urn (or the number of fleas on the first dog) at time t. The state of the system is thus represented by the ordered pair $(i, n-i)$. From the rule described in words, the transition probability is modified as follows

$$w(i,i+1) = \mathbb{P}(Y_{t+1} = i+1|Y_t = i) = \frac{n-i}{n}, \; i = 0,\dots,n-1$$
$$w(i,i-1) = \mathbb{P}(Y_{t+1} = i-1|Y_t = i) = \frac{i}{n}. \; i = 1,\dots,n. \tag{6.9}$$

Note that in all the other cases $w(i,j) = \mathbb{P}(Y_{t+1} = j|Y_t = i) = 0$ and that one has $\sum_j w(i,j) = 1$. Most of the mathematical complications of the model are a consequence of the fact that the randomly chosen flea is forced to change dog. This implies that, at each transition, if i was even it must change into an odd number, and vice versa. Thus, the parity is conserved after two transitions, and this constraint is not cancelled by the time evolution. Therefore, the probability of finding an even number oscillates from a positive number (for even steps) to zero (for odd steps), and this holds forever. By means of a slight modification of the model, this periodicity can be avoided. It is enough to let the selected flea return to the dog from which it came. Assume that a flea on the first dog is selected and then hops on to the second dog. The dog occupation vector passes from the initial state $(i, n-i)$, to the intermediate state $(i-1, n-i)$; eventually, there is the transition from $(i-1, n-i)$ to $(i-1, n-i+1)$, the final state. Using the physical jargon introduced above, this transition is equivalent to the destruction of a flea on the first dog, followed by the creation of a flea on the second. In statistical jargon this may describe the fact that an element changes category. Now, if the flea can also come back to the original dog, instead, one has the transition $(i, n-i) \to (i-1, n-i) \to (i, n-i)$, that is the final state does coincide with the initial one. If the creation probability is symmetric, Equation (6.9) becomes

$$w(i,i+1) = \mathbb{P}(Y_{t+1} = i+1|Y_t = i) = \frac{n-i}{n}\frac{1}{2}, \; i = 0,\dots,n-1$$
$$w(i,i) = \mathbb{P}(Y_{t+1} = i|Y_t = i) = \frac{1}{2}, \; i = 0,\dots,n$$
$$w(i,i-1) = \mathbb{P}(Y_{t+1} = i-1|Y_t = i) = \frac{i}{n}\frac{1}{2}, \; i = 1,\dots,n. \tag{6.10}$$

Again, all the other cases are forbidden transitions and their probability is 0. This process is a simplification of the previous dice process. Equation (6.10) describes n coins on the table, a coin is picked at random and then tossed, so that $Y_t = i$ denotes that, after t moves, there are i coins displaying heads and $n-i$ coins displaying tails. Repeating the previous considerations, one can show that the equilibrium distribution of the chain is the binomial distribution of parameters $n, 1/2$, that is

$$\pi_i = \binom{n}{i}\left(\frac{1}{2}\right)^n, \tag{6.11}$$

with $i = 0,\dots,n$. In order to prove these intuitive properties, and to discover many new results, the formal notion of Markov chain and the minimal set of theorems needed in this book are introduced in the following sections.

6.3 Finite Markov chains

As discussed in Chapter 4, a stochastic process is a family of random variables $\{X_i \in S, i \in I\}$, where S is called the state space, and I is the set of indices. The state space can be either discrete (finite or numerable) or continuous; also the set of indices can be either discrete or continuous. For a fixed value of the index, X_i is a random variable, and $\mathbb{P}(X_i \in A)$ is the probability that the ith value of X belongs to some subset $A \subseteq S$. In all the cases considered so far, the state space was the range of the individual random variables, that is the label set $\{1,\dots,g\}$; the set of indices was also finite, each index being attached to an element chosen from a finite population. Until now, several examples of exchangeable processes have been considered (the hypergeometric process, the multinomial process and the generalized Pólya process), whose interpretations are sampling, or removing (or allocating) individuals from (or into) categories. Now, our attention is focused on the time evolution of the system as a whole; X_t denotes the *state* of the system at (discrete) time t. For a while, let us abandon any interpretation of the state (individual/statistical description of n elements in g categories, and so on), and just consider an abstract finite state space S:

1. A family of random variables $\{X_i \in S, i \in I\}$, where S is discrete and finite, and I is a subset of non-negative integers is called a *finite Markov chain* if

$$\begin{aligned}&\mathbb{P}(X_0 = x_0, X_1 = x_1, \dots, X_s = x_s)\\ &\quad = \mathbb{P}(X_0 = x_0)\mathbb{P}(X_1 = x_1|X_0 = x_0)\cdots\mathbb{P}(X_s = x_s|X_{s-1} = x_{s-1}).\end{aligned} \tag{6.12}$$

   In other words, the typical Markovian property is that the predictive probability $\mathbb{P}(X_s = x_s|X_{s-1} = x_{s-1},\dots,X_0 = x_0)$ simplifies to

$$\mathbb{P}(X_s = x_s|X_{s-1} = x_{s-1},\dots,X_0 = x_0) = \mathbb{P}(X_s = x_s|X_{s-1} = x_{s-1}). \tag{6.13}$$

   This means that the probability of the new state depends only on the actual state, and not on the past history: the future depends only on the present, not on the past.
2. If $\mathbb{P}(X_{s+1} = j|X_s = i)$ depends only on i,j and not on s, one can write

$$\mathbb{P}(X_{s+1} = j|X_s = i) = w(i,j); \tag{6.14}$$

in this case, the Markov chain is called *homogeneous*. The set of numbers $\{w(i,j)\}$, where $i,j \in \{1,\ldots,m\}$, and $m = \#S$, can be represented as a square matrix

$$\mathbb{W} = \{w(i,j)\}_{i,j=1,\ldots,m}, \tag{6.15}$$

called the (one-step) *transition matrix*. Each row of the matrix represents the probability of reaching the states $j = 1,2,\ldots,m$ starting from the same state i. Then $w(i,j) = \mathbb{P}(X_{s+1} = j|X_s = i)$ is non-negative, and, for any i, $\sum_{j=1}^{m} w(i,j) = 1$. Hence, $\mathbb{W}$ is also called a *stochastic matrix*.

3. The probability of any sequence is given in terms of $p(x_0) = \mathbb{P}(X_0 = x_0)$ and $\mathbb{W}$. Indeed, one has

$$\mathbb{P}(X_0 = x_0, X_1 = x_1 \ldots, X_s = x_s) = p(x_0)w(x_0,x_1)\cdots w(x_{s-1},x_s), \tag{6.16}$$

and the terms $w(x_{t-1},x_t)$ are the corresponding entries in $\mathbb{W}$. Given that all the sequences $X_0 = x_0, X_1 = x_1, \ldots, X_s = x_s$ are the constituents of any event depending on the first s steps, the probability space is solvable in principle and the stochastic process is fully determined. But shortcuts are possible.

4. Consider the probability of reaching state j from state i in r steps, that is

$$w^{(r)}(i,j) = \mathbb{P}(X_r = j|X_0 = i). \tag{6.17}$$

If one introduces all the possible states after one step, one of which necessarily occurs, it follows by the total probability theorem that

$$w^{(r)}(i,j) = \sum_{k=1}^{m} w(i,k)w^{(r-1)}(k,j), \tag{6.18}$$

which is a particular instance of the so-called *Chapman–Kolmogorov equation*. From matrix theory, one recognizes (6.18) as the formula of matrix multiplication. Therefore, one has

$$\mathbb{W}^{(r)} = \mathbb{W} \times \mathbb{W}^{(r-1)} = \mathbb{W} \times \mathbb{W} \times \mathbb{W}^{(r-2)} = \ldots = \mathbb{W}^r, \tag{6.19}$$

where the first equation has been iterated $r-2$ times. In words, the r-step transition probabilities $w^{(r)}(i,j)$ are the entries of $\mathbb{W}^r$, the rth power of the one-step transition probability matrix $\mathbb{W}$.

5. Consider the probability that the chain state is j after r steps: conditioned to the initial state i, $w^{(r)}(i,j) = \mathbb{P}(X_r = j|X_0 = i)$. If the initial condition is probabilistic, and $P_i^{(0)} = \mathbb{P}(X_0 = i)$, as a consequence of the total probability theorem, one finds that

$$P_j^{(r)} = \mathbb{P}(X_r = j) = \sum_{k=1}^{m} P_k^{(0)} w^{(r)}(k,j). \tag{6.20}$$

From matrix theory, one again recognizes (6.20) as the formula for the multiplication of a row vector with m components by an $m \times m$ matrix. In other words, if $\mathbf{P}^{(r)} = (P_1^{(r)}, \ldots, P_m^{(r)})$ represents the equations $\mathbb{P}(X_r = 1), \ldots, \mathbb{P}(X_r = m)$, then one can write

$$\mathbf{P}^{(r)} = \mathbf{P}^{(0)} \times \mathbb{W}^r. \tag{6.21}$$

The simplest case of (6.20) is

$$P_j^{(r+1)} = \sum_{k=1}^{m} P_k^{(r)} w(k,j); \tag{6.22}$$

its matrix form is

$$\mathbf{P}^{(r+1)} = \mathbf{P}^{(r)} \times \mathbb{W}. \tag{6.23}$$

Equation (6.22) is called the *Markov equation.*

6.3.1 The Ehrenfest urn: periodic case

All the previous abstract formulae can be clearly interpreted in a simple, well-known example, the original Ehrenfest model. Consider the case $n = 4$; the state space is $\{0, 1, 2, 3, 4\}$, denoting the number of balls in the right urn. From (6.9), the transition matrix $\mathbb{W}$ is the following 5×5 matrix:

$$\mathbb{W} = \begin{pmatrix} 0 & 1 & 0 & 0 & 0 \\ 1/4 & 0 & 3/4 & 0 & 0 \\ 0 & 1/2 & 0 & 1/2 & 0 \\ 0 & 0 & 3/4 & 0 & 1/4 \\ 0 & 0 & 0 & 1 & 0 \end{pmatrix}. \tag{6.24}$$

All rows of the matrix sum to 1, because some final state has to be chosen. The diagonal entries $w(i,i)$ are all zero, because the state must change. The row representing the transition starting from $k = 0$ is *deterministic*, because, for sure, a ball is drawn from the second urn, and added to the first one. A similar situation is found for $k = 4$, where, with probability 1, a ball is drawn from the first urn and replaced in the second one. There are many 0s because only states that differ by 1 communicate in a single step. If one starts with $k = 0$, and wants to reach $k = 4$, the shortest possibility is the sequence $k_1 = 1, k_2 = 2, k_3 = 3, k_4 = 4$, obtained by always choosing a ball in the second urn. From Equation (6.9) one has

$$\mathbb{P}(k_1 = 1, k_2 = 2, k_3 = 3, k_4 = 4 | k_0 = 0) = \frac{4}{4}\frac{3}{4}\frac{2}{4}\frac{1}{4} = \frac{3}{32}. \tag{6.25}$$

Considering the four-step transition matrix $\mathbb{W}^{(4)} = \mathbb{W} \times \mathbb{W} \times \mathbb{W} \times \mathbb{W} = \mathbb{W}^4$, one finds that

$$\mathbb{W}^{(4)} = \begin{pmatrix} 5/32 & 0 & 3/4 & 0 & 3/32 \\ 0 & 17/32 & 0 & 15/32 & 0 \\ 1/8 & 0 & 3/8 & 0 & 1/8 \\ 0 & 15/32 & 0 & 17/32 & 0 \\ 3/32 & 0 & 3/4 & 0 & 5/32 \end{pmatrix}, \tag{6.26}$$

and $w^{(4)}(0,4)$ is just $3/32$, as there is only a single path leading from 0 to 4 in four steps. Having considered the two most distant states, it is true that all the states communicate with each other; this means that, for each couple i,j, there exists a number of steps for which $w^{(s)}(i,j) > 0$. Its minimum value is $s(i,j) = \min\{r : w^{(r)}(i,j) > 0)$. A Markov chain with this property is said to be *irreducible*, and all the states are *persistent*. However, unfortunately, also in $\mathbb{W}^{(4)}$ there are a lot of 0s: the number of 0s initially decreases from 17 in $\mathbb{W}^{(1)} = \mathbb{W}$, to 14 in $\mathbb{W}^{(2)}$; it is 13 in $\mathbb{W}^{(3)}$, it reaches its minimum 12 in $\mathbb{W}^{(4)}$; but, in $\mathbb{W}^{(5)}$it climbs to 13 again, and eventually it indefinitely oscillates $12, 13, 12, 13, \ldots$. While the initial filling is commonsensical (it is a kinematical truism that the longer is the step the greater is the number of reachable states), the residual oscillating number of 0s is due to the (dynamical) fact that the selected ball must change urn. In other terms, while destruction is random, creation (given the destruction) is sure, so that if $w^{(r)}(i,j) > 0$, one has $w^{(r+1)}(i,j) = 0$. Alternatively, considering the probability of returning to the starting state after r steps, $w_{ii}^{(r)} = 0$ if r is odd. This chain has *period* equal to 2. However, for large r, $\mathbb{W}^{(2r)}$ and $\mathbb{W}^{(2r+2)}$ become closer and closer. Within the 3-digit precision, $\mathbb{W}^{(10)}$ and $\mathbb{W}^{(12)}$ are given by the following limiting matrix:

$$\mathbb{W}^{(even)} = \begin{pmatrix} 0.125 & 0 & 0.750 & 0 & 0.125 \\ 0 & 0.500 & 0 & 0.500 & 0 \\ 0.125 & 0 & 0.750 & 0 & 0.125 \\ 0 & 0.500 & 0 & 0.500 & 0 \\ 0.125 & 0 & 0.750 & 0 & 0.125 \end{pmatrix}. \tag{6.27}$$

For odd values, the limiting matrix is obtained within the same precision for $\mathbb{W}^{(11)}$ and $\mathbb{W}^{(13)}$:

$$\mathbb{W}^{(odd)} = \begin{pmatrix} 0 & 0.500 & 0 & 0.500 & 0 \\ 0.125 & 0 & 0.750 & 0 & 0.125 \\ 0 & 0.500 & 0 & 0.500 & 0 \\ 0.125 & 0 & 0.750 & 0 & 0.125 \\ 0 & 0.500 & 0 & 0.500 & 0 \end{pmatrix}. \tag{6.28}$$

Considering the binomial distribution

$$\mathrm{Bin}(k;4,1/2) = \binom{4}{k}\frac{1}{2^4}, \tag{6.29}$$

that is $(1/16, 4/16, 6/16, 4/16, 1/16)$, one can see that $w^{(even)}(i,j) = 2\,\mathrm{Bin}\,(j;4,1/2)$ for j even, and 0 for j odd, whereas $w^{(odd)}(i,j) = 2\ \mathrm{Bin}(j;4,1/2)$ for j odd, and 0 for j even. For instance, if the column of $\mathbb{W}^{(odd)}$ is considered, corresponding to the final state 1, one can see that all non-zero entries are the same. After an odd number of steps, state 1 can only be reached from states $0,2,4$. If the odd number of steps is large, the limiting probability $1/2$ is the same for all even starting states. Hence, within the same parity all the initial states are forgotten, but the chain remembers the initial parity forever, so that $\mathbf{P}^{(t)}$, conditioned on any initial state x_0, forever oscillates and cannot reach any limit (see Exercises 6.1, 6.2 and 6.7).

6.3.2 The Ehrenfest urn: aperiodic case

If, instead of Equation (6.9), one uses the transition matrix corresponding to Equation (6.10), one obtains the following 5×5 matrix:

$$\mathbb{W} = \begin{pmatrix} 1/2 & 1/2 & 0 & 0 & 0 \\ 1/8 & 1/2 & 3/8 & 0 & 0 \\ 0 & 1/4 & 1/2 & 1/4 & 0 \\ 0 & 0 & 3/8 & 1/2 & 1/8 \\ 0 & 0 & 0 & 1/2 & 1/2 \end{pmatrix}. \tag{6.30}$$

The diagonal terms are all $1/2$, corresponding to the probability of returning to the starting urn. For instance, if one starts from $k = 0$, and wishes to reach $k = 4$, the shortest possibility is the sequence $k_1 = 1, k_2 = 2, k_3 = 3, k_4 = 4$ again, obtained by always choosing a ball in the second urn. From (6.9), one finds that

$$\mathbb{P}(k_1 = 1, k_2 = 2, k_3 = 3, k_4 = 4 | k_0 = 0) = \frac{4}{8}\frac{3}{8}\frac{2}{8}\frac{1}{8} = \frac{3}{512} = 5.86 \times 10^{-3}. \tag{6.31}$$

Considering the four-step transition matrix: $\mathbb{W}^{(4)} = \mathbb{W} \times \mathbb{W} \times \mathbb{W} \times \mathbb{W} = \mathbb{W}^4$, one has

$$\mathbb{W}^{(4)} = \begin{pmatrix} 166 & 406 & 328 & 93.8 & 5.86 \\ 102 & 330 & 375 & 170 & 23.4 \\ 54.7 & 250 & 391 & 250 & 54.7 \\ 23.4 & 170 & 375 & 330 & 102 \\ 5.86 & 93.8 & 328 & 406 & 166 \end{pmatrix} \times 10^{-3}, \tag{6.32}$$

and $w^{(4)}(0,4)$ is just 5.86×10^{-3}. However for large r, $\mathbb{W}^{(r)}$ and $\mathbb{W}^{(r+1)}$ become closer and closer to the following limiting matrix

$$\mathbb{W}^{(\infty)} = \begin{pmatrix} 0.0625 & 0.250 & 0.375 & 0.250 & 0.0625 \\ 0.0625 & 0.250 & 0.375 & 0.250 & 0.0625 \\ 0.0625 & 0.250 & 0.375 & 0.250 & 0.0625 \\ 0.0625 & 0.250 & 0.375 & 0.250 & 0.0625 \\ 0.0625 & 0.250 & 0.375 & 0.250 & 0.0625 \end{pmatrix}. \tag{6.33}$$

Within 3-digit precision, this is obtained for $r = 30$. In $\mathbb{W}^{(\infty)}$ all rows are the same, and thus all columns have a constant value. In this case each row (in decimal form) is equal to the binomial distribution (6.29), that is (in rational form) $(1/16, 4/16, 6/16, 4/16, 1/16)$. Hence, if one is going to observe the system after a large number of steps, one notices that the probability of finding the system in the jth state is

$$w^{(\infty)}(i,j) = \pi_j = \binom{4}{j} \frac{1}{2^4}, \tag{6.34}$$

whatever the initial state i was. Formally, one has

$$\lim_{r \to \infty} \mathbb{P}(X_r = j | X_0 = i) = \lim_{r \to \infty} w^{(r)}(i,j) = \pi_j. \tag{6.35}$$

This means that, in the long run, the probability of a state only depends on the dynamics (the transition matrix), which completely deletes any influence of the initial conditions. If the limiting matrix exists, it means that for large r, $\mathbf{P}^{(r)}$ and $\mathbf{P}^{(r+1)}$ are very close to the limiting distribution π given by Equation 6.29. One has that $r = 30$, for a 10^{-3} (3-digit) precision, it becomes $r = 38$ for a 4-digit precision, $r = 46$ for a 5-digit precision, ... Is there justification to write $\mathbb{W}^{(\infty)}$? Under what conditions can one say that $\mathbb{W}^{(\infty)}$ (and thus π_j) exists? And, if this is possible, how can π_j be found? In the following section, these questions will find an answer.

6.4 Convergence to a limiting distribution

The destruction–creation mechanism described at the beginning of the chapter defines irreducible and aperiodic Markov chains. The more complicated periodic case was described in order to underline the difference between a fully probabilistic dynamics (the aperiodic case) and a dynamics still retaining some deterministic features (see also Exercise 6.1). It is now time to discuss the limiting behaviour of an irreducible aperiodic Markov chain, such as the aperiodic Ehrenfest urn, in a general way.

In order to prove that $\mathbf{P}^{(r)}$ converges to a limiting distribution, that is, following Cauchy, that

$$\lim_{r\to\infty} |\mathbf{P}^{(r+s)} - \mathbf{P}^{(r)}| = 0, \tag{6.36}$$

one needs a suitable definition of distance between two probability distributions $\boldsymbol{\mu} = (\mu_1, \ldots, \mu_m)$ and $\boldsymbol{\nu} = (\nu_1, \ldots, \nu_m)$, with $\sum_{i=1}^m \mu_i = \sum_{i=1}^m \nu_i = 1$. In other words, the symbol $|\cdot|$ informally used in Equation (6.36) must be defined.

Definition (Distance between probability distributions) Given two probability distributions over m states $\boldsymbol{\mu} = (\mu_1, \ldots, \mu_m)$ and $\boldsymbol{\nu} = (\nu_1, \ldots, \nu_m)$, with $\sum_{i=1}^m \mu_i = \sum_{i=1}^m \nu_i = 1$, the distance between them is defined as follows

$$d(\boldsymbol{\mu}, \boldsymbol{\nu}) = \frac{1}{2}\sum_{i=1}^{m} |\mu_i - \nu_i| = \frac{1}{2}\sum{}^{+}(\mu_i - \nu_i) - \frac{1}{2}\sum{}^{-}(\mu_i - \nu_i), \tag{6.37}$$

where $\sum^+$ ($\sum^-$) denotes the sum over positive (negative) terms.

Remark (Axioms of metric space) A set M equipped with a map $d : M \times M \to \mathbb{R}_0^+$ with the following properties

1. $d(x,y) = 0 \iff x = y$,
2. (symmetry) $d(x,y) = d(y,x)$,
3. (triangular inequality) $d(x,z) \le d(x,y) + d(y,z)$,

is called a *metric space*, and d is called a *distance*. The reader can verify that definition (6.37) satisfies the axioms of distance and that $d(\boldsymbol{\mu}, \boldsymbol{\nu}) \le 1$.

Remark (A property of the distance) Given that $\sum_{i=1}^m (\mu_i - \nu_i) = 0$, one has that $\sum^+(\mu_i - \nu_i) = \sum^+(\nu_i - \mu_i)$; this yields

$$d(\boldsymbol{\mu}, \boldsymbol{\nu}) = \sum{}^{+}(\mu_i - \nu_i). \tag{6.38}$$

Example (Aperiodic Ehrenfest urn) Considering the five states of the above aperiodic Ehrenfest model, let $\boldsymbol{\mu} = (1/3, 1/3, 1/3, 0, 0)$, $\boldsymbol{\nu} = (0, 0, 0, 1/2, 1/2)$ and $\boldsymbol{\xi} = (1/3, 0, 1/3, 0, 1/3)$ be three possible distributions. The first distribution means that the chain is equiprobably in the set $\{0, 1, 2\}$, the second one means that the chain is equiprobably in the set $\{3, 4\}$, the third distribution means that the chain is equiprobably in the set of even-numbered states $\{0, 2, 4\}$. Now applying Equation (6.38), one has $d(\boldsymbol{\mu}, \boldsymbol{\nu}) = \sum_i^+ (\mu_i - \nu_i) = \sum_{i=1}^3 (\mu_i - \nu_i) = 1/3 + 1/3 + 1/3 = 1$. Also $d(\boldsymbol{\nu}, \boldsymbol{\mu})$ has the same value: $d(\boldsymbol{\nu}, \boldsymbol{\mu}) = \sum_{i=3}^4 (\nu_i - \mu_i) = 1/2 + 1/2 = 1$, that is $d(\boldsymbol{\mu}, \boldsymbol{\nu}) = d(\boldsymbol{\nu}, \boldsymbol{\mu})$ as a consequence of the distance symmetry. Further, one has $d(\boldsymbol{\mu}, \boldsymbol{\xi}) = \mu_2 - \xi_2 = 1/3$, whereas $d(\boldsymbol{\nu}, \boldsymbol{\xi}) = \nu_4 - \xi_4 + \nu_5 - \xi_5 = 1/2 + 1/2 - 1/3 = 2/3$, so that $d(\boldsymbol{\mu}, \boldsymbol{\nu}) = 1 \le d(\boldsymbol{\mu}, \boldsymbol{\xi}) + d(\boldsymbol{\nu}, \boldsymbol{\xi}) = 1$, which satisfies the triangular inequality. Moreover, $0 \le d(\boldsymbol{\mu},\boldsymbol{\nu}) \le 1$; the distance is zero if and only if $\boldsymbol{\mu} = \boldsymbol{\nu}$ and it reaches its maximum 1 for *orthogonal* distributions, such as $(1/3, 1/3, 1/3, 0, 0)$ and $(0, 0, 0, 1/2, 1/2)$, which are positive on non-overlapping domains.

6.4.1 An abstract theorem

Theorem If $\mathbb{W} = \{w(i,j)\}$ is an $m \times m$ positive stochastic matrix, then $d(\boldsymbol{\mu} \times \mathbb{W}, \boldsymbol{\nu} \times \mathbb{W}) \le d(\boldsymbol{\mu},\boldsymbol{\nu})$.

Remark (Interpretation of the theorem) Assume that $\mathbb{W}$ represents the Markovian evolution of some system, for instance, it may be given by Equation (6.32) in Section 6.3.2, with no zero entries, i.e. $w(i,j) \ge \delta > 0$. If such a positive matrix is applied to two different initial probability distributions $\boldsymbol{\mu}$ and $\boldsymbol{\nu}$, the distance between the outcomes is reduced after four steps. As this operation can be indefinitely repeated, it is not difficult to understand what happens in the long run.

Proof From Markov Equation (6.22), one has that

$$\mu'_j = \sum_i \mu_i w(i,j), \tag{6.39}$$

and that

$$\nu'_j = \sum_i \nu_i w(i,j), \tag{6.40}$$

so that the initial distance $d(\boldsymbol{\mu},\boldsymbol{\nu}) = \sum_i{}^+ (\mu_i - \nu_i)$ transforms into

$$\begin{aligned} d(\boldsymbol{\mu} \times \mathbb{W}, \boldsymbol{\nu} \times \mathbb{W}) &= \sum_j{}^+ \left(\mu'_j - \nu'_j\right) = \sum_j{}^+ \left(\sum_{i=1}^m \mu_i w(i,j) - \sum_{i=1}^m \nu_i w(i,j)\right) \\ &= \sum_j{}^+ \sum_{i=1}^m (\mu_i - \nu_i) w(i,j). \end{aligned} \tag{6.41}$$

Consider the terms in $\sum_j{}^+$, where by definition $\mu'_j - \nu'_j > 0$, that is $\boldsymbol{\mu}'$ dominates $\boldsymbol{\nu}'$, and call them $\boldsymbol{\mu}'$-dominating states. From Equation (6.22), one has

$$\mu'_j - \nu'_j = \sum_{i=1}^m (\mu_i - \nu_i) w(i,j). \tag{6.42}$$

Some of the terms $\mu_i - \nu_i$ will be positive, and they are all states i where $\boldsymbol{\mu}$ was dominating $\boldsymbol{\nu}$ ($\boldsymbol{\mu}$-dominating states), whereas the remaining terms are negative or vanish. Therefore, one further has

$$\mu'_j - \nu'_j = \sum_{i=1}^{m} (\mu_i - \nu_i) w(i,j) \leq \sum_i{}^+ (\mu_i - \nu_i) w(i,j). \tag{6.43}$$

Each term of $\sum_i^+$ denotes the fraction of the $\boldsymbol{\mu}$-dominance which is transferred to a $\boldsymbol{\mu}'$-dominating state. Then summing on all $\boldsymbol{\mu}'$-dominating states, one arrives at

$$\sum_j{}^+ (\mu'_j - \nu'_j) \leq \sum_j{}^+ \sum_i{}^+ (\mu_i - \nu_i) w(i,j). \tag{6.44}$$

Inverting the summation order, one finds that the term

$$\sum_i{}^+ (\mu_i - \nu_i) \sum_j{}^+ w(i,j) \tag{6.45}$$

takes into account how each 'piece' of the old distance $\mu_i - \nu_i > 0$ is transferred to the new $\boldsymbol{\mu}'$-dominating states. Therefore, one gets that

$$\begin{aligned} d(\boldsymbol{\mu} \times \mathbb{W}, \boldsymbol{\nu} \times \mathbb{W}) &\leq \sum_i{}^+ (\mu_i - \nu_i) \sum_j{}^+ w(i,j) \leq \sum_i{}^+ (\mu_i - \nu_i) \sum_{j=1}^{m} w(i,j) \\ &= \sum_i{}^+ (\mu_i - \nu_i) = d(\boldsymbol{\mu}, \boldsymbol{\nu}). \end{aligned} \tag{6.46}$$

Given that $\sum_j{}^+ w(i,j)$ differs from $\sum_{j=1}^{m} w(i,j) = 1$ in at least one term, one has that $\sum_j{}^+ w(i,j) \leq 1 - \delta$, and thus

$$d(\boldsymbol{\mu} \times \mathbb{W}, \boldsymbol{\nu} \times \mathbb{W}) \leq (1 - \delta) d(\boldsymbol{\mu}, \boldsymbol{\nu}), \tag{6.47}$$

with $\delta = \min_{i,j}(w(i,j))$. □

Remark (Conceptual core) This is the conceptual core of the previous theorem: Markov evolution mixes all the states, so that, if some $\boldsymbol{\mu}$-dominance is present at the beginning, given that all $w(i,j)$ are positive, the quantity $\sum_i{}^+ (\mu_i - \nu_i)$ diffuses not only towards $\boldsymbol{\mu}'$-dominated states, but also towards $\boldsymbol{\nu}'$-dominated states, and thus the dominating mass is reduced.

6.4.2 Applications to finite Markov chains

Considering the seven dice chain, or the aperiodic Ehrenfest model, $\mathbb{W}^{(n)}$ is a positive stochastic matrix, where in general n stands for the size of the population (of dice, or of balls). Intuitively, after n steps all the states have positive probability of being reached from any initial state. The transition matrix for the dice chain is 792 × 792 and it was not explicitly written, but from Equation (6.6) one can see that $\mathbb{P}(\mathbf{Y}_{t+1} = \mathbf{n}|\mathbf{Y}_t = \mathbf{n}) = 1/6 > 0$ implies no periodicity at all. The periodic case is different, where for no r, it is true that $\mathbb{W}^{(r)}$ is positive. The rest of the book is devoted to irreducible aperiodic chains of the kind described in Section 6.1.4, for which there always exists an r such that $\mathbb{W}^{(r)}$ is positive (an alternative definition is: $\mathbb{W}$ is *regular* if $\mathbb{W}^{(r)}$ is positive for some r). If $\mathbb{W}^{(r)}$ is positive, also $\mathbb{W}^{(r+1)} = \mathbb{W} \times \mathbb{W}^{(r)}$ is positive. Moreover, if $\mathbb{W}^{(r)}$ is positive, also $\mathbf{P}^{(r)} = \mathbf{P}^{(0)} \times \mathbb{W}^{(r)}$ is positive for all states, whatever $\mathbf{P}^{(0)}$ may be.

Theorem (Main theorem for irreducible aperiodic Markov chains (Sinai)) For any irreducible aperiodic Markov chain, that is a Markov chain whose transition matrix is regular, there exists a unique probability distribution $\boldsymbol{\pi} = (\pi_1, \ldots, \pi_m)$, $\pi_j > 0$, such that:

1. π_j is the long-term limit of the probability $\mathbb{P}(X_r = j) = P_j^{(r)}$:

$$\pi_j = \lim_{r\to\infty} \mathbb{P}(X_r = j); \tag{6.48}$$

2. $\boldsymbol{\pi}$ is the (unique) invariant distribution of the chain, meaning that

$$\boldsymbol{\pi} \times \mathbb{W} = \boldsymbol{\pi}; \tag{6.49}$$

3. π_j is the long-term limit of the rows of the r-step transition matrix:

$$\lim_{r\to\infty} w^{(r)}(i,j) = \pi_j. \tag{6.50}$$

Proof For concrete systems governed by a regular transition matrix, one can consider $\mathbf{P}^{(0)}, \mathbf{P}^{(1)} = \mathbf{P}^{(0)} \times \mathbb{W}, \ldots, \mathbf{P}^{(n)} = \mathbf{P}^{(0)} \times \mathbb{W}^{(n)}$, and show that the sequence $\{\mathbf{P}^{(n)}\}_{n\in\mathbb{N}}$ is a Cauchy one, meaning that

$$\lim_{n\to\infty} d(\mathbf{P}^{(n)}, \mathbf{P}^{(n+p)}) = 0. \tag{6.51}$$

This equation clarifies the meaning of Equation (6.36). Now $\mathbf{P}^{(n)}$ and $\mathbf{P}^{(n+p)}$ stand for $\boldsymbol{\mu}$ and $\boldsymbol{\nu}$ in the theorem discussed above. The matrix being regular, assume that $\mathbb{W}^{(r)}$ is positive; it plays the role of $\mathbb{W}$ in the previous theorem. As a consequence of Equations (6.21) and (6.47), one can write

$$d(\mathbf{P}^{(n-r)} \times \mathbb{W}^{(r)}, \mathbf{P}^{(n+p-r)} \times \mathbb{W}^{(r)}) \leq (1-\delta) d(\mathbf{P}^{(n-r)}, \mathbf{P}^{(n+p-r)}), \tag{6.52}$$

where $\delta = \min_{i,j}(w^{(r)}(i,j))$; in other words, the time evolution reduces the distance with respect to r steps before. This consideration can be iterated, leading to

$$d(\mathbf{P}^{(n)}, \mathbf{P}^{(n+p)}) \leq (1-\delta)^2 d(\mathbf{P}^{(n-2r)}, \mathbf{P}^{(n+p-2r)}), \tag{6.53}$$

and so on. If $n = mr + s$, $s = 1,2,\ldots,m-1$, then one finds

$$d(\mathbf{P}^{(n)},\mathbf{P}^{(n+p)}) \le (1-\delta)^m d(\mathbf{P}^{(n-mr)},\mathbf{P}^{(n+p-mr)}) \le (1-\delta)^m, \tag{6.54}$$

as one always has that $d(\boldsymbol{\mu},\boldsymbol{\nu}) < 1$. Hence, for any $\varepsilon > 0$ there is an n large enough so that $(1-\delta)^m < \varepsilon$ for some m. Therefore, the sequence $\{\mathbf{P}^{(n)}\}_{n\in\mathbb{N}}$ is a Cauchy sequence and the limit

$$\lim_{r\to\infty} \mathbf{P}^{(r)} = \boldsymbol{\pi}, \tag{6.55}$$

exists;[1] in other words, the mixing property of the transition matrix produces a limiting distribution in the long run.

With Equation (6.55) point 1 of the theorem is proven. Now, consider the following chain of equalities

$$\boldsymbol{\pi} \times \mathbb{W} = \lim_{t\to\infty} \mathbf{P}^{(t)} \times \mathbb{W} = \lim_{t\to\infty} \mathbf{P}^{(0)} \times \mathbb{W}^{(t)} \times \mathbb{W} = \lim_{t\to\infty} \mathbf{P}^{(0)} \times \mathbb{W}^{(t+1)} = \boldsymbol{\pi}; \tag{6.56}$$

this means that

$$\boldsymbol{\pi} \times \mathbb{W} = \boldsymbol{\pi}, \tag{6.57}$$

which proves one part of point 2. A probability distribution satisfiying Equation (6.57) is said to be an *invariant* distribution of the chain. Therefore, *the limiting distribution is just the invariant distribution* of the chain. That $\boldsymbol{\pi}$ is unique is given by the fact that, if $\boldsymbol{\pi}$ and $\boldsymbol{\pi}'$ are both invariant, they do not change by multiplication by $\mathbb{W}^{(r)}$; hence, as a consequence of the abstract theorem $d(\boldsymbol{\pi} \times \mathbb{W}^{(r)}, \boldsymbol{\pi}' \times \mathbb{W}^{(r)}) = d(\boldsymbol{\pi},\boldsymbol{\pi}') \le (1-\delta) d(\boldsymbol{\pi},\boldsymbol{\pi}')$, then $d(\boldsymbol{\pi},\boldsymbol{\pi}') = 0$, and as a consequence of the distance properties, one eventually gets $\boldsymbol{\pi} = \boldsymbol{\pi}'$. This proves the uniqueness of the invariant distribution in the case under scrutiny and completes the proof of point 2.

It is often useful and instructive to consider the initial distribution as degenerate, or deterministic, that is $\mathbf{P}^{(0)} = (0,\ldots,0,1,0,\ldots,0) = \delta(x-x_0)$, in order to observe how the probability mass spreads as a consequence of the Markov dynamics. The mass is initially concentrated on a single state, then diffuses on all the available states as time goes by. In this case, one can write

$$\mathbf{P}^{(r)} = \mathbf{P}^{(0)} \times \mathbb{W}^{(r)} = w^{(r)}(i,j), \tag{6.58}$$

[1] Strictly speaking, one needs the *completeness* for the metric space of probability distributions on m states. Completeness means that every Cauchy distribution has a limit. This is true if the distributions are real valued.

where i is the index of the only initial state. Therefore, as a consequence of Equation (6.55), one immediately has that

$$\lim_{r\to\infty} w^{(r)}(i,j) = \pi_j; \tag{6.59}$$

this shows that the equilibrium distribution does not depend on the initial one (there is a complete memory loss) and completes the proof of point 3 and of the theorem. □

Remark (Speed of convergence) Regarding the speed of convergence of $\mathbf{P}^{(0)} = \delta(x - x_0)$ to π, the following considerations are possible. Given that $\boldsymbol{\pi} = \boldsymbol{\pi} \times \mathbb{W}^{(n)}$, and $d(\mathbf{P}^{(n)}, \boldsymbol{\pi}) = d(\mathbf{P}^{(0)} \times \mathbb{W}^{(n)}, \boldsymbol{\pi} \times \mathbb{W}^{(n)}) \leq \ldots \leq (1-\delta)^m$, where m is the integer part of n/r, one knows that the distance diminishes each r steps, but one desires a formula containing any step n. Now, assume that $r = 4$, as in the example of Section 6.3.2, then one has $d(\mathbf{P}^{(31)}, \pi) \leq (1-\delta)^7$ as $31 = 7\cdot 4 + 3$, and $7 \geq 31/4 - 1 = 6.75$. As a function of n, $m+1$ is 1 for $n = 1,2,3$; it is 2 for $n = 4,5,6,7$. Therefore $m \geq (n/r) - 1$ and given that $(1-\delta) < 1$, it increases if the exponent decreases. Then $(1-\delta)^m \leq (1-\delta)^{(n/r)-1} = (1-\delta)^{-1}(1-\delta)^{n/r} = (1-\delta)^{-1}\gamma^n$, where $\gamma = (1-\delta)^{1/r} < 1$. Finally, one finds

$$d(\mathbf{P}^{(n)}, \boldsymbol{\pi}) \leq (1-\delta)^{-1}\gamma^n, \tag{6.60}$$

and the speed of convergence is exponential, or it is faster than any power of n. See Exercise 6.3 for further details.

6.5 The invariant distribution

6.5.1 *On the existence of the invariant distribution*

The invariant distribution exists (and is unique) not only for any regular (irreducible and aperiodic) Markov chain, but also in the irreducible and periodic case. For regular chains, it is also the equilibrium distribution, that is $\lim_{r\to\infty} \mathbb{P}(X_r = j|X_0 = x_0) = \pi_j$ for any initial state x_0. This is not possible for periodic chains, where the parity of the initial state is there forever. However, if there exists a probability distribution on all the m states $\boldsymbol{\pi}$ satisfying the Markov Equation (6.22), one can conclude that, given that $\mathbf{P}^{(r)} = \boldsymbol{\pi}$, also $\mathbf{P}^{(r+1)} = \boldsymbol{\pi} \times \mathbb{W} = \boldsymbol{\pi}$, and this is true for all the future steps. For the aperiodic case this would be an artificial way to prepare the initial state: if the initial state is such that $\mathbf{P}^{(0)} = \boldsymbol{\pi}$, then the Markov chain would be a stationary process (see Exercise 6.1). In order to find the invariant distribution π, one should check if the set of m

linear equations

$$\begin{cases} \pi_1 = \sum_{i=1}^{m} \pi_i w(i,1), \\ \dots \\ \pi_j = \sum_{i=1}^{m} \pi_i w(i,j), \\ \dots \\ \pi_m = \sum_{i=1}^{m} \pi_i w(i,m) \end{cases} \tag{6.61}$$

has a solution for $\pi_j > 0$, with $\sum_{i=1}^{m} \pi_j = 1$. In matrix terms, it is the solution of the equation $\boldsymbol{\pi} = \boldsymbol{\pi} \times \mathbb{W}$, where $\boldsymbol{\pi}$ is the left eigenvector of $\mathbb{W}$ corresponding to the eigenvalue 1. In very simple examples (see the exercises), this route can be followed, but in our applications to physical and economic systems the state space is very large, and the search for the invariant distribution will follow another method. Regarding the meaning of Equation (6.61), writing the Markov equation explicitly for the jth state, one has that

$$\mathbb{P}(X_{s+1} = j) = \sum_{i=1}^{m} \mathbb{P}(X_s = i) w(i,j) \tag{6.62}$$

decribes the decomposition of the probability mass for the jth state at step $s+1$ as a sum of all contributions at the previous step. If there exists a distribution $\mathbb{P}(X_s = i)$, $i = 1, \dots, m$, such that the 'traffic' $w(i,j)$ leaves it invariant, $\boldsymbol{\pi}$ is found. This is just a mathematical property, and nothing is said about the fact that a specific system, prepared in some state, will sooner or later be described by $\boldsymbol{\pi}$.

In order to prove the existence of the invariant distribution, one can simply consider an irreducible Markov chain described by the one-step transition probability matrix $\mathbb{W}$ and by the initial probability distribution $\mathbf{P}^{(0)}$. As mentioned above, the hypothesis of aperiodicity is not necessary. For instance, if $\mathbf{P}^{(0)}$ has all its mass on x_0, $\mathbf{P}^{(1)}$ will be positive on the nearest neighbours of x_0 (the states with which x_0 communicates), and, as time goes by, all entries of $\mathbf{P}^{(s)}$ will become positive (if the matrix is regular, that is the chain is aperiodic), or they will have an oscillatory presence of zero and positive entries, if the matrix is periodic. What would happen if one uses this abstract mean:

$$\overline{\mathbf{P}}^{(0)}(s) = \frac{1}{s} \sum_{r=0}^{s-1} \mathbf{P}^{(r)} = \frac{1}{s} \{\mathbf{P}^{(0)} + \dots + \mathbf{P}^{(s-1)}\} \tag{6.63}$$

as initial probability? This is the time average of the first s probability distributions on the states. An intuitive meaning of this distribution is given below. However, at the second step, from Markov Equation (6.22), one gets

$$\overline{\mathbf{P}}^{(1)}(s) = \overline{\mathbf{P}}^{(0)}(s) \times \mathbb{W} = \frac{1}{s} \{\mathbf{P}^{(1)} + \dots + \mathbf{P}^{(s)}\}, \tag{6.64}$$

so that

$$\overline{\mathbf{P}}^{(1)}(s) - \overline{\mathbf{P}}^{(0)}(s) = \frac{1}{s}\{\mathbf{P}^{(s)} - \mathbf{P}^{(0)}\}, \tag{6.65}$$

and

$$|\overline{\mathbf{P}}^{(1)}(s) - \overline{\mathbf{P}}^{(0)}(s)| = \frac{1}{s}|\mathbf{P}^{(s)} - \mathbf{P}^{(0)}| \leq \frac{2}{s}. \tag{6.66}$$

This difference can be made as small as desired, provided that $s \to \infty$. In other words, $\lim_{s\to\infty} \overline{\mathbf{P}}^{(0)}(s)$ does not vary during the time evolution. Therefore, the invariant distribution

$$\pi = \lim_{s\to\infty} \frac{1}{s}\{\mathbf{P}^{(1)} + \ldots + \mathbf{P}^{(s)}\} \tag{6.67}$$

does exist for periodic chains also.

Now, consider the periodic Ehrenfest urn; if $w^{(r)}(i,j)$ is positive, then $w^{(r+1)}(i,j)$ is zero, and, for large s, if r is even, $w^{(r)}(i,j) \simeq w^{(\text{even})}(i,j)$, while $w^{(r+1)}(i,j) \simeq w^{(\text{odd})}(i,j)$. Moreover, introducing the limiting matrix of the aperiodic case,

$$w^{(\infty)}(i,j) = \frac{1}{2}(w^{(\text{odd})}(i,j) + w^{(\text{even})}(i,j)), \tag{6.68}$$

and given that one of them is zero, either $w^{(\text{odd})}(i,j)$ is 0 or it is equal to $2w_{ij}^{(\infty)}$, and the same is true for $w^{(\text{even})}$.

Assume that one considers a periodic and an aperiodic chain with the same space state and both starting from the same initial state, say $x_0 = 0$, and focus on the jth state. Consider

$$\overline{P_j}^{(0)}(s) = \frac{1}{s}\{P_j^{(0)} + \ldots + P_j^{(s-1)}\}. \tag{6.69}$$

If $j \neq 0$, then one starts with $P_j^{(0)} = 0$, but, after some steps, this probability will be positive, as all states communicate with each other in both cases (both chains are assumed to be irreducible). Note that starting with $x_0 = 0$ amounts to setting $P_j^{(t)} = w^{(t)}(0,j)$. Omitting the initial steps, whose weight goes to zero if s is large, and considering two consecutive steps, in the aperiodic case, both $w^{(t)}(0,j)$ and $w^{(t+1)}(0,j)$ are positive and converge to $w^{(\infty)}(0,j)$, whereas only one of them is positive in the periodic case, but the positive term converges to $2w^{(\infty)}(0,j)$. Then in the sum

$$\frac{1}{s}\{P_j^{(0)} + \ldots + P_j^{(s-1)}\}, \tag{6.70}$$

the correlation between time and values is lost, and the invariant distribution exists for both the aperiodic and the periodic case.

6.5.2 The frequency meaning of the invariant distribution

An intuitive meaning of the invariant distribution is as follows. One is interested in the time-fraction in which the system visits any given state as time goes by. As usual (see Section 4.1.8), one can define $I_j^{(r)}$ as the indicator function $\mathbb{I}_{\{X_r=j\}}$ of the event $X_r = j$. Therefore the random variable

$$J_{j,s} = \frac{1}{s}\sum_{r=0}^{s-1} I_j^{(r)} \tag{6.71}$$

denotes the relative frequency of visits to the state j after s steps. The expected value of $I_j^{(r)}$ is

$$\mathbb{E}(I_j^{(r)}) = \mathbb{P}(X_r = j) = P_j^{(r)}. \tag{6.72}$$

Therefore,

$$\mathbb{E}(J_{j,s}) = \frac{1}{s}\sum_{r=0}^{s-1} \mathbb{E}(I_j^{(r)}) = \frac{1}{s}\{P_j^{(0)} + \ldots + P_j^{(s-1)}\} \tag{6.73}$$

is just the expected relative frequency of visits to the jth state after s steps. If there exists

$$\pi_j = \lim_{s\to\infty} \frac{1}{s}\{P_j^{(0)} + \ldots + P_j^{(s-1)}\}, \tag{6.74}$$

then the expected relative frequency of visits is such that

$$\lim_{s\to\infty} \mathbb{E}(J_{j,s}) = \pi_j. \tag{6.75}$$

This link between a probability value π_j (which is a measure) and the relative frequency $J_{j,s}$ (that is a random variable) can be made stronger by also considering the variance of $J_{j,s}$. Indeed, it is possible to prove that $\sqrt{\mathbb{V}(J_{j,s})}/\mathbb{E}(J_{j,s})$ vanishes for large s, and this is the basis of the (weak) *law of large numbers* for irreducible Markov chains and the consequent *ergodic theorem*.

6.5.3 The law of large numbers for irreducible Markov chains

From Exercise 5.2, the variance of the sum of s random variables is given by

$$\mathbb{V}\left(\sum_{i=1}^{s} X_i\right) = \sum_{i=1}^{s} \mathbb{V}(X_i) + 2\sum_{i=1}^{s}\sum_{j=i+1}^{s} \mathbb{C}(X_i, X_j), \tag{6.76}$$

a formula consisting of s diagonal terms and $s(s-1)$ out-of-diagonal terms for a grand-total of s^2 terms.

If the random variables X_i were independent (or uncorrelated), one would have $\mathbb{C}(X_i, X_j) = 0$, and only s terms would survive in Equation (6.76); if they were exchangeable, then $\mathbb{C}(X_i, X_j)$ would not depend on i, j and they would be positive; as a consequence, all the s^2 terms would be positive. When one is interested in the variance of $\sum_{i=1}^{s} X_i/s$, one must divide Equation (6.76) by s^2 and this is the reason the variance of $\sum_{i=1}^{s} X_i/s$ vanishes for independent variables (self-averaging), and does not vanish for exchangeable variables (lack of self-averaging). In the Markovian case, correlations decrease as a function of $|j-i|$, so that the variance of $\sum_{i=1}^{s} X_i/s$ vanishes for $s \to \infty$.

Now, consider the variance of $J_{j,s}$ defined in Equation (6.71). The following calculations are valid only for an aperiodic irreducible Markov chain, but the result can be extended to the periodic case. Following Penrose [1], one has

$$\mathbb{V}(J_{j,s}) = \frac{1}{s^2}\mathbb{V}\left(\sum_{r=1}^{s} I_j^{(r)}\right), \tag{6.77}$$

and

$$\mathbb{V}\left(\sum_{r=1}^{s} I_j^{(r)}\right) = \sum_{r=1}^{s} \mathbb{V}(I_j^{(r)}) + \sum_{r=1}^{s}\sum_{u=r+1}^{s} \mathbb{C}(I_j^{(r)}, I_j^{(u)})$$
$$+ \sum_{r=1}^{s}\sum_{u=1}^{r-1} \mathbb{C}(I_j^{(r)}, I_j^{(u)}) = v_1 + v_2 + v_3, \tag{6.78}$$

where the out-of-diagonal terms are divided into future ($u > r$) and past ($u < r$) covariances. Regarding v_1 one finds that (see Exercise 4.4)

$$\mathbb{V}(I_j^{(r)}) = P_j^{(r)}[1 - P_j^{(r)}] \leq 1/4; \tag{6.79}$$

therefore,

$$v_1 \leq s/4. \tag{6.80}$$

Regarding v_2, one can use the result of Exercise 5.3 to find that

$$\mathbb{C}(I_j^{(r)}, I_j^{(u)}) = \mathbb{P}(X_r = j, X_u = j) - \mathbb{P}(X_r = j)\mathbb{P}(X_u = j)$$
$$= P_j^{(r)}[w^{(u-r)}(j,j) - P_j^{(u)}], \tag{6.81}$$

which is the covariance of the rth variable and the uth, with $r < u \leq s$, distant $u - r$ steps. Further,

$$v_2 = \sum_{r=1}^{s} \sum_{u=r+1}^{s} P_j^{(r)} [w^{(u-r)}(j,j) - P_j^{(u)}] = \sum_{r=1}^{s} P_j^{(r)} \sum_{u=r+1}^{s} [w^{(u-r)}(j,j) - P_j^{(u)}]. \tag{6.82}$$

The correlation $\mathbb{C}(I_j^{(r)}, I_j^{(u)})$ can also be negative; therefore, given that $P_j^{(r)} \leq 1$,

$$|v_2| \leq \sum_{r=1}^{s} \sum_{u=r+1}^{s} |w^{(u-r)}(j,j) - P_j^{(u)}|. \tag{6.83}$$

Note that

$$|w^{(u-r)}(j,j) - P_j^{(u)}| \leq |w^{(u-r)}(j,j) - \pi_j| + |\pi_j - P_j^{(u)}|, \tag{6.84}$$

and both $w^{(u-r)}(j,j)$ and $P_j^{(u)}$ converge to π_j for large u and $u - r$. Therefore, when s is large, most of the $s(s-1)/2$ terms in the double sum will vanish. Eventually, using the inequality in (6.84), one finds

$$|v_2| \leq \sum_{r=1}^{s} \sum_{u=r+1}^{s} |w^{(u-r)}(j,j) - \pi_j| + \sum_{r=1}^{s} \sum_{u=r+1}^{s} |\pi_j - P_j^{(u)}|. \tag{6.85}$$

For any r, the first summand on the right-hand side of inequality (6.85) sums $s - r$ terms $|w^{(u-r)}(j,j) - \pi_j| = |w^{(t)}(j,j) - \pi_j|$, where $t = u - r$ varies in $1, \ldots, s - r$. If one also adds the remaining r terms, a new upper bound is obtained:

$$\begin{aligned} \sum_{r=1}^{s} \sum_{u=r+1}^{s} |w^{(u-r)}(j,j) - \pi_j| &= \sum_{r=1}^{s} \sum_{t=1}^{s-r} |w^{(t)}(j,j) - \pi_j| \\ &\leq \sum_{r=1}^{s} \sum_{t=1}^{s} |w^{(t)}(j,j) - \pi_j| = s \sum_{t=1}^{s} |w^{(t)}(j,j) - \pi_j|, \end{aligned} \tag{6.86}$$

as the first sum does not depend on r. Symmetrically adding r terms in the second summand, the final new upper bound for v_2 is

$$|v_2| \leq s \sum_{t=1}^{s} [|w^{(t)}(j,j) - \pi_j| + |\pi_j - P_j^{(t)}|]. \tag{6.87}$$

From the limiting properties of $w^{(t)}(j,j)$ and $P_j^{(t)}$, it turns out that for any $\varepsilon > 0$ there exists a k such that for $t > k$ one has[2]

$$|w_{jj}^{(t)} - \pi_j| + |\pi_j - P_j^{(t)}| < \varepsilon. \tag{6.88}$$

But in Equation (6.87) all values of t between 1 and s are present, also small values, for which $w^{(t)}(j,j)$ and $P_j^{(t)}$ can be very different from π_j. However, the sum in Equation (6.87) can be split into two parts

$$\frac{|v_2|}{s} \leq \sum_{t=1}^{k} [|w^{(t)}(j,j) - \pi_j| + |\pi_j - P_j^{(t)}|] + \sum_{t=k+1}^{s} [|w^{(t)}(j,j) - \pi_j| + |\pi_j - P_j^{(t)}|]. \tag{6.89}$$

In the first part, each summand is not greater than 2, whereas, in the second, each term is not greater than ε; it follows that

$$|v_2| \leq s\sum_{t=1}^{k} 2 + s \sum_{t=k+1}^{s} \varepsilon \leq 2sk + s^2\varepsilon. \tag{6.90}$$

Returning to the variance of the relative frequency of visit, this is given by

$$\mathbb{V}(J_{j,s}) = \frac{1}{s^2}(v_1 + v_2 + v_3), \tag{6.91}$$

where $v_3 = v_2$ and

$$|\mathbb{V}(J_{j,s})| \leq \left(\frac{|v_1|}{s^2} + 2\frac{|v_2|}{s^2}\right). \tag{6.92}$$

From Equation (6.80), it is immediately apparent that

$$\lim_{s\to\infty} \frac{|v_1|}{s^2} = \lim_{s\to\infty} \frac{v_1}{s^2} \leq \lim_{s\to\infty} \frac{1}{4s} = 0. \tag{6.93}$$

To show that also

$$\lim_{s\to\infty} \frac{|v_2|}{s^2} = 0, \tag{6.94}$$

it is sufficient to note that, from Equation (6.90), one has

$$\frac{|v_2|}{s^2} \leq \frac{2k}{s} + \varepsilon. \tag{6.95}$$

[2] It is assumed that s is large enough, as one is interested in $s \to \infty$.

Then, if $s \geq 2k/\varepsilon$ in addition to $s \geq k$, one gets

$$\frac{|v_2|}{s^2} \leq \frac{2k}{s} + \varepsilon \leq \varepsilon + \varepsilon = 2\varepsilon, \tag{6.96}$$

which suffices to prove (6.94). Therefore, from Equation (6.92), one finds that

$$\lim_{s \to \infty} \mathbb{V}(J_{j,s}) = 0. \tag{6.97}$$

In order to see this more clearly, consider the previous aperiodic model, setting $j = 0$ and $P_0^{(0)} = 1$, $\mathbf{P}^{(0)} = (1,0,0,0,0)$, in which case $w^{(t)}(j,j) = P_j^{(t)}$; if one chooses $\varepsilon = 0.01$, then one finds that $|w^{(t)}(j,j) - \pi_j| + |\pi_j - P_j^{(t)}| = 2|w^{(t)}(j,j) - \pi_j| < 0.01$ for $t = k > 14$. Writing down the upper limit, one finds that $|v_2|/s^2 \leq 28/s + 0.01$. If one requires that $|v_2|/s^2 \leq 2\varepsilon$, one needs $28/s < 0.01$, that is $s > 2800$.

Using the Bienaymé–Chebyshev inequality (see Exercise 6.4),

$$\mathbb{P}(|J_{j,s} - \mathbb{E}(J_{j,s})| \geq \delta) \leq \frac{\mathbb{V}(J_{j,s})}{\delta^2}, \tag{6.98}$$

and recalling Equation (6.75) yields

$$\lim_{s \to \infty} \mathbb{P}(|J_{j,s} - \pi_j| \geq \delta) = 0. \tag{6.99}$$

This is the weak law of large numbers for irreducible Markov chains, stating that, given a difference $\delta > 0$ between the observed relative frequency $J_{i,s}$ and the value π_i, there exists $\varepsilon > 0$ and a number of steps $s(\delta, \varepsilon)$ such that the probability of exceeding δ is less than ε. Considering all states together, the vector $\mathbf{J}_s = (J_{1,s}, \ldots, J_{m,s})$, with $\sum_{i=1}^{m} J_{i,s} = 1$, represents the normalized histogram of the relative visiting frequencies after s steps. One expects that $\lim_{s \to \infty} d(\mathbf{J}_s, \boldsymbol{\pi}) \to 0$. If one considers a function X of the state i such that $X(i) = x_i$, then X can be decomposed in normal form, that is

$$X_t = \sum_{i=1}^{n} x_i I_i^{(t)}. \tag{6.100}$$

Its time average is the random variable

$$\overline{X}_s = \frac{1}{s} \sum_{t=1}^{s} X_t = \frac{1}{s} \sum_{t=1}^{s} \sum_{i=1}^{m} x_i I_i^{(t)} = \frac{1}{s} \sum_{i=1}^{m} x_i \sum_{t=1}^{s} I_i^{(t)} = \sum_{i=1}^{m} x_i J_{i,s}, \tag{6.101}$$

where the sum on steps was transformed into a sum on values weighted by their relative frequency. Due to the fact that $d(\mathbf{J}_s, \boldsymbol{\pi}) \to 0$ for $s \to \infty$, one expects that

$\overline{X}_s = \sum_{i=1}^{m} x_i J_{i,s}$ will converge to

$$\mathbb{E}_\pi(X) = \sum_{i=1}^{m} x_i \pi_i. \tag{6.102}$$

This is stronger than claiming that, for $s \to \infty$, one has $\mathbb{E}(\overline{X}_s) \to \mathbb{E}_\pi(X)$. This property means that, repeating many times $i = 1,\ldots,n$ a very long sequence $X_0^{(i)}, X_1^{(i)}, \ldots, X_s^{(i)}$, the time averages $\overline{X}_s^{(i)}$ are distributed around $\mathbb{E}_\pi(X)$. The (weak) ergodic theorem states that

$$\lim_{s\to\infty} \mathbb{P}(|\overline{X}_s - \mathbb{E}_\pi(X)| \geq \delta) = 0. \tag{6.103}$$

Equation (6.103) means that $\overline{X}_s$ converges to its limiting expectation value $\mathbb{E}_\pi(X)$ with a variance which is vanishing for large s. This theorem will not be proved here. As a final remark, note that equations such as (6.99) and (6.103) lie at the foundation of the *Markov chain Monte Carlo* simulation method which will be used in the following sections.

6.6 Reversibility

Is there any property which allows an exact evaluation of the invariant distribution π? In order to answer this question, consider the Markov equation once more,

$$\mathbb{P}(X_{s+1} = j) = \sum_{i=1}^{m} \mathbb{P}(X_s = i) w(i,j), \tag{6.104}$$

and given that

$$\sum_{i=1}^{m} w(j,i) = 1, \tag{6.105}$$

one finds

$$\mathbb{P}(X_s = j) = \sum_{i=1}^{m} \mathbb{P}(X_s = j) w(j,i). \tag{6.106}$$

Subtracting this equation from Equation (6.104), one gets the so-called discrete *master equation*

$$\mathbb{P}(X_{s+1} = j) - \mathbb{P}(X_s = j) = \sum_{i=1}^{m} [\mathbb{P}(X_s = i) w(i,j) - \mathbb{P}(X_s = j) w(j,i)]. \tag{6.107}$$

Equation (6.107) can be written in a shorter form, omitting the term $i = j$, which vanishes

$$P_j^{(s+1)} - P_j^{(s)} = \sum_{i \neq j} \left[P_i^{(s)} w(i,j) - P_j^{(s)} w(j,i) \right]. \qquad (6.108)$$

The master equation can be understood as follows: the difference $\Delta P_j^{(s)} = P_j^{(s+1)} - P_j^{(s)}$ is the increment of the probability mass for the jth state, whereas $P_i^{(s)} w(i,j) - P_j^{(s)} w(j,i)$ is the net traffic (the net flux) along the road connecting the states i and j. Note that $P_i^{(s)} w(i,j)$ is the ith state mass flux entering the state j, and $P_j^{(s)} w(j,i)$ is jth state mass flux leaving towards state i. If, for any couple i,j, the total flux is zero

$$P_i^{(s)} w(i,j) = P_j^{(s)} w(j,i), \qquad (6.109)$$

then $P_j^{(s+1)} = P_j^{(s)} = \pi_j$. In other words, if a distribution satisfies the *detailed balance conditions* (6.109), it is the invariant distribution. Note that Equation (6.109) is not a necessary condition for the invariant distribution: for any π_j, the global balance is zero, whereas detailed balance is a property valid for each pair of states. The master equation and the balance equations may seem intuitive in the *ensemble interpretation* of probability (see Exercise 6.1). In this framework, $P_i^{(s)} w(i,j)$ is proportional to the number of systems passing from i to j, and $P_j^{(s)} w(j,i)$ is proportional to the number of systems passing from j to i. If they are equal for all couples, the number of systems in the jth state does not change in time.

One may wonder whether there is some simple property which can help in showing that the invariant distribution satisfies detailed balance. This property exists, and it is called *reversibility*. The intuitive meaning is the following. Consider the sequence of events $S_1 = \ldots, i_{t-1}, j_t, k_{t+1}, \ldots$, where j_t is a shortcut for $X_t = j$, as a *movie*; the movie is reversible if the observer is unable to say whether the movie is being run forward or backward. The reversed sequence is $S_2 = \ldots, k_{t-1}, j_t, i_{t+1}, \ldots$ Under what conditions can one assume that the two sequences are *equiprobable*? If so, a filmgoer could not discern whether the movie is correctly projected, or whether it is rewinding. One can show that, if a forward movie has the scene $i_t j_{t+1}$, then after a lag $s > t$ depending on the origin of reversal, rewinding shows $j_{s-t-1} i_{s-t}$. Referring to the example of Section 2.4.1, the scene could be 'agent A is a bull at time 0 and becomes a bear at time 1'. At time step 10, the movie is stopped and reversed, and for a new observer the scene at time step 10 is the initial one. This filmgoer will see that 'agent A is a bear at time 9 and passes to a bull at time 10'.

6.6.1 Backward movies

If the direct movie is a homogeneous irreducible Markov chain, what can be said of the reversed one? Consider the sequence $S = \ldots, i_{t-1}, j_t, k_{t+1}, \ldots$, where, as before,

j_t is a shortcut for $X_t = j$. Starting from a generic t, consider

$$P(i_{t-1}|j_t, k_{t+1}, \ldots) = \mathbb{P}(X_{t-1} = i|X_t = j, X_{t+1} = k, \ldots), \tag{6.110}$$

which is the predictive probability of the backward movie at time step t. One can show that:

1. the reversed predictive probability still has the Markov property, that is

$$P(i_{t-1}|j_t, k_{t+1}, \ldots) = P(i_{t-1}|j_t). \tag{6.111}$$

   Indeed, it is sufficient to consider $P(i_{t-1}|j_t, k_{t+1})$; from Bayes' rule one finds:

$$P(i_{t-1}|j_t, k_{t+1}) = P(i_{t-1}, j_t, k_{t+1})/P(j_t, k_{t+1}). \tag{6.112}$$

   If the direct chain is homogeneous, then

$$\frac{P(i_{t-1}, j_t, k_{t+1})}{P(j_t, k_{t+1})} = \frac{P(i_{t-1})w(i,j)w(j,k)}{P(j_t)w(j,k)} = \frac{P(i_{t-1})w(i,j)}{P(j_t)}, \tag{6.113}$$

   which does not depend on k_{t+1}, and therefore the reversed process of a homogeneous Markov chain is still a Markov chain:

$$P(i_{t-1}|j_t, k_{t+1}) = P(i_{t-1}|j_t) = \frac{P(i_{t-1})w(i,j)}{P(j_t)}; \tag{6.114}$$

2. from Equation (6.114), if the direct chain is stationary, that is $P(i_{t-1}) = \pi_i$, and $P(j_t) = \pi_j$, then the reversed chain is homogeneous, and

$$P(i_{t-1}|j_t) = u(j,i) = \frac{\pi_i w(i,j)}{\pi_j}; \tag{6.115}$$

3. finally, the reversed chain has the same transition matrix as the direct chain if and only if $P(i_{t-1}|j_t) = u(j,i)$ is equal to $P(i_{t+1}|j_t) = w(j,i)$, that is

$$w(j,i) = u(j,i) = \frac{\pi_i w(i,j)}{\pi_j}, \tag{6.116}$$

   which is again the *detailed balance condition*. The condition $P(i_{t-1}|j_t) = P(i_{t+1}|j_t)$ expresses reversibility in the clearest way: given that $X_t = j$, if the movie is run forward one sees $X_{t+1} = i$ with the same probability as one sees $X_{t-1} = i$, if the movie is run backward.

The above considerations are a proof of the following

Theorem (reversible Markov chains) A homogeneous irreducible Markov chain is reversible if and only if it is stationary and the stationary distribution satisfies the detailed balance conditions.

Concerning search of the invariant distribution, one can conclude that if the chain cannot be reversible, then one cannot make use of Equation (6.109). For instance, for a moment jump to Exercise 6.1, case a). The proposed transition matrix is such that the state b is always followed by c; hence, reversing the movie one will always see the opposite transition $c \to b$, and the chain, even if stationary, is not reversible. The case of the seven dice game (in Section 6.1.4) is different: there is no logical way to distinguish a die displaying 1 which, when cast, becomes a 4 from the reversed transition, in which a 4 becomes a 1. In the case of dice, after a number of steps sufficient to forget the initial conditions, the matrix being regular, as a consequence of Sinai's theorem, the chain becomes closer and closer to the stationary chain, which is described by the equilibrium distribution; therefore, one can try to see if there exists a $\pi(\mathbf{n})$ such that:

$$\pi(\mathbf{n})w(\mathbf{n},\mathbf{n}_i^j) = \pi(\mathbf{n}_i^j)w(\mathbf{n}_i^j,\mathbf{n}),\ j \neq i \tag{6.117}$$

where $w(\mathbf{n},\mathbf{n}_i^j)$ is given by Equation (6.5). Replacing (6.5) into Equation (6.117) and simplifying one gets

$$\pi(\mathbf{n})n_i = \pi(\mathbf{n}_i^j)(n_j + 1). \tag{6.118}$$

Given that $\mathbf{n}_i^j = (n_1,\ldots,n_i - 1,\ldots,n_j + 1,\ldots,n_g)$, and that $\pi(\mathbf{n}_i^j)/\pi(\mathbf{n}) = n_i/(n_j+1)$, one finds that $\pi(\mathbf{n}) \sim \left(\prod_{i=1}^g n_i!\right)^{-1}$ satisfies (6.117). The normalization condition $\sum_{\mathbf{n}\in S_g^n} \pi(\mathbf{n}) = 1$ leads to the symmetric multinomial distribution

$$\pi(\mathbf{n}) = \frac{n!}{\prod_{i=1}^g n_i!} g^{-n}, \tag{6.119}$$

written for any value of n and g. This distribution is *regular*, that is positive for all possible $\mathbf{n} \in S_g^n$. The same distribution describes a single experiment, where n 'fair' dice are cast and the frequency distribution of the $g = 6$ results is observed. In this case, one could ask 'what is the meaning of the number $\pi(\mathbf{n})$?', and the usual discussion on the meaning of probability would start (see Chapter 1). On the contrary, in the case of the dice game, one could record the normalized histogram of the visiting frequencies, and compare it with $\pi(\mathbf{n})$. Distinguishing between the histogram and $\pi(\mathbf{n})$, by the law of large numbers for irreducible Markov chains, one can determine how long the game has to go on so that the probability that they are closer than the fixed distance is as close to 1 as desired. Therefore, as discussed above, the chain contains a natural 'frequency interpretation' of the equilibrium probability $\pi(\mathbf{n})$. However, this is not the only interpretation physicists give to $\pi(\mathbf{n})$. If an observer is going to study the system after a long time during which the system was left running, the limit $\lim_{s\to\infty} \mathbb{P}(X_s = j) = \pi_j$ means that the probability distribution to be used is $\mathbb{P}(X_s = j) = \pi_j$, whatever past information could have been available. In the 'ensemble interpretation', π_j is the fraction of (fictitious) systems which at time s are in state j. Further, if the observer knows that π_j is also a measure of the relative visiting frequency for state j, π_j can also be considered as the fraction of 'favourable cases' of finding state j over all the 'possible cases'. Nowadays, the 'ensemble interpretation' can be realized in practice, by setting up a large number of parallel computer simulations running the same Markov chain, and with the same initial condition (and different pseudo-random number sequences). The coincidence of ensemble averages and time (or ergodic) averages is an old problem in *ergodic theory*, which finds a satisfactory solution within the framework of finite Markov chains.

6.6.2 Stationarity, reversibility and equilibrium

Any irreducible chain has an invariant distribution $\boldsymbol{\pi}$, and, if it is aperiodic, sooner or later, it will be described by $\boldsymbol{\pi}$. For an aperiodic 'concrete' chain, starting from the initial state x_0, with $\mathbf{P}^{(0)} = \delta(x - x_0)$, there will be a *relaxation time* during which $\mathbf{P}^{(t)}$ moves step-by-step towards $\boldsymbol{\pi}$ until the chain reaches stationarity. As an example, consider a system where 1000 fleas are present on dog A and 0 fleas on dog B. For a long initial period, it is far more likely to observe a flea flux from A to B than the converse. This is why the Ehrenfests introduced their model, to clarify the meaning of Boltzmann's H-theorem and of irreversibility. During the initial period, the Ehrenfest chain is non-stationary and cannot be reversible. The most probable event is always to choose a flea from dog A, as in the event $X_1 = 999, X_2 = 998, \ldots, X_{10} = 990$, in any case a decreasing X_t, so that the backward movie shows a curiously increasing number of fleas on dog A, which contains all of them at the end of the movie. The situation is quite different after the chain has reached stationarity. Assume that this happens when $t \geq 2000$; then, one is entitled to believe that $\mathbf{P}^{(t)}$ is the binomial distribution. However, imagine that the observer measures $O_1 = X_{2000} = 500, X_{2001} = 501, \ldots, X_{2499} = 1000$, and compare this measure with the opposite $O_2 = X_{2000} = 1000, X_{2001} = 999, \ldots, X_{2499} = 500$. Which of these sequences is more likely? In the periodic case, which is simpler, one finds that

$$\mathbb{P}(O_1) = \pi_{500}\frac{500}{1000}\frac{499}{1000}\cdots\frac{2}{1000}\frac{1}{1000} = \frac{1000!}{500!500!}2^{-1000}\frac{500!}{1000^{500}},$$
$$\mathbb{P}(O_2) = \pi_{1000}\frac{1000}{1000}\frac{999}{1000}\cdots\frac{502}{1000}\frac{501}{1000} = 2^{-1000}\frac{1000!}{500!}\frac{1}{1000^{500}}, \tag{6.120}$$

that is $\mathbb{P}(O_1) = \mathbb{P}(O_2)$! This is an alternative way of understanding reversibility; the probability of a sequence is the same as the probability of the reversed one. While in the irreversible conditions at the beginning of time evolution, conditioned on $X_0 = 1000$, it was trivial to find that the probability of O_2 is zero, here, one has that $\mathbb{P}(O_1) = \mathbb{P}(O_2)$; this is due to the fact that for the starting equilibrium probabilities $\pi_{500} \gg \pi_{1000}$. In other words, it is far more probable to find the system around 500 than around 1000, but the transition probabilities are such that $w^{(500)}(500,1000) \ll w^{(500)}(1000,500)$. In fact, the detailed balance condition holds true, and $\pi_{500}w^{(500)}(500,1000) = \pi_{1000}w^{(500)}(1000,500)$. As a conclusion, one cannot decide if one is watching either the backward or the forward movie. Therefore, for reversibility to eventually occur, one has to wait until the chain becomes stationary.

6.6.3 Markov chains for individual descriptions

The individual description is a sequence as long as the size of the population n, and there is nothing in principle against the description of time evolution in terms of individual descriptions. A transition matrix $g^n \times g^n$ is difficult to handle. But the possibility of studying such a complicated Markov chain is given solely by the relationship between the invariant distribution, detailed balance conditions and reversibility. Returning to the previous example of seven dice (see Section 6.1.4): if $\mathbf{X}_0 = (2,6,2,1,1,5,4)$, and $\mathbf{X}_1 = (2,6,1,1,1,5,4)$, one is able to calculate the transition probability $\mathbb{P}(\mathbf{X}_1|\mathbf{X}_0)$ and then to build the Markov chain $\mathbf{X}_0, \mathbf{X}_1, \ldots, \mathbf{X}_n, \ldots$ For instance, a move requires that die number 3 is chosen, with uniform probability, and that casting the die results in a 1, with probability equal to p_1. To be more precise, one can introduce the following symbols for a move: let $\mathbf{x}_{j,k}^{(i)}|\mathbf{x}$ denote that the ith die changes from the jth category to the kth category, all the rest being unchanged. This transition corresponds to the frequency change $\mathbf{n}_j^k|\mathbf{n}$. Then

$$\mathbb{P}(\mathbf{x}_{j,k}^{(i)}|\mathbf{x}) = \frac{1}{n} p_k, \tag{6.121}$$

and a little thought leads us to deduce that the inverse transition has probability

$$\mathbb{P}(\mathbf{x}|\mathbf{x}_{j,k}^{(i)}) = \frac{1}{n} p_j. \tag{6.122}$$

In the example, $p_j = p_k = 1/6$, but one can consider the more general non-uniform case. One can search for the invariant distribution of the chain looking for a $\pi(\mathbf{x})$ which possibly satisfies the detailed balance conditions (6.109). In this case, this leads to

$$\pi(\mathbf{x})P(\mathbf{x}_{j,k}^{(i)}|\mathbf{x}) = \pi(\mathbf{x}_{j,k}^{(i)})\mathbb{P}(\mathbf{x}|\mathbf{x}_{j,k}^{(i)}), \tag{6.123}$$

that is $\pi(\mathbf{x})p_k = \pi(\mathbf{x}^{(i)}_{j,k})p_j$, or

$$\frac{\pi(\mathbf{x}^{(i)}_{j,k})}{\pi(\mathbf{x})} = \frac{p_k}{p_j}. \tag{6.124}$$

Recalling that $n_k(\mathbf{x}^{(i)}_{j,k}) = n_k(\mathbf{x}) + 1$, while $n_j(\mathbf{x}^{(i)}_{j,k}) = n_j(\mathbf{x}) - 1$, a distribution

$$\pi(\mathbf{x}) \propto \prod_{i=1}^{g} p_i^{n_i} \tag{6.125}$$

satisfies (6.124). This is the multinomial distribution on all the possible individual sequences. This distribution, which in the uniform case ($\forall i, p_i = 1/g$) makes all the individual descriptions equiprobable, was discussed in an abstract way in Chapter 3.

6.7 Exercises

6.1 Consider the following chain in the three cases: a) $\beta = 0$, b) $\beta = 1$, c) $0 < \beta < 1$.

$$\mathbb{W} = \begin{array}{cccc} \textit{state} & a & b & c \\ a & \beta & 1-\beta & 0 \\ b & 0 & 0 & 1 \\ c & 1 & 0 & 0 \end{array}.$$

For each case, making use of common sense also:
1. determine if the Markov chain is irreducible or not;
2. determine if the Markov chain is periodic or not;
3. find (if it exists) the invariant distribution;
4. find (if it exists) the limiting frequency distribution;
5. find (if it exists) the equilibrium distribution;
6. does detailed balance hold true?

6.2 Consider the periodic Ehrenfest urn with $n = 4$ balls or fleas; one can write the transition matrix the following way:

$$\mathbb{W} = \begin{array}{cccccc} \textit{state} & \textit{0} & \textit{1} & \textit{2} & \textit{3} & \textit{4} \\ \textit{0} & 0 & 1 & 0 & 0 & 0 \\ \textit{1} & 1/4 & 0 & 3/4 & 0 & 0 \\ \textit{2} & 0 & 1/2 & 0 & 1/2 & 0 \\ \textit{3} & 0 & 0 & 3/4 & 0 & 1/4 \\ \textit{4} & 0 & 0 & 0 & 1 & 0 \end{array};$$

by simply rearranging the list of states, the matrix becomes

$$\widetilde{\mathbb{W}} = \begin{array}{cccccc} \textit{state} & \textit{0} & \textit{2} & \textit{4} & \textit{1} & \textit{3} \\ \textit{0} & 0 & 0 & 0 & 1 & 0 \\ \textit{2} & 0 & 0 & 0 & 1/2 & 1/2 \\ \textit{4} & 0 & 0 & 0 & 0 & 1 \\ \textit{1} & 1/4 & 3/4 & 0 & 0 & 0 \\ \textit{3} & 0 & 3/4 & 1/4 & 0 & 0 \end{array}.$$

This rearrangement demonstrates that, after one step, all the states are obliged to change parity.

Write $\mathbb{W}^{(4)}$ in the same way and comment on the form of the matrix. Write $\mathbb{W}^{(odd)}$ and $\mathbb{W}^{(even)}$ in the same form, and show that if the initial distribution is such that $\mathbf{P}^{(0)}(\text{odd}) = \mathbf{P}^{(0)}(\text{even}) = 1/2$, then $\lim_{n\to\infty} \mathbf{P}^{(n)} = \boldsymbol{\pi}$, where $\boldsymbol{\pi}$ is the binomial distribution.

6.3 Consider Sinai's theorem for an aperiodic Markov chain described by (6.10) and the aperiodic Ehrenfest urn with $n = 4$ balls or fleas. Assume that the initial state is $x_0 = 0$, that is $\mathbf{P}^{(0)} = (1,0,0,0,0)$. The invariant distribution is $\boldsymbol{\pi} = (1,4,6,4,1)/16$. Write a simple computer program and compute the exact distance $\left(\sum |P_i^{(t)} - \pi_i|\right)/2 = d(\mathbf{P}^{(t)}, \boldsymbol{\pi})$ as a function of $t = 0,1,\ldots,30$. Then noticing that $\min(w^{(4)}(i,j)) = \delta = 0.0059$, as after 4 steps the matrix is regular, determine $(1-\delta)^{(t/4)-1}$. The graph in Fig. 6.2 will be produced,

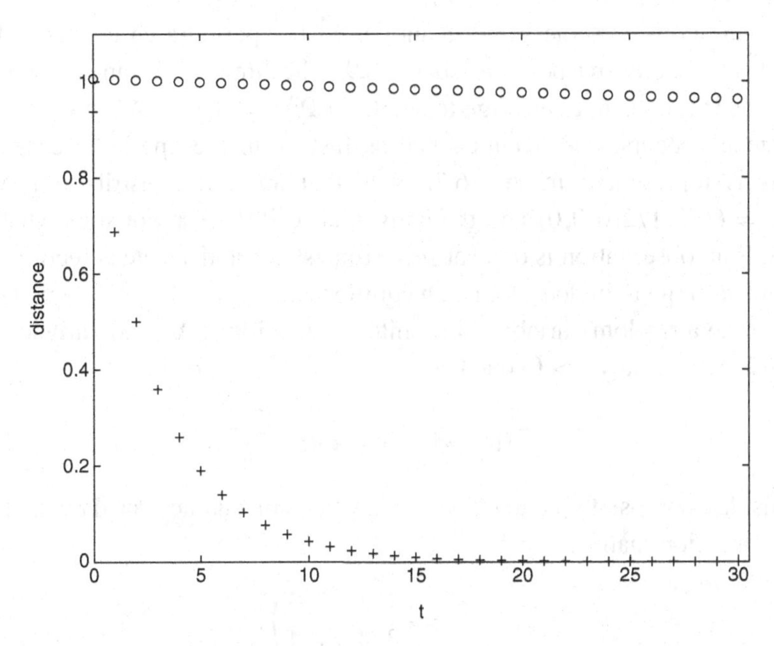

Figure 6.2. Sinai's distance bound (open circles) and effective distance (crosses) for the aperiodic Ehrenfest urn model with $n = 4$ balls.

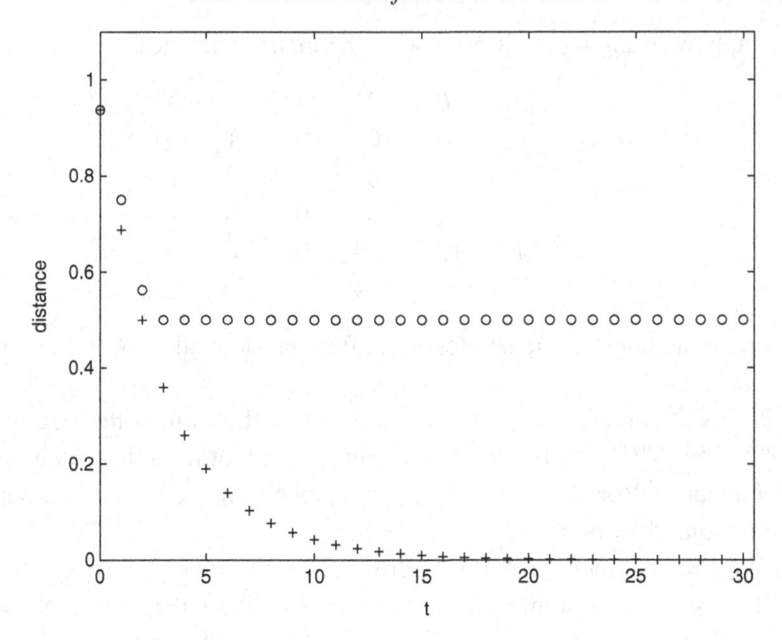

Figure 6.3. Effective distance for the periodic Ehrenfest urn model with $n = 4$ balls starting from $\mathbf{P}^{(0)} = (1, 0, 0, 0, 0)$ (open circles) and for the aperiodic Ehrenfest urn model starting from the same state (crosses).

where the true distance vanishes much faster than the bound fixed by Sinai's theorem. To test the exponential convergence of $d(\mathbf{P}^{(t)}, \boldsymbol{\pi})$, show numerically that $d(\mathbf{P}^{(t+1)}, \boldsymbol{\pi})/d(\mathbf{P}^{(t)}, \boldsymbol{\pi}) \simeq 0.75$.

If one considers the same quantities in the periodic case, that is $\mathbf{P}^{(0)} = (1, 0, 0, 0, 0)$ and the periodic chain (6.9), the limiting distance is not 0 but $1/2$, as $\mathbf{P}^{(t)}$ does not converge to $\boldsymbol{\pi}$. But if $\mathbf{P}^{(0)} = (1/2, 1/2, 0, 0, 0)$, then convergence occurs, and occurs twice as fast as in the aperiodic case. Why? This is represented in Fig. 6.3. Note that an initial distribution such as $\mathbf{P}^{(0)} = (1/2, 1/2, 0, 0, 0)$ and the convergence $\mathbf{P}^{(t)} \to \boldsymbol{\pi}$ are somewhat artificial. If an observation is done at some time step s and a state is recorded, then the restarting chain does not reach equilibrium.

6.4 Let X be a random variable with finite expectation $\mathbb{E}(X)$ and variance $\mathbb{V}(X)$. Prove that, for any $\delta > 0$, one has

$$\mathbb{P}(|X - \mathbb{E}(X)| \geq \delta) \leq \frac{\mathbb{V}(X)}{\delta^2}. \tag{6.126}$$

6.5 Consider a two-state homogeneous Markov chain characterized by the following transition matrix:

$$\mathbb{W} = \begin{pmatrix} 1/3 & 2/3 \\ 3/4 & 1/4 \end{pmatrix}.$$

Find the invariant distribution.

6.6 Evaluate $\mathbb{W}^{(t)}$ for the transition probability matrix of Exercise 6.5 and for $t = 2, 5, 10$.

6.7 Consider a two-state homogeneous Markov chain characterized by the following transition matrix:

$$\mathbb{W} = \begin{pmatrix} 0 & 1 \\ 1 & 0 \end{pmatrix}.$$

What happens to the powers $\mathbb{W}^t$?

6.8 Summary

In this chapter, the following basic properties for irreducible Markov chains were proved:

1. irreducibility leads to the existence of a (unique) invariant distribution $\boldsymbol{\pi}$, whose meaning is the limiting relative frequency of visiting times for any state (6.99); the ergodic theorem (6.103) is a consequence of this result;
2. aperiodicity further implies that the invariant distribution $\boldsymbol{\pi}$ is the equilibrium distribution, that is $\lim_{s\to\infty} \mathbb{P}(X_s = j|X_0 = x_0) = \pi_j$, and the speed of convergence is geometric (exponential);
3. the main task in studying an irreducible Markov chain is to find the invariant distribution $\boldsymbol{\pi}$. One way is solving the eigenvalue problem discussed in Section 6.5.1. However, given that the solution of the linear system (6.61) is practically impossible in several applications to large physical and economic systems, the search for an invariant distribution must follow some other route;
4. the invariant distribution $\boldsymbol{\pi}$ can be evaluated by means of a Monte Carlo computer simulation. As long as one sees that (hopefully) the histogram of relative visiting frequencies stabilizes, one can apply Equation (6.99); this procedure is affected by all the difficulties related to a frequency evaluation of probability. Also the Monte Carlo method soon becomes impractical for large systems;
5. given that a homogeneous irreducible Markov chain is reversible if and only if it is stationary and the stationary distribution satisfies the detailed balance conditions, imposing the detailed balance condition may help in finding the invariant distribution. If the chain is irreducible, this is the unique invariant measure; moreover, if the chain is aperiodic, it is also the equilibrium distribution.

Further reading

The Ehrenfest urn model was introduced by Paul and Tatiana Ehrenfest in order to justify Boltzmann's ideas on statistical equilibrium.

P. Ehrenfest and T. Ehrenfest, *Begriffliche Grundlagen der statistischen Auffassung in der Mechanik*, in F. Klein and C. Müller (eds), *Encyclopädie der Mathematischen Wissenschaften mit Einschluß ihrer Anwendungen* **4**, 3–90, Teubner, Leipzig (1911). English translation in P. Ehrenfest and T. Ehrenfest, *The Conceptual Foundations of the Statistical Approach in Mechanics*, Dover, New York (1990).

Marc Kac discusses this model in his book on the applications of probability theory in physics:

M. Kac, *Probability and Related Topics in Physical Science*, Interscience, London (1959).

and it is now a standard subject in books on the application of probability theory and stochastic processes to statistical physics.

H. Wio, *An Introduction to Stochastic Processes and Nonequilibrium Statistical Physics*, World Scientific, Singapore (1994).

Markov chains are named after Russian mathematician Andrey Andreyevich Markov, who extended the weak law of large numbers and the *central limit theorem* (informally discussed below in Section 10.8) to the class of dependent random-variable sequences now carrying his name.

A.A. Markov, *Rasprostranenie predel'nyh teorem ischisleniya veroyatnostej na summu velichin svyazannyh v cep'*, Zapiski Akademii Nauk po Fiziko-matematicheskomu otdeleniyu, VIII seriya, tom 25(3) (1908). English translation: A.A. Markov, *Extension of the Limit Theorems of Probability Theory to a Sum of Variables Connected in a Chain*, in R.A. Howard (ed.), *Dynamic Probabilistic Systems, volume 1: Markov Chains*, Wiley, New York (1971).

According to Basharin *et al.*,

G.P. Basharin, A.N. Langville and V.A. Naumov, *The Life and Work of A. A. Markov*, in A.N. Langville and W.J. Stewart (eds.), *Proceedings of the 2003 conference on The Numerical Solution of Markov Chains*, Linear Algebra and its Applications, **386**, 3–26 (2004).

it was Bernstein

S. N. Bernstein, *Sur l'extension du thèoréme du calcul de probabilités aux sommes de quantités dépendantes*, Mathematische Annalen, **97**, 1–59 (1926).

who used the name 'Markov chains' for the first time, to denote the processes described in this chapter.

The reader can find an alternative and complete account on the theory of finite Markov chains in the classic book by Kemeny and Snell

J.G. Kemeny and J.L. Snell, *Finite Markov Chains*, Van Nostrand, Princeton, NJ (1960).

Hoel, Port and Stone present an exhaustive discussion on discrete Markov chains in the first two chapters of their book on stochastic processes

P.G. Hoel, S.C. Port and C.J. Stone, *Introduction to Stochastic Processes*, Houghton Mifflin, Boston, MA (1972).

Our discussion of reversibility was inspired by

P. Suppes, *Weak and Strong Reversibility of Causal Processes*, in M.C. Gallavotti, P. Suppes and D. Costantini (eds.), *Stochastic Causality*, CSLI Publications, Stanford, CA (2001).

Finally, those interested in the extension of this theory to continuous state spaces can read the book by Meyn and Tweedie with profit

S.P. Meyn and R.L. Tweedie, *Markov Chains and Stochastic Stability*, Springer, Berlin (1993). S.P. Meyn and R.L. Tweedie, *Markov Chains and Stochastic Stability*, 2nd edition, Cambridge University Press, Cambridge, UK (2009).

References

[1] O. Penrose, *Foundations of Statistical Mechanics*, Pergamon Press, Oxford (1970).

7

The Ehrenfest–Brillouin model

Before studying this chapter, the reader is advised to (re-)read Section 6.1.4. In fact, the present chapter is devoted to a generalization of that example where random 'destructions' are followed by 'creations' whose probability is no longer uniform over all the categories, but follows a rule due to L. Brillouin and directly related to the Pólya distribution discussed in Chapter 5.

After reading this chapter, you should be able to:

- use random destructions (*à la* Ehrenfest) and Pólya distributed creations (*à la* Brillouin) to study the kinematics of a system of n elements moving within g categories;
- define unary, binary, ..., m-ary moves;
- write the appropriate transition probabilities for these moves;
- use detailed balance to find the invariant distribution for the Ehrenfest–Brillouin Markov chain;
- discuss some applications to economics, finance and physics illustrating the generality of the Ehrenfest–Brillouin approach.

7.1 Merging Ehrenfest-like destructions and Brillouin-like creations

Statistical physics studies the macroscopic properties of physical systems at the human scale in terms of the properties of microscopic constituents. In 1996, Aoki explicitly used such a point of view in economics, in order to describe macro variables in terms of large collections of interacting microeconomic entities (agents, firms, and so on). These entities are supposed to change their state unceasingly, ruled by a Markov-chain probabilistic dynamics. All these situations are characterized by a dynamical mechanism destroying an entity in a category and re-creating an entity in another category. From an ontological point of view, in physics and economics, initial and final entities are usually the same, so that the destruction–creation mechanism can be interpreted just as a category change; this is the case of the velocity change for a particle after a collision, or of the strategy variation of an agent after a decision, and so on. On the contrary, in genetics, the destruction–creation mechanism can be interpreted as a death–birth process, where the two entities are different, and even belong to different generations. In any case, in general, the elementary destruction–creation event depends on the present state of the whole system. The state of the system, coinciding with the joint description of its microscopic constituents, evolves as a consequence of these microscopic changes. If these features hold, the evolution is Markovian.

The Ehrenfest urn model can be seen as the starting point of this chapter. Once this model is considered as a 'destruction–creation mechanism', as in the aperiodic version of Equation (6.5), it can be generalized in many ways, conserving the original destruction mechanism, and using a more general creation term. If the 'creation probability' is a function of the present occupation numbers, it can represent correlations between agents, and the full transition probability still does not depend on the time step leading to a homogeneous Markov chain. The 'creation probability' introduced below is called 'the Brillouin term' for historical reasons. Both the Ehrenfests and Brillouin were physicists, and they essentially considered particles, cells and energy levels. However, nothing changes but the interpretation of quantities, when agents and strategies or firms and workers are considered.

As usual, consider a population made up of n entities (agents, particles, etc.) and g categories (cells, strategies, and so on). The state of the system is described by the occupation number vector

$$\mathbf{n} = (n_1, \ldots, n_i, \ldots, n_g), \tag{7.1}$$

with $n_i \geq 0$, and $\sum_{i=1}^{g} n_i = n$. Therefore, the state space is the set of g-ples summing up to n; this is denoted by

$$S_g^n = \{\mathbf{n},\ n_i \geq 0,\ \sum n_i = n\}. \tag{7.2}$$

The dynamical discrete-time evolution is given by the sequence of random variables

$$\mathbf{Y}_0 = \mathbf{n}(0), \mathbf{Y}_1 = \mathbf{n}(1), \ldots, \mathbf{Y}_t = \mathbf{n}(t), \ldots, \tag{7.3}$$

where $\mathbf{n}(t)$ belongs to S_g^n. It is assumed that Equation (7.3) describes the realization of a homogeneous Markov chain, whose one-step transition probability is a matrix whose entries are given by

$$w\left(\mathbf{n}, \mathbf{n}'\right) = \mathbb{P}(\mathbf{Y}_{t+1} = \mathbf{n}' | \mathbf{Y}_t = \mathbf{n}), \tag{7.4}$$

with $\mathbf{n}', \mathbf{n} \in S_g^n$, not depending on time explicitly. How many entries has this stochastic matrix?

7.2 Unary moves

The most elementary event considered here is: 'an entity changes its state from category i to category k'. Therefore, the state of the system changes from $\mathbf{Y}_t = \mathbf{n}$ to $\mathbf{Y}_{t+1} = \mathbf{n}_i^k$, where $\mathbf{n} = (n_1,\ldots,n_i,\ldots,n_k,\ldots,n_g)$ denotes the initial state and $\mathbf{n}_i^k = (n_1,\ldots,n_i - 1,\ldots,n_k + 1,\ldots,n_g)$ the final state, and the notation described in Section 6.1.1 is used. This transition can be split into two distinct parts. The first (the so-called 'Ehrenfest's term') is the destruction of an entity belonging to the ith category, with probability

$$\mathbb{P}(\mathbf{n}_i|\mathbf{n}) = \frac{n_i}{n}. \tag{7.5}$$

The second part (the so-called 'Brillouin's term') is the creation of a particle in the kth category given the vector $\mathbf{n}_i = (n_1, n_i - 1,\ldots,n_k,\ldots,n_g)$, whose probability is:

$$\mathbb{P}(\mathbf{n}_i^k|\mathbf{n}_i) = \frac{\alpha_k + n_k - \delta_{k,i}}{\alpha + n - 1}, \tag{7.6}$$

with $\alpha = \sum_{i=1}^{g} \alpha_i$, and $\boldsymbol{\alpha} = (\alpha_1,\ldots,\alpha_g)$ is a vector of parameters; $\delta_{k,i}$ is the usual Kronecker's delta equal to 1 for $k = i$ and zero otherwise. The meaning of α_i is related to the probability of accommodation on the category i if it is empty. Two cases are interesting: all $\alpha_i > 0$, and all α_i negative integers. In this second case, the size of the population is limited by the inequality $n \leq |\alpha|$, and categories with occupation numbers $n_i = |\alpha_i|$ cannot accommodate further elements. The resulting transition probability is:

$$\mathbb{P}(\mathbf{n}_i^k|\mathbf{n}) = \mathbb{P}(\mathbf{n}_i|\mathbf{n})\mathbb{P}(\mathbf{n}_i^k|\mathbf{n}_i) = \frac{n_i}{n}\frac{\alpha_k + n_k - \delta_{k,i}}{\alpha + n - 1}. \tag{7.7}$$

When all the α_is are positive, if one starts from a given occupation vector $\mathbf{n}$, by repeated applications of (7.7), each possible vector of S_g^n can be reached with positive probability. The cardinality of the state space is

$$\#S_g^n = \binom{n+g-1}{n}. \tag{7.8}$$

Moreover, all these states are persistent: in fact, given that $\alpha_i > 0$, an empty category can be occupied again. For negative integer α_is, Equation (7.7) is understood to return zero if $\alpha_k + n_k - \delta_{k,i} \leq 0$. The set of possible persistent states reduces to

$$S_g^n(|\alpha|) = \left\{\mathbf{n} : \sum_{i=1}^{g} n_i = n,\ n_i \leq |\alpha_i|\right\}, \tag{7.9}$$

and starting from some initial state $\mathbf{n} \in S_g^n(|\alpha|)$, all the states belonging to $S_g^n(|\alpha|)$ can be reached with positive probability.[1] Therefore, the transition probability (7.7) describes an irreducible Markov chain. As a consequence of the discussion in Chapter 6, there exists a unique invariant measure $\pi(\mathbf{n})$ on the set of persistent states. In addition, Equation (7.7) does not exclude the case $\mathbf{n}' = \mathbf{n}_i^k = \mathbf{n}$, that is the case $k = j$. Therefore, in all cases, the chain is aperiodic, due to the possibility for the selected element to immediately return into its original category. In other words, the invariant measure $\pi(\mathbf{n})$ is also the equilibrium distribution on the ergodic set and

$$\lim_{t\to\infty} \mathbb{P}(\mathbf{Y}_t = \mathbf{n}|\mathbf{Y}_0 = \mathbf{n}(0)) = \pi(\mathbf{n}), \tag{7.10}$$

irrespective of the initial state $\mathbf{n}_0 = \mathbf{n}(0)$.

At a first glance, here, nothing is 'irreversible' in the transition mechanism. Then, the search for the invariant distribution can be done by means of detailed balance conditions. Indeed, using Equations (6.109) and (7.7) for $i \neq j$, one finds that

$$\frac{\pi(\mathbf{n}_i^k)}{\pi(\mathbf{n})} = \frac{n_i}{n_k + 1} \frac{\alpha_k + n_k}{\alpha_j + n_j - 1}, \tag{7.11}$$

a relationship which is satisfied by the generalized g-dimensional Pólya distribution (see Chapter 5)

$$\pi(\mathbf{n}) = \text{Pólya}(\mathbf{n}; \boldsymbol{\alpha}) = \frac{n!}{\alpha^{[n]}} \prod_{i=1}^{g} \frac{\alpha_i^{[n_i]}}{n_i!}, \tag{7.12}$$

with $\sum_{i=1}^{g} n_i = n$ and $\sum_{i=1}^{g} \alpha_i = \alpha$. If $\{\alpha_i > 0\}$, then creations are positively correlated with occupation numbers $\mathbf{n}$. The equilibrium probability (7.12) can be related to a Pólya urn scheme with initial composition $\boldsymbol{\alpha}$ and unitary prize. A simple subcase of Equation (7.12) is the so-called *Bose–Einstein distribution*, which is the symmetric g-dimensional Pólya distribution with $\alpha_i = 1$, and $\alpha = g$. In this case $\alpha_i^{[n_i]}$ becomes $n_i!$, and

$$\frac{n!}{\alpha^{[n]}} = \frac{n!}{g^{[n]}}, \tag{7.13}$$

so that one eventually finds

$$\pi_{\text{BE}}(\mathbf{n}) = \text{Pólya}(\mathbf{n}; \mathbf{1}) = \binom{n+g-1}{n}^{-1}, \tag{7.14}$$

[1] For instance, if one starts from $\mathbf{n} = (n, 0, \ldots, 0), n \leq |\alpha|$, and $n > |\alpha_1|$ from Equation (7.7), it turns out that n_1 is forced to decrease until the value $n_1 = |\alpha_1|$ is reached, and n_1 will never become greater than $|\alpha_1|$ again. Therefore, all the states with $n \leq |\alpha|$ and with some $n_i > |\alpha_i|$ are not persistent (they are called transient states). Hence, also in this unusual case after a finite number of steps the dynamics is restricted to the set $S_g^n(|\alpha|)$ of Equation (7.9).

where $\boldsymbol{\alpha} = \mathbf{1} = (1, \ldots, 1)$ is a vector containing g values equal to 1. In other words, if $\boldsymbol{\alpha} = \mathbf{1}$, all the occupation vectors have the same equilibrium probability, that is π_{BE} does not depend on $\mathbf{n}$. When $\{\alpha_i < 0\}$ and α_i are negative integers, creations are negatively correlated with the occupation numbers $\mathbf{n}$. The equilibrium probability (7.12) becomes the g-dimensional hypergeometric distribution, that can be re-written as

$$\pi(\mathbf{n}) = \mathrm{HD}(\mathbf{n}; |\boldsymbol{\alpha}|) = \frac{\prod_{i=1}^{g} \binom{|\alpha_i|}{n_i}}{\binom{|\alpha|}{n}}, \tag{7.15}$$

with $\sum_{i=1}^{g} n_i = n$, and $n_i \leq |\alpha_i|$. A remarkable example of Equation (7.15) is the so-called *Fermi–Dirac distribution*, which is obtained for $\alpha_i = -1$, and $\alpha = -g$. Given that $n_i \in \{0, 1\}$, the numerator is 1, and one finds that

$$\pi_{\mathrm{FD}}(\mathbf{n}) = \text{Pólya}(\mathbf{n}; -\mathbf{1}) = \binom{g}{n}^{-1}, \tag{7.16}$$

which is uniform on all occupation vectors with $n_i \in \{0, 1\}$, as also π_{FD} does not depend on $\mathbf{n}$. Note that both π_{BE} and π_{FD} are the only uniform distributions on the occupation vectors in the two possible domains of $\boldsymbol{\alpha}$. They are at the basis of the statistical description of quantum particles.

The correlation is large and positive for small $\alpha > 0$, it is large and negative for small (integer) $\alpha < 0$. The smallest possible value of $\alpha < 0$ is $\alpha = -g$, where no more than one unit can occupy each category (this is known in physics as *Pauli's exclusion principle*). Correlations vanish for $|\alpha| \to \infty$ from both sides, where the creation probability becomes $p_k = \alpha_k/\alpha$ independent of $\mathbf{n}$. In this limit, Equation (7.12) becomes

$$\pi(\mathbf{n}) = \text{Pólya}(\mathbf{n}; \pm\infty) = \mathrm{MD}(\mathbf{n}; \mathbf{p}) = \frac{n!}{\prod_{i=1}^{g} n_i!} \prod_{i=1}^{g} p_i^{n_i}, \tag{7.17}$$

which is the multinomial distribution; if $p_1 = p_2 = \ldots = p_n = 1/g$, one gets the symmetric multinomial distribution

$$\pi_{\mathrm{MB}}(\mathbf{n}) = \frac{n!}{\prod_{i=1}^{g} n_i!} g^{-n}; \tag{7.18}$$

in statistical physics, this is known as the *Maxwell–Boltzmann distribution*. The case $|\alpha| \to \infty$ coincides with an independent accommodation of the moving element; it was discussed in Sections 6.1.2 and 6.1.4. The equilibrium probability (7.18) is such that all the individual descriptions $\mathbf{X}$ are equiprobable,[2] and this distribution lies at the foundation of classical statistical mechanics.

It is interesting to observe that all the above distributions can be discussed in a unified way within the same probabilistic scheme, without introducing notions such as '(in-)distinguishable' particles or elements, which are sometimes used even in probability (by Feller for instance) and in economics. The different behaviour of the microscopic entities is due to the vector parameter $\boldsymbol{\alpha}$ leading to different kinds of inter-element correlation.

In statistical physics, the cases of Equations (7.14), (7.16) and (7.18) exhaust all the possibilities. In economics, the so-called 'parastatistics' are possible, which seem forbidden in physics. This issue will be further discussed in Chapter 8.

7.3 From fleas to ants

The Ehrenfest–Brillouin model is more powerful than the Ehrenfest urn models due to probabilistic creations depending on the state of the system. Ehrenfest's fleas are randomly selected, and they must change dog. The creation is then deterministic (it happens with probability 1). In the aperiodic version, the selected flea chooses the new dog tossing a coin. Here, creation is probabilistic, but independent of the state of the system. What is the meaning of a creation probability with the form of Equation (7.6)? Consider the case $k \neq i$, just to avoid the term $\delta_{k,i}$. Introducing $p_k = \alpha_k/\alpha$, the probability of occupying cell k given that the present occupation vector is $\mathbf{Y}^{(n-1)} = \mathbf{n}_i$, with $\sum_i y_i^{(n-1)} = n-1$, can be re-written as

$$\mathbb{P}(\mathbf{n}_i^k|\mathbf{n}_i) = \frac{\alpha p_k + n_k}{\alpha + n - 1} = \frac{\alpha}{\alpha + n - 1} p_k + \frac{n-1}{\alpha + n - 1}\frac{n_k}{n-1}. \tag{7.19}$$

In selecting a category, the moving element uses two probability distributions, an 'initial' (or theoretical) distribution $\{p_k\}_{k=1}^g$ and an 'empirical' distribution $\{n_k/(n-1)\}_{k=1}^g$, given by the relative accommodation frequency of the $n-1$ other elements. Equation (7.19) can be regarded as a weighted average (mixture) of the two distributions, where α is the weight of the initial distribution, and $n-1$ is the weight of the empirical one. The theoretical and empirical distributions are the extremes of a simplex, able to represent both independence ($\alpha \gg n$) and strong correlation ($\alpha \ll n$). This mixture can be seen as a randomization of two strategies, following Kirman's ideas (see Further reading). One can interpret the moving elements as economic agents and the categories as strategies followed by the agents. Assume that the choice is performed in two stages: first the choice of the distribution, followed by the choice of the category given the chosen distributions. Focusing on

[2] This means that the distribution on the set of the g^n possible individual descriptions is *uniform*.

a category/strategy, say the kth strategy, one can write for the creation term (7.19)

$$\mathbb{P}(C=k)=\mathbb{P}(C=k|\text{Theor})\mathbb{P}(\text{Theor})+\mathbb{P}(C=k|\text{Emp})\mathbb{P}(\text{Emp}), \tag{7.20}$$

where $C=k$ denotes a creation in the kth category given $\mathbf{n}_i$

$$\mathbb{P}(C=k|\text{Theor})=p_k, \tag{7.21}$$

and

$$\mathbb{P}(C=k|\text{Emp})=\frac{n_k}{n-1}, \tag{7.22}$$

whereas

$$\mathbb{P}(\text{Theor})=\frac{\alpha}{\alpha+n-1}, \tag{7.23}$$

and

$$\mathbb{P}(\text{Emp})=\frac{n-1}{\alpha+n-1}. \tag{7.24}$$

If a moving agent chooses the theoretical distribution (behaving as a 'fundamentalist'), she is not influenced by her colleagues. In this case, one has a self-modification of behaviour. If a moving agent chooses the empirical distribution (behaving as a 'chartist'), then he chooses a colleague at random and he selects her strategy. This is exactly the ant foraging behaviour model introduced by Kirman in the very special case of two symmetric sources of food. The 'strange' behaviour of the ants is such that one can find almost all of them at the same food source for a long time, and the change from one source to the other is similar to an avalanche. On average, the two sources are exploited in the same way, and the equilibrium distribution is U-shaped. This behaviour is just the opposite of the Ehrenfest case, with fleas hopping onto two dogs, where the equilibrium distribution is bell-shaped. The generalization presented here admits g sources of food, and a general initial probability distribution $(p_1,\ldots,p_k,\ldots,p_g)$ that may reflect objective differences ('fundamentals') between the sources. The original Kirman's model can be formulated as a birth–death Markov chain by means of the following general transition probabilities, which work in the case of fleas, as well. Focusing again on a single category and setting $w(k,h)=\mathbb{P}(Y_{t+1}=h|Y_t=k)$, one has

$$w(k,k+1)=\frac{n-k}{n}\frac{(\alpha/2)+k}{\alpha+n-1}, \tag{7.25}$$

$$w(k,k-1)=\frac{k}{n}\frac{(\alpha/2)+n-k}{\alpha+n-1}, \tag{7.26}$$

and

$$w(k,k) = 1 - w(k,k+1) - w(k,k-1), \tag{7.27}$$

with all the other transitions forbidden. The 'normal' behaviour of fleas is obtained by letting $\alpha \to \infty$ (see Equation (6.10)), whereas the 'strange' ant behaviour is obtained setting $\alpha = 1$. Following the discussion in the previous section, the equilibrium distribution for fleas is the binomial distribution of parameter $p = 1/2$; in other words, one has

$$\pi_{\text{fleas}}(k) = \text{Bin}(k;n,1/2), \tag{7.28}$$

whereas the equilibrium distribution for Kirman's ants is

$$\pi_{\text{ants}}(k) = \text{Pólya}(k,n-k;1/2,1/2), \tag{7.29}$$

a U-shaped distribution. In this case, in the continuous limit $k \gg 1$, $n \gg 1$, the fraction $x = k/n$ is distributed according to the Beta distribution, whose probability density function Beta $(x;1/2,1/2)$ is proportional to $1/\left(\sqrt{x(1-x)}\right)$. Note that $\mathbb{E}(Y) = n/2$ in both cases (they are both symmetric distributions), whereas the variance depends on α and is given by

$$\mathbb{V}(Y) = \frac{n}{4}\left(\frac{\alpha+n}{\alpha+1}\right), \tag{7.30}$$

which is proportional to n^2 for ants (and to n for fleas). Indeed, the ant model is non-self-averaging (see Chapter 5).

Given that both ants and (aperiodic) fleas share the same probabilistic engine (Equations (7.25), (7.26) and (7.27)), their behaviour can be simulated with the same program (code), where the weight of the initial distribution α can be continuously changed from large values $\alpha \gg 1$ (see also Exercise 5.5 and Fig. A5.5) to Kirman's value $\alpha = 1$ (see Exercise 7.2 below).

Another important difference between the cases $\alpha \to \infty$ (independent accommodations) and $\alpha = 1$ (strong herding behaviour) is given by the rate of approach to equilibrium. While fleas run very fast towards the equilibrium 'fifty–fifty' situation, and after that oscillate around the expected value, ants stay for a long time at one food source, and they suddenly rush to the other source, as an avalanche; moreover, they are seldom found in the expected 'fifty–fifty' situation. Later, it will be shown that the number of steps needed for the expected value to reach its equilibrium value is roughly $n(\alpha+n-1)/\alpha$, which gives exactly n for fleas, but n^2 for ants.

In Kirman's original formulation the weight of the theoretical distribution is δ and the empirical weight is $1-\delta$, independent of the population size. In his framework, in order to get the beta distribution, it is necessary to assume that $n \to \infty$, $\delta \to 0$ and

$n\delta \to \alpha$. On the contrary, in the Equations (7.25), (7.26) and (7.27), the empirical weight depends on the size of population from the beginning. It is quite reasonable to imagine that the weight of the 'herding behaviour' is just the size of the herd. This weight must be compared to α, the weight of 'self-modification'.

7.4 More complicated moves

Unary moves are the smallest and simplest changes that can take place in a population. As briefly discussed above, the marginal Ehrenfest–Brillouin model belongs to the family of birth–death Markov chains, admitting a continuous limit both in space and in time (see also Section 10.8 below). However, more complicated moves are needed in many applications. For instance, in physics, interparticle collisions cause a change of category (state) of two particles. A binary collision is represented by two particles in the initial categories (also called *cells* in this case) (i,j) and ending in the final categories (k,l). A binary transition can be represented by $\mathbf{n}_{ij}^{kl}|\mathbf{n}$, where

$$\mathbf{n} = (n_1,\ldots,n_i,\ldots,n_j,\ldots,n_k,\ldots,n_l,\ldots,n_g), \tag{7.31}$$

and

$$\mathbf{n}_{ij}^{kl} = (n_1,\ldots,n_i-1,\ldots,n_j-1,\ldots,n_k+1,\ldots,n_l+1,\ldots,n_g). \tag{7.32}$$

A generalization of Equations (7.7) to binary moves is the following, for i,j,k,l all different:

$$\mathbb{P}(\mathbf{n}_{ij}^{kl}|\mathbf{n}) = P(\mathbf{n}_{ij}|\mathbf{n})P(\mathbf{n}_{ij}^{kl}|\mathbf{n}_{ij}) = 4\frac{n_i}{n}\frac{n_j}{n-1}\frac{(\alpha_k+n_k)(\alpha_l+n_l)}{(\alpha+n-2)(\alpha+n-1)}. \tag{7.33}$$

Neglecting the coefficient 4, the first two terms give the probability of selecting a particle in the ith cell, followed by the extraction of a particle in cell j in a hypergeometric process of parameter $\mathbf{n}$, where $\mathbf{n}$ is the initial occupation vector. The third term is an accommodation Pólya process with weights $\alpha_k + n_k$ for the kth cell when the total weight is $\alpha + n - 2$, and the fourth is an accommodation Pólya process with weights $\alpha_l + n_l$ and total weight given by $\alpha + n - 1$. This mechanism is just the same as for unary moves, considering that each destruction or creation step is conditioned to the current state of the population and that the population diminishes during destructions, and increases during creations. The factor 4 is due to the four paths connecting $\mathbf{n}$ to $\mathbf{n}_{ij}^{kl}$. In fact, denote by $D_1 = i$, $D_2 = j$, $C_1 = k$, $C_2 = l$ the sequence of the two extractions followed by the two accommodations. There are two destruction sequences connecting $\mathbf{n}$ to $\mathbf{n}_{ij}$, that is $D_1 = i, D_2 = j$ and $D_1 = j, D_2 = i$, and there are two creation sequences from $\mathbf{n}_{ij}$ to $\mathbf{n}_{ij}^{kl}$, that is $C_1 = k$, $C_2 = l$ and $C_1 = l$, $C_2 = k$. Due to the exchangeability of destructions $\{D_r\}$ and creations $\{C_r\}$, the four paths D_1, D_2, C_1, C_2 are equiprobable. The coefficient is not 4 if some index is repeated, but these subtleties can be omitted. The simplest route to the invariant distribution is using the detailed balance conditions, and showing that the invariant distribution of (7.33) is still the Pólya distribution (7.12). However, a more refined approach is possible, based on a deep property of the Pólya distribution.

7.5 Pólya distribution structures

Consider ten 'fair' coins on a table, which were 'honestly' tossed, but they are covered and the result of the tosses is not known. Then a coin is removed from the table; whether it was 'heads' or 'tails' is unknown, all coins being on a par for being removed. What can be said about the probability distribution of the remaining coins? In order to formally discuss this problem, note that one has a population of ten elements which can belong to two categories; focusing on the occupation vector $\mathbf{n} = (n_{\mathrm{H}}, n_{\mathrm{T}})$, with $n = 10 = n_{\mathrm{H}} + n_{\mathrm{T}}$, and defining $k = n_{\mathrm{H}}$, one has that

$$\mathbb{P}_{10}(k) = \binom{10}{k} 2^{-10},\ k = 0, 1, \ldots, 10. \tag{7.34}$$

Now an element is selected at random, and the probability $\mathbb{P}_9(k)$ is desired, for $k = 0, 1, \ldots, 9$. Consider the case in which on the table there are 2 heads (and 7 tails), whose probability is $\mathbb{P}_9(2)$. This event is consistent with two hypotheses: either there were 3 heads on the table (an event with probability $\mathbb{P}_{10}(3)$) and a head was removed (with conditional probability $3/10$), or there were 2 heads on the table (an event with probability $\mathbb{P}_{10}(2)$) and a tail was removed (with conditional probability $8/10$). Then, one finds that

$$\mathbb{P}_9(2) = (3/10)\mathbb{P}_{10}(3) + (8/19)\mathbb{P}_{10}(2); \tag{7.35}$$

this result can be generalized to any number n of coins, leading to the following equation

$$\mathbb{P}_{n-1}(k) = \mathbb{P}_n(k+1)\frac{k+1}{n} + \mathbb{P}_n(k)\frac{n-k}{n}. \tag{7.36}$$

One can directly verify that if $\mathbb{P}_n(k) = \mathrm{Bin}(k; n, p)$, then one has $\mathbb{P}_{n-1}(k) = \mathrm{Bin}(k; n-1, p)$. In other words, hypergeometric sampling does conserve the structure of the probability distribution. Returning to the problem discussed above, its solution is

$$\mathbb{P}_9(2) = \binom{9}{2} 2^{-9}. \tag{7.37}$$

This depends on the probabilistic rule for removing coins. For instance, if one knows that the coin removed was showing 'heads', then $\mathbb{P}_9(2)$ would just be equal to $\mathbb{P}_{10}(3)$. But, if the coin is randomly removed and not observed, it is just as if, initially, there were only 9 coins on the table. This result can be generalized to any exchangeable dichotomous distribution, where the focus is on successes.

7.5.1 Random destructions: the general case

Assume that n elements are described by a dichotomous exchangeable distribution. Then, from Equation (4.59), one can write

$$\mathbb{P}_n(k) = \binom{n}{k}\mathbb{P}(\mathbf{X}(k,n)), \tag{7.38}$$

where $\mathbf{X}(k,n)$ denotes an individual description with k successes in n trials. Then, a random destruction is performed, leading to

$$\mathbb{P}_{n-1}(k) = \mathbb{P}_n(k+1)\mathbb{P}(k|k+1) + \mathbb{P}_n(k)\mathbb{P}(k|k) = \mathbb{P}_n(k+1)\frac{k+1}{n} + \mathbb{P}_n(k)\frac{n-k}{n}. \tag{7.39}$$

Here, one can reason as in the case of the coins. The situation with k successes in $n-1$ trials can be reached either from $k+1$ successes in n trials by randomly removing one success or from k successes in n trials by removing a failure. Using Equations (7.38) and (7.39), one is led to

$$\begin{aligned}\mathbb{P}_{n-1}(k) &= \mathbb{P}_n(k+1)\frac{k+1}{n} + \mathbb{P}_n(k)\frac{n-k}{n} \\ &= \frac{k+1}{n}\binom{n}{k+1}\mathbb{P}(\mathbf{X}(k+1,n)) + \frac{n-k}{n}\binom{n}{k}\mathbb{P}(\mathbf{X}(k,n)).\end{aligned} \tag{7.40}$$

Further observing that

$$\frac{k+1}{n}\binom{n}{k+1} = \frac{n-k}{n}\binom{n}{k} = \binom{n-1}{k}, \tag{7.41}$$

eventually yields

$$\mathbb{P}_{n-1}(k) = \binom{n-1}{k}[\mathbb{P}(\mathbf{X}(k+1,n)) + \mathbb{P}(\mathbf{X}(k,n))] = \binom{n-1}{k}\mathbb{P}(\mathbf{X}(k,n-1)), \tag{7.42}$$

meaning that the resulting distribution is still exchangeable, and it is known in terms of $\mathbb{P}_n(k)$. The last equality

$$\mathbb{P}(\mathbf{X}(k,n-1)) = \mathbb{P}(\mathbf{X}(k,n)) + \mathbb{P}(\mathbf{X}(k+1,n)) \tag{7.43}$$

is simply the second compatibility condition of Kolmogorov (4.49). Indeed, as mentioned above, the situation with k successes in $n-1$ trials can be reached in two ways: either from $k+1$ successes in n trials or from k successes in n trials.

If the exchangeable distribution is the Pólya one, from

$$\mathbb{P}(\mathbf{X}(k,n)) = \frac{\alpha_1^{[k]}(\alpha-\alpha_1)^{[n-k]}}{\alpha^{[n]}}, \tag{7.44}$$

after a random destruction, one finds

$$\mathbb{P}(\mathbf{X}(k,n-1)) = \frac{\alpha_1^{[k]}(\alpha-\alpha_1)^{[n-k-1]}}{\alpha^{[n-1]}}, \tag{7.45}$$

or (which is the same): if a random destruction occurs in a population described by Pólya$(k,n-k;\alpha_1,\alpha-\alpha_1)$, the remaining population is described by Pólya$(k,n-1-k;\alpha_1,\alpha-\alpha_1)$. This result is valid for any number of steps $m \le n$, and it is also a property of the multivariate Pólya distribution.

Returning to the search of the invariant distribution for the Ehrenfest–Brillouin process with more complicated moves: if the chain is described by Pólya$(k,n-k;\alpha_1,\alpha-\alpha_1)$, then, after a sequence of $m \le n$ hypergeometric destructions, the chain will be described by a Pólya$(k,n-m-k;\alpha_1,\alpha-\alpha_1)$ and, in the multivariate case, if $\mathbf{m}$ denotes the vector of destructions, one will pass from Pólya$(\mathbf{n};\boldsymbol{\alpha})$ to Pólya$(\mathbf{n}-\mathbf{m};\boldsymbol{\alpha})$.

7.5.2 Pólya creations

Now, consider the case in which an element is added to a population of n elements. In the case of creation probabilities following the Pólya scheme, the analogue of Equation (7.39) is given by:

$$\mathbb{P}_{n+1}(k) = \mathbb{P}_n(k-1)\frac{\alpha_1+k-1}{\alpha+n} + \mathbb{P}_n(k)\frac{\alpha-\alpha_1+n-k}{\alpha+n}. \tag{7.46}$$

The reader can check directly that, if $\mathbb{P}_n(k)$ is Pólya$(k,n-k;\alpha_1,\alpha-\alpha_1)$, then $\mathbb{P}_{n+1}(k)$ is Pólya$(k,n+1-k;\alpha_1,\alpha-\alpha_1)$. Again this result is valid for any number of steps, and also for the multivariate Pólya distribution. Then applying Equation (7.46) m times in sequence to Pólya$(\mathbf{n}-\mathbf{m};\boldsymbol{\alpha})$, one recovers Pólya$(\mathbf{n};\boldsymbol{\alpha})$, that is the starting distribution. Summarizing all the previous discussion: if the population is described by a distribution Pólya$(\mathbf{n};\boldsymbol{\alpha})$, then after a sequence of $m \le n$ hypergeometric destructions the shrunken population is described by a distribution which is Pólya$(\mathbf{n}-\mathbf{m};\boldsymbol{\alpha})$, and after a sequence of m Pólya creations, the restored population is described by the same initial distribution, that is Pólya$(\mathbf{n};\boldsymbol{\alpha})$; this means that Pólya$(\mathbf{n};\boldsymbol{\alpha})$ is the invariant distribution of the m-ary chain.

7.5.3 Occupation numbers as random variables

In the general case let the initial state be $\mathbf{n}$. If m elements are selected without replacement and destroyed, one classifies them by category and indicates this event by $\mathbf{D}^{(m)} = (d_1, \ldots, d_g)$, with $\sum_{i=1}^{g} d_i = m$. After the destruction, the m units are redistributed following the vector $\mathbf{C}^{(m)} = (c_1, \ldots, c_g)$, again with $\sum_{i=1}^{g} c_i = m$. The resulting process can be written as

$$\mathbf{Y}_{t+1} = \mathbf{Y}_t + \mathbf{I}_{t+1}, \tag{7.47}$$

where

$$\mathbf{I}_{t+1} = -\mathbf{D}_{t+1} + \mathbf{C}_{t+1} \tag{7.48}$$

is the frequency vector of the increment, and, as always in this book, the sum of two vectors follows the usual rules in terms of components. Destructions always precede creations. The probabilistic structure of the process is completely specified, as

$$\mathbb{P}(\mathbf{D}_{t+1} = \mathbf{d}|\mathbf{Y}_t = \mathbf{n}) = \frac{\prod_{i=1}^{g} \binom{n_i}{d_i}}{\binom{n}{m}}, \tag{7.49}$$

and

$$\mathbb{P}(\mathbf{C}_{t+1} = \mathbf{c}|\mathbf{Y}_t = \mathbf{n}, \mathbf{D}_{t+1} = \mathbf{d}) = \mathbb{P}(\mathbf{C}_{t+1} = \mathbf{c}|\mathbf{n} - \mathbf{d})$$
$$= \frac{\prod_{i=1}^{g} \binom{\alpha_i + n_i - d_i + c_i - 1}{c_i}}{\binom{\alpha + n - 1}{m}}. \tag{7.50}$$

The transition probability

$$\mathbb{P}(\mathbf{Y}_{t+1} = \mathbf{n}'|\mathbf{Y}_t = \mathbf{n}) = \mathbb{P}(\mathbf{I}_{t+1} = \mathbf{n}' - \mathbf{n}|\mathbf{Y}_t = \mathbf{n}) \tag{7.51}$$

is the sum over all paths of the kind

$$B_m = \left\{(\mathbf{d}, \mathbf{c}) : -\mathbf{d} + \mathbf{c} = \mathbf{n}' - \mathbf{n}\right\}; \tag{7.52}$$

therefore, one has

$$W_m(\mathbf{n}, \mathbf{n}') = \mathbb{P}\left(\mathbf{Y}_{t+1} = \mathbf{n}'|\mathbf{Y}_t = \mathbf{n}\right) = \sum_{B_m} \mathbb{P}(\mathbf{c}|\mathbf{n} - \mathbf{d})\mathbb{P}(\mathbf{d}|\mathbf{n}). \tag{7.53}$$

The quite general transition matrix $W_m(\mathbf{n}, \mathbf{n}')$ given by Equation (7.53) defines Markov chains with the Pólya$(\mathbf{n}; \boldsymbol{\alpha})$ as invariant distribution. Note that

Equation (7.53) encompasses Equations (7.7) and (7.33). If the only constraint is $\sum_{i=1}^{g} n_i = n$, given that for $m = 1$ the Markov chains defined by Equation (7.53) are irreducible and aperiodic, a fortiori the same holds true for any $m \leq n$; thus the invariant distribution Pólya$(\mathbf{n};\boldsymbol{\alpha})$ is also the limiting equilibrium distribution. Moreover, one expects that the rate with which equilibrium is reached is an increasing function of m. In order to study the rate of approach to equilibrium, one can consider the marginal description for a fixed category, whose initial weight is β:

$$Y_{t+1} = Y_t - D_{t+1} + C_{t+1} = Y_t + I_{t+1}. \tag{7.54}$$

Suppose that $Y_t = k$, and merge the other $n-k$ elements into a single category, whose initial weight is $\alpha - \beta$. The destruction move chooses m objects from the occupation vector $(k, n-k)$ without replacement, and d units are removed from the fixed category, whereas $m-d$ are taken away from the remaining part of the system. The resulting occupation vector is $(k-d, n-k-m+d)$. The creation consists in m extractions from a Pólya urn with initial composition $(\beta+k-d, \alpha-\beta+n-k-m+d)$, and c elements are created in the category with initial weight β, whereas $m-c$ are created in the category with initial weight $\alpha-\beta$. Given the initial state $(k, n-k)$, the expected value of D is a function of the starting state $Y_t = k$:

$$\mathbb{E}(D|Y_t = k) = m\frac{k}{n}. \tag{7.55}$$

As for creations, adding to the evidence the destroyed state $(d, m-d)$ one gets:

$$\mathbb{E}(C|Y_t = k, D_{t+1} = d) = m\frac{\beta+k-d}{\alpha+n-m}. \tag{7.56}$$

To eliminate d, one can use the following equation

$$\mathbb{E}(C|Y_t = k) = \mathbb{E}(\mathbb{E}(C|Y_t = k, D_{t+1} = d)) = m\frac{\beta+k-\mathbb{E}(D|Y_t = k)}{\alpha+n-m}, \tag{7.57}$$

leading to

$$\mathbb{E}(I_{t+1}|Y_t = k) = -\frac{m\alpha}{n(\alpha+n-m)}\left(k - \frac{n\beta}{\alpha}\right). \tag{7.58}$$

The average increment vanishes if $k = (n\beta)/\alpha = \mu$, which is the equilibrium value for k, and it is mean-reverting: it reduces the deviations away from the equilibrium expected value. One can further see that

$$r = \frac{m\alpha}{n(\alpha+n-m)} \tag{7.59}$$

is the rate of approach to equilibrium, depending on the size n, the total initial weight α and the number of changes m. Setting $m = 1$ and $\alpha \to \infty$, one obtains

$r = 1/n$, the value for the unary Ehrenfest aperiodic model. The rate is a linear function of m if $m \ll n$, and grows to 1 for $m = n$. Note that for $m = n$, and $r = 1$, the destruction completely empties the system, so that, all at once, the subsequent creation probability is already the equilibrium one. In this limiting case Y_{t+1} does not depend on Y_t, and one expects a vanishing correlation $\mathbb{C}(Y_t, Y_{t+1}) = 0$. Therefore, one can derive the equation

$$\mathbb{E}(I_{t+1}|Y_t = k) = -r(k - \mu), \tag{7.60}$$

or equivalently, adding $k - \mu$ to both sides and taking out of the expected value what is known

$$\mathbb{E}(Y_{t+1}|Y_t = k) - \mu = (1 - r)(k - \mu). \tag{7.61}$$

The expected increment acts as a restoring force, making the new expected position closer to the equilibrium value. Iterating Equation (7.61), one finds

$$\mathbb{E}(Y_{t+s}|Y_t = k) - \mu = (1 - r)^s (k - \mu). \tag{7.62}$$

Introducing the 'centred variable' $U_t = Y_t - \mu$, whose equilibrium value is $\mu_U = 0$, and letting $q = 1 - r$, the evolution Equation (7.62) reduces to

$$\mathbb{E}(U_{t+s}|U_t = u) = q^s u, \tag{7.63}$$

and taking the expectation on both sides

$$\mathbb{E}(U_{t+s}) = q^s \mathbb{E}(U_t). \tag{7.64}$$

All these equations show that the *regression* of Y_{t+1} on Y_t, that is, by definition, the expected value of Y_{t+1} conditioned on Y_t, is linear, proportional to Y_t.

This is a simple example (restricted to the first moment) of the general geometrical approach to equilibrium for all Markov chains (see Chapter 6). Note that $I_{t+1} = Y_{t+1} - Y_t = U_{t+1} - U_t$; therefore, passing to the second centred moments, which are the same for Y and U, one gets

$$\mathbb{C}(Y_{t+1}, Y_t) = \mathbb{C}(U_{t+1}, U_t) = \mathbb{E}(U_{t+1}U_t) - \mathbb{E}(U_{t+1})\mathbb{E}(U_t). \tag{7.65}$$

Now, one has that

$$\mathbb{E}(U_{t+1})\mathbb{E}(U_t) = q\mathbb{E}^2(U_t), \tag{7.66}$$

as $\mathbb{E}(U_{t+1}) = q\mathbb{E}(U_t)$. Moreover, one can write

$$\mathbb{E}(U_{t+1}U_t) = \mathbb{E}(\mathbb{E}(U_{t+1}U_t|U_t = u)) = q\mathbb{E}(U_t^2), \tag{7.67}$$

as a consequence of Equation (7.63). Finally, replacing Equations (7.66) and (7.67) into Equation (7.65), one finds that

$$\mathbb{C}(Y_{t+1}, Y_t) = \mathbb{C}(U_{t+1}, U_t) = q[\mathbb{E}(U_t^2) - \mathbb{E}^2(U_t)] = q\mathbb{V}(U_t). \tag{7.68}$$

This is another way of showing the linear regression, as Equation (7.61) can be re-written in the following way:

$$\mathbb{E}(U_{t+1}|U_t = u) = qu, \tag{7.69}$$

that is the regression of U_{t+1} on U_t is linear, and the angular coefficient b of the regression line of U_{t+1} on U_t is a constant

$$b = \frac{\mathbb{C}(U_{t+1}, U_t)}{\mathbb{V}(U_t)} = q = 1 - r, \tag{7.70}$$

and the equation of the regression line is

$$\mathbb{E}(U_{t+1}|U_t = u) - \mathbb{E}(U_{t+1}) = q(u - \mathbb{E}(U_t)), \tag{7.71}$$

obtained from $\mathbb{E}(U_{t+1}) = q\mathbb{E}(U_t)$, subtracted from both sides of Equation (7.69). Although both $\mathbb{C}(U_{t+1}, U_t)$ and $\mathbb{V}(U_t)$ are difficult to calculate in the general case, their ratio is known also far from equilibrium.

When the chain reaches equilibrium, it becomes stationary and both $\mathbb{E}(U_{t+1})$ and $\mathbb{E}(U_t)$ vanish, and $\mathbb{E}(U_t^2)$ converges to $\mathbb{V}(X_t) \to \sigma^2$, where

$$\sigma^2 = n\frac{\beta}{\alpha}\left(1 - \frac{\beta}{\alpha}\right)\left(\frac{\alpha + n}{\alpha + 1}\right), \tag{7.72}$$

and

$$\mathbb{C}(Y_{t+1}, Y_t) = \mathbb{C}(U_{t+1}, U_t) = (1 - r)\sigma^2 = q\sigma^2, \tag{7.73}$$

and, iterating Equation (7.73), one finds

$$\mathbb{C}(Y_{t+s}, Y_t) = (1 - r)^s\sigma^2. \tag{7.74}$$

Then the Bravais–Pearson autocorrelation function of the process once it attains stationarity is

$$\rho(Y_{t+s}, Y_t) = \rho(s) = (1 - r)^s. \tag{7.75}$$

The variance of increments is

$$\mathbb{V}(I_{t+1}) = \mathbb{E}(U_{t+1}^2) + \mathbb{E}(U_t^2) - 2\mathbb{C}(U_{t+1}, U_t) = 2\sigma^2 - 2(1 - r)\sigma^2 = 2r\sigma^2. \tag{7.76}$$

Suppose two categories are considered, and call K and L the respective occupation variables; the two categories evolve in time against the rest of the system, which simulates a thermostat. In the thermodynamic limit, when the rest of the system is very large, K and L become uncorrelated and even independent (see Section 7.6); this means that for any s:

$$\begin{aligned}\mathbb{C}(K_{t+s}, K_t) &= (1-r)^s \sigma_K^2,\\ \mathbb{C}(L_{t+s}, L_t) &= (1-r)^s \sigma_L^2,\\ \mathbb{C}(L_{t+s}, K_t) &= 0 \text{ for any } s.\end{aligned} \tag{7.77}$$

Then considering the random variable $V = K - L$, one has

$$\mathbb{V}(V_t) = \sigma_K^2 + \sigma_L^2 = \sigma_V^2, \tag{7.78}$$

and

$$\mathbb{C}(V_{t+s}, V_t) = \mathbb{C}(K_{t+s}, K_t) + \mathbb{C}(L_{t+s}, L_t) = (1-r)^s(\sigma_K^2 + \sigma_L^2) = (1-r)^s \sigma_V^2. \tag{7.79}$$

These equations lead to the following Bravais–Pearson autocorrelation function

$$\rho(V_{t+s}, V_t) = (1-r)^s. \tag{7.80}$$

Summarizing all the previous results, Equation (7.47), with probability conditions (7.49), (7.50) and (7.53), defines a *random walk* on the set of states $\mathbf{n} \in S_n^g$, whose degree of time correlation is triggered by $m \leq n$, *ceteris paribus*. If m increases, the Bravais–Pearson autocorrelation function $\rho(\mathbf{Y}_{t+1}, \mathbf{Y}_t)$ diminishes, and, in the case $m = n$, it vanishes because the variables $\mathbf{Y}_t$ become independent. This feature is important in applications, as this model is very flexible and can represent empirical time series with very different time-correlation behaviour (see Exercise 7.1, as well).

7.6 An application to stock price dynamics

The previous model can be applied to price increments (or returns) in a stock market. Consider a stock market with n agents, labelled from 1 to n, trading a single risky asset, whose log-price at time step t is denoted by the random variable $X(t)$. During each time period, an agent may choose either to buy, or to sell or not to trade. The quantity of shares requested by agent i is represented by the random variable Φ_i, which can assume three values $+1$ (*bullish* behaviour: agent i wants to buy hoping that the price will increase), 0 (*neutral* behaviour, agent i does not want either to sell or to buy) and -1 (*bearish* behaviour: agent i wants to sell fearing that the price will decrease). The aggregate excess demand for the risky asset at time t is then given by

$$D(t) = \sum_{i=1}^{n} \Phi_i(t). \tag{7.81}$$

Assume that the price log-return is proportional to $D(t)$:

$$\Delta X(t) = X(t+1) - X(t) = \frac{1}{\eta} D(t) = \frac{1}{\eta} \sum_{i=1}^{n} \Phi_i(t), \tag{7.82}$$

where η is the excess demand needed to move the return of one unit. In the following $\eta = 1$ for the sake of simplicity. In order to evaluate the distribution of returns, one needs the joint distribution of $\{\Phi_i(t)\}_{i=1,\ldots,n}$. There are three strategies (bullish, bearish, neutral) and n agents. The state of the system is denoted by the occupation vector $\mathbf{n} = (n_+, n_-, n_0)$. This allows us to treat the number of active agents $n_+ + n_-$ as a random variable. The excess demand is given by $D(t) = n_+(t) - n_-(t)$. The parameters $\alpha_+ = a$, $\alpha_- = b$ and $\alpha_0 = c = \theta - a - b$, with $\theta = a + b + c$, associated to the three strategies, determine the transition probabilities of a Markov chain. In this framework, the financial interpretation should be clear. At each time step, a fixed number m of agents has the possibility of changing strategy. The fraction m/n is the percentage of agents that, at each step, are involved in strategy changes. For positive initial weights the moving agent tends to join the majority (this is known as *herding* behaviour); for negative weights, the selected agent tends to behave at odds with the current majority's behaviour (this is known as *contrarian* behaviour); for very large weights agents are not influenced by their environment.

If the probabilistic dynamics follows Equation (7.47), with the usual probability conditions (7.49), (7.50) and (7.53), the equilibrium distribution is the three-dimensional Pólya distribution, that is ($i \in \{+, -, 0\}$),

$$\pi(\mathbf{n}) = \pi(n_+, n_-, n_0) = \frac{n!}{\theta^{[n]}} \prod_i \frac{\alpha_i^{[n_i]}}{n_i!}, \tag{7.83}$$

with the constraint $n_+ + n_- + n_0 = n$. As mentioned before, the effective demand $\sum_{i=1}^{n} \Phi_i(t) = n_+(t) - n_-(t)$ is a function of $\mathbf{n}$. Now the 'thermodynamic limit' can be introduced. In the limit $\alpha_0 = c \to \infty$, $n \to \infty$, with $\chi = n/\theta$ kept constant, the Pólya distribution factorizes, leading to

$$\pi(n_+, n_-, n_0 = n - n_+ - n_-) \to \mathbb{P}(n_+)\mathbb{P}(n_-), \tag{7.84}$$

where

$$\mathbb{P}(n_+) = \frac{a^{[n_+]}}{n_+!}\left(\frac{1}{1+\chi}\right)^a\left(\frac{\chi}{1+\chi}\right)^{n_+},$$
$$\mathbb{P}(n_-) = \frac{b^{[n_-]}}{n_-!}\left(\frac{1}{1+\chi}\right)^b\left(\frac{\chi}{1+\chi}\right)^{n_-}, \tag{7.85}$$

and one recognizes the negative binomial distribution discussed in Section 5.3.3. The proof of Equations (7.84) and (7.85) is discussed in Section 5.3.3 also. The thermodynamic limit corresponds to increasing the number of agents and the initial propensity to be 'neutral', conserving the average number of 'bulls' and 'bears'

$$\mathbb{E}(n_+) = n\frac{a}{\theta} \to a\chi; \quad \mathbb{E}(n_-) \to b\chi, \tag{7.86}$$

surrounded by a 'reservoir' of neutral agents, that can provide new active agents, or absorb them. In order to get the moments for the equilibrium distribution of the excess demand note that

$$\mathbb{E}(n_+) = a\chi,$$
$$\mathbb{V}(n_+) = a\chi(1+\chi),$$
$$\mathbb{K}^*(n_+) = \frac{1}{a}\left(6 + \frac{1}{\chi(\chi+1)}\right), \tag{7.87}$$

where $\mathbb{K}^*(\cdot)$ denotes the excess kurtosis for the negative binomial distribution, which is large for small a. The same formulae hold true for n_- with a replaced by b. Therefore, one has

$$\mathbb{E}(\Delta X) = \mathbb{E}(n_+ - n_-) = (a-b)\chi. \tag{7.88}$$

Given the independence of n_+ and n_-, one further gets:

$$\mathbb{V}(\Delta X) = \mathbb{V}(n_+ - n_-) = (a+b)\chi(1+\chi), \tag{7.89}$$

and

$$\mathbb{K}^*(\Delta X) = \frac{\mathbb{V}(n_+)^2\mathbb{K}^*(n_+) + \mathbb{V}(n_-)^2\mathbb{K}^*(n_-)}{(\mathbb{V}(n_+) + \mathbb{V}(n_-))^2}$$
$$= \frac{1}{a+b}\left(6 + \frac{1}{\chi(\chi+1)}\right), \tag{7.90}$$

where the last equation is explained later. In general, assuming that the skewness of the empirical distribution is negligible, one has three equations connecting the moments of the excess demand distribution to the three parameters a, b and χ that specify the model. Then, one can estimate a, b, χ from the mean (7.88), the standard deviation (7.89) and the (excess) kurtosis (7.90) of the empirical distribution of the data. Moreover, one can determine the 'intensity of the market' m/n (the fraction of active agents) from the daily autocorrelation.

Example: the Italian stock exchange between 1973 and 1998

In Table 7.1, $\widehat{\mu}$ (the mean), $\widehat{\sigma}$ (the standard deviation), $\widehat{\kappa}$ (the kurtosis) of the daily percentage returns are reported for the Italian stock exchange as well as daily (d), weekly (w) and monthly (m) autocorrelation estimates, in the period June 1973 to April 1998, divided into a further six subperiods. Note that periods I, III and V correspond to (long) stable motion within a price band, whereas II, IV and VI correspond to (short) transitions to a new band. In Table 7.2, the estimated values are given for the model parameters a, b, χ and $f = m/n$. Using the estimated values of a, b and χ, one gets the exact equilibrium distribution for 'bulls' and 'bears'. The exact equilibrium distribution of returns is the distribution of $\Delta X = n_+ - n_-$, where n_+ and n_- are independent random variables distributed according to Equation (7.85). This distribution does not depend on m, or on the intensity $f = m/n$. However, the latter quantities are essential for the time evolution of the process. The parameter m can be derived from the daily autocorrelation (d) in the six periods. The rate of convergence to equilibrium does depend on f and on $\chi = n/\theta$. Indeed, from Equation (7.59), in the present case, one can write

$$r = \frac{f}{1 + \chi(1 - f)}, \tag{7.91}$$

Table 7.1. *Italian Stock Exchange: descriptive statistics of the daily percentage returns for the market index for the period June 1973 to April 1998. Source: Gardini* et al., *1999 (see Further reading below).*

	I	II	III	IV	V	VI
	6.73–8.80	8.80–6.81	6.81–5.85	5.85–5.86	5.86–6.97	6.97–4.98
n. of obs	1870	206	1028	266	2892	187
$\widehat{\mu}$	−0.02	0.42	0.00	0.43	−0.00	025
$\widehat{\sigma}$	1.23	1.94	1.50	1.40	1.21	1.50
$\widehat{\kappa}$	6.84	6.02	13.75	6.24	14.56	9.30
d autoc	0.14	0.09	0.17	0.06	0.15	−0.08
w autoc	0.02	−0.08	−0.01	−0.05	0.00	−0.03
m autoc	0.02	0.21	0.03	−0.10	−0.03	−0.08

Table 7.2. *Parameter estimates for the model described in the text, based on the data presented in Table 7.1*

Period	(Observed)				(Estimated)			
	$\widehat{\mu}$	$\widehat{\sigma}$	$\widehat{\kappa}$	d	a	b	χ	f
I	−0.02	1.23	6.84	0.14	0.47	0.50	0.84	0.92
II	0.42	1.94	6.02	0.09	0.67	0.38	1.47	1.00
III	0.00	1.50	13.75	0.17	0.23	0.23	1.79	0.93
IV	0.43	1.40	6.24	0.06	0.75	0.30	0.96	0.97
V	0.00	1.21	14.56	0.15	0.22	0.22	1.41	0.93
VI	0.25	1.50	9.30	−0.08	0.43	0.25	1.39	1.00

leading to

$$f = \frac{r + r\chi}{1 + r\chi}. \tag{7.92}$$

The rate is estimated from the empirical value of the daily autocorrelation, based on Equation (7.75); one can write $r = 1 - \mathrm{d}$, and obtain an estimate of f. The weekly and monthly autocorrelations of the daily returns can be estimated from Equation (7.75) for $s = 5$ and $s = 20$ (one must use the number of weekly and monthly working days), and then compared to the last two rows of Table 7.1.

Excess kurtosis for the sum of independent random variables

The kurtosis of a random variable X is defined as

$$\mathbb{K}(X) = \frac{\mathbb{E}[(X - \mathbb{E}(X))^4]}{\mathbb{V}^2(X)}. \tag{7.93}$$

The kurtosis for a normally distributed random variable, whose probability density is

$$p(x) = N(x; \mu, \sigma) = \frac{1}{\sqrt{2\pi}\sigma} \exp\left(-\frac{(x-\mu)^2}{2\sigma^2}\right), \tag{7.94}$$

is given by

$$\mathbb{K}_N(X) = 3. \tag{7.95}$$

This value is taken as reference for the excess kurtosis, defined as

$$\mathbb{K}^*(X) = \mathbb{K}(X) - 3. \tag{7.96}$$

Random variables are classified as *leptokurtic* (with positive excess kurtosis), *mesokurtic* (with zero excess kurtosis) and *platykurtic* (with negative excess kurtosis). In particular, the tails of a leptokurtic random variable are *heavier* than the tails of the normal distribution, meaning that extreme events are more likely to occur than in the normal case. Now, one can prove that, for two independent random variables X and Y, with $Z = X - Y$, Equation (7.90) holds true.

Without loss of generality, one can assume that $\mathbb{E}(X) = \mathbb{E}(Y) = 0$, moreover, one has

$$(X - Y)^4 = X^4 - 4X^3Y + 6X^2Y^2 - 4XY^3 + Y^4, \tag{7.97}$$

and, while averaging, the terms with odd powers vanish, so that one gets

$$\mathbb{E}[(X - Y)^4] = \mathbb{E}(X^4) + 6\mathbb{V}(X)\mathbb{V}(Y) + \mathbb{E}(Y^4). \tag{7.98}$$

Further, one has that

$$\mathbb{V}[(X - Y)]^2 = (\mathbb{V}(X) + \mathbb{V}(Y))^2. \tag{7.99}$$

Now, from Equation (7.93), one immediately has that $\mathbb{E}(X^4) = \mathbb{K}(X)\mathbb{V}^2(X)$, and $\mathbb{E}(Y^4) = \mathbb{K}(Y)\mathbb{V}^2(Y)$. Putting all the above results together in Equation (7.93) yields

$$\mathbb{K}(Z) = \frac{\mathbb{V}^2(X)\mathbb{K}(X) + 6\mathbb{V}(X)\mathbb{V}(Y) + \mathbb{V}^2(Y)\mathbb{K}(Y)}{(\mathbb{V}(X) + \mathbb{V}(Y))^2}. \tag{7.100}$$

Replacing Equation (7.100) into Equation (7.96) eventually leads to

$$\begin{aligned}
\mathbb{K}^*(Z) &= \frac{\mathbb{V}^2(X)\mathbb{K}(X) + 6\mathbb{V}(X)\mathbb{V}(Y) + \mathbb{V}^2(Y)\mathbb{K}(Y)}{(\mathbb{V}(Y) + \mathbb{V}(Y))^2} - 3 \\
&= \frac{\mathbb{V}^2(X)(\mathbb{K}(X) - 3) + \mathbb{V}^2(Y)(\mathbb{K}(Y) - 3)}{(\mathbb{V}(X) + \mathbb{V}(Y))^2} \\
&= \frac{\mathbb{V}^2(X)\mathbb{K}^*(X) + \mathbb{V}^2(Y)\mathbb{K}^*(Y)}{(\mathbb{V}(X) + \mathbb{V}(Y))^2}.
\end{aligned} \tag{7.101}$$

7.7 Exogenous constraints and the most probable occupation vector

All the previous discussion in this chapter was exact; both the *invariant distribution*, 'solving' a great deal of the problems on the probabilistic dynamics, and the *approach to equilibrium*, describing the transitory period towards stationarity, were studied without approximations. This was possible because there were no external constraints to the motion of the system. A counterexample immediately comes from physics, for which the original Ehrenfest–Brillouin model was developed. Suppose that two particles collide, and that their final states are different from the initial ones. The transition from the occupation state $\mathbf{n}$ to the occupation state $\mathbf{n}_{ij}^{kl}$ is associated with the events $D_1 = i, D_2 = j$, representing the initial categories of the two selected particles, and $C_1 = k, C_2 = l$, representing the final categories of the colliding particles. Now, further assume that the collision is *elastic*, meaning that the total energy of the two particles is the same before and after the collision. In other words, the transition can occur only if

$$\varepsilon(i) + \varepsilon(j) = \varepsilon(l) + \varepsilon(k), \tag{7.102}$$

where $\varepsilon(i)$ denotes the energy of a particle in the ith category, otherwise it is forbidden. In this case, it is useful to call the categories *cells* and divide the g cells into energy levels, grouping cells of equal energy. The parameters of the system are n, the number of particles, and the set $\{g_i, \varepsilon_i\}_{i=1,\dots,d}$, where $g = \sum_{i=1}^{d} g_i$ is the decomposition of the g cells according to their energy. If cells are re-ordered with respect to their energy in the following way $(1,\dots,g_1,g_1+1,\dots,g_1+g_2,g_1+g_2+1,\dots,\sum_{i=1}^{d} g_i)$, with $g_0 = 0$, the variable

$$N_i = \sum_{j=1+\sum_{k=1}^{i} g_{k-1}}^{\sum_{k=1}^{i} g_k} n_j \tag{7.103}$$

represents the occupation number of a level. The level-occupation vector $\mathbf{N} = (N_1,\dots,N_d)$ is the so-called *macrostate* of the system, whereas the original cell-occupation vector (with cells numbered from 1 to g, as usual) $\mathbf{n} = (n_1,\dots,n_g)$ is the *microstate* of the system. The same procedure can be used for the initial weights, leading to

$$\beta_i = \sum_{j=1+\sum_{k=1}^{i} g_{k-1}}^{\sum_{k=1}^{i} g_k} \alpha_j. \tag{7.104}$$

If the system starts from a given microstate $\mathbf{n}$ and the energy is conserved, with unary moves, it cannot leave the initial macrostate $\mathbf{N}$. On the contrary, binary moves can mimic elastic collisions, and the transition probability $\mathbb{P}(\mathbf{n}_{ij}^{kl}|\mathbf{n})$ still has the form of Equation (7.33), with a slight modification due to the existence of forbidden transitions. In fact, the destruction term is given by the usual term $2(n_i/n)[n_j/(n-1)]$ for $i \neq j$, or by $(n_i/n)[(n_i-1)/(n-1)]$ for $i = j$; if creation concerns cells not involved in the previous destructions, the creation term is proportional to

$2(\alpha_k + n_k)(\alpha_l + n_l)$ for $l \neq k$, or to $(\alpha_k + n_k)(\alpha_k + n_k + 1)$ for $l = k$. These cumbersome notes are useful in order to write a program simulating the dynamics: at any stage of the move, one must use the current occupation number, possibly modified by previous partial moves. In order to normalize the creation probability, one must consider all final states reachable from $\mathbf{n}_{ij}$ adding two particles whose total energy is $\varepsilon(l) + \varepsilon(k) = \varepsilon(i) + \varepsilon(j)$. Defining the term

$$Q(\mathbf{n}_{ij}^{kl}|\mathbf{n}_{ij}) = (2 - \delta_{lk})(\alpha_k + n'_k)(\alpha_l + n'_l + \delta_{lk}), \tag{7.105}$$

where n'_k and n'_l are the occupation numbers of $\mathbf{n}_{ij}$, then one has

$$\mathbb{P}(\mathbf{n}_{ij}^{kl}|\mathbf{n}_{ij}) = \frac{Q(\mathbf{n}_{ij}^{kl}|\mathbf{n}_{ij})}{\sum_{k',l'} Q(\mathbf{n}_{ij}^{k'l'}|\mathbf{n}_{ij})}, \tag{7.106}$$

and, finally, the transition probability for the binary move is

$$\begin{aligned}\mathbb{P}(\mathbf{n}_{ij}^{kl}|\mathbf{n}) &= \mathbb{P}(\mathbf{n}_{ij}|\mathbf{n})\mathbb{P}(\mathbf{n}_{ij}^{kl}|\mathbf{n}_{ij}) \\ &= (2 - \delta_{ij})\frac{n_i}{n}\frac{n_j - \delta_{ij}}{n-1}\frac{Q(\mathbf{n}_{ij}^{kl}|\mathbf{n}_{ij})}{D}\delta[\varepsilon(l) + \varepsilon(k) - (\varepsilon(i) + \varepsilon(j))],\end{aligned} \tag{7.107}$$

where

$$D = \sum_{k',l'} Q(\mathbf{n}_{ij}^{k'l'}|\mathbf{n}_{ij}), \tag{7.108}$$

and the sum is on the set

$$A(\mathbf{n}, i, j) = \{k', l' : \varepsilon(l') + \varepsilon(k') = \varepsilon(i) + \varepsilon(j)\}. \tag{7.109}$$

Therefore, not all the 'first neighbours' of the initial microstate $\mathbf{n}$ can be reached in a single step, as the motion is confined to the so-called *constant energy surface*. Nevertheless, it is remarkable that, if $\mathbb{P}(\mathbf{n}_{ij}^{kl}|\mathbf{n}) > 0$, then also the inverse transition is possible, that is $\mathbb{P}(\mathbf{n}|\mathbf{n}_{ij}^{kl}) > 0$. This is called *microscopic reversibility*. If one considers all the states reachable from the initial state $\mathbf{Y}_0 = \mathbf{n}(0)$ by means of repeated applications of the transition probability (7.107), for all these states $\mathbf{Y}_t = \mathbf{n}(t)$, one finds that $\sum_{j=1}^{g} n_j(t)\varepsilon(j) = \sum_{j=1}^{g} n_j(0)\varepsilon(j) = E(\mathbf{n}(0))$ is a constant of motion. One can assume that all the states compatible with the constraint, i.e. the *energy surface* or *energy shell* $\{\mathbf{n} : \sum_{j=1}^{g} n_j\varepsilon(j) = E(\mathbf{n}(0))\}$, can be reached by means of binary collisions (if this is not the case, in principle, nothing prevents introduction of m-ary collisions). In any case, the energy surface is a subset of the unconstrained state space S_g^n, which strongly depends on the values $\{g_i, \varepsilon_i\}_{i=1,2,\ldots,d}$. Therefore, the transition matrix (7.107) rules a multidimensional random walk on the energy surface which is an irreducible and aperiodic Markov chain.

In order to discuss the inverse transition, one can start from

$$\mathbf{n}_{ij}^{kl} = \mathbf{n} + \mathbf{k} + \mathbf{l} - \mathbf{i} - \mathbf{j}, \tag{7.110}$$

where **k**, **l**, **i** and **j** indicate vectors which are zero everywhere except for the positions k, l, i and j, respectively, where they are equal to 1. Now, consider the path $D_1 = l, D_2 = k$, leading to $\mathbf{n}_{ij} = \mathbf{n} - \mathbf{i} - \mathbf{j}$, and the path $C_1 = j, C_2 = i$, leading to **n**. Then, one has

$$\mathbb{P}(\mathbf{n}_{ij}|\mathbf{n}_{ij}^{kl}) = (2 - \delta_{lk}) \frac{n_k}{n} \frac{n_l - \delta_{lk}}{n-1}, \tag{7.111}$$

whereas

$$\mathbb{P}(\mathbf{n}_{ij}^{ij}|\mathbf{n}_{ij}) = \frac{Q(\mathbf{n}_{ij}^{ij}|\mathbf{n}_{ij})}{D}, \tag{7.112}$$

where $\mathbf{n}_{ij}^{ij} = \mathbf{n}$ and the denominator D coincides with the one in Equation (7.107). In words, any path connecting **n** to $\mathbf{n}_{ij}^{kl}$ passes through the intermediate state $\mathbf{n}_{ij}$ in both directions. In both cases the destructions end in $\mathbf{n}_{ij}$, which is the starting state for creations, all constrained to belong to the set $A(\mathbf{n}, i, j)$. Note that if one sets $\mathbf{n}_{ij}^{kl} = \mathbf{n}'$, then $\mathbf{n}_{ij} = \mathbf{n}'_{kl}$, and $A(\mathbf{n}, i, j) = A'(\mathbf{n}', k, l)$, as $\varepsilon(l) + \varepsilon(k) = \varepsilon(i) + \varepsilon(j)$. Given that the multiplicity factors are equal, one gets from Equation (7.107) and from the previous discussion

$$\frac{\mathbb{P}(\mathbf{n}_{ij}^{kl}|\mathbf{n})}{\mathbb{P}(\mathbf{n}|\mathbf{n}_{ij}^{kl})} = \frac{n_i n_j (\alpha_k + n_k)(\alpha_l + n_l)}{(n_k + 1)(n_l + 1)(\alpha_i + n_i - 1)(\alpha_j + n_j - 1)}, \tag{7.113}$$

for the case $i \neq j \neq k \neq l$. Indeed, there exists a probability distribution which satisfies the detailed balance conditions, such that

$$\frac{\pi(\mathbf{n}_{ij}^{kl})}{\pi(\mathbf{n})} = \frac{\mathbb{P}(\mathbf{n}_{ij}^{kl}|\mathbf{n})}{\mathbb{P}(\mathbf{n}|\mathbf{n}_{ij}^{kl})}, \tag{7.114}$$

where the right-hand side is given by Equation (7.113), and it is the following:

$$\pi(\mathbf{n}) \propto \prod_{j=1}^{g} \frac{\alpha_j^{[n_j]}}{n_j!}. \tag{7.115}$$

It is the usual generalized Pólya distribution now restricted to (or conditional on) all the states belonging to the energy surface. Note that one can check that Equation (7.115) works for all the possible cases, including those situations in which initial and final cells coincide. This is straightforward but boring.

These last equations can be compared to the examples discussed at the beginning of the chapter. If $\forall j$, one has $\alpha_j = 1$ (the Bose–Einstein case), then one finds $\pi(\mathbf{n}_{ij}^{kl}) = \pi(\mathbf{n})$ and one gets the uniform distribution *restricted* to the energy shell; if $\forall j$, $\alpha_j = -1$ (the Fermi–Dirac case), then one has $\pi(\mathbf{n}_{ij}^{kl}) = \pi(\mathbf{n})$ for all $\mathbf{n} : n_i \in \{0, 1\}$; finally, the case in which $\forall j$, one has $\alpha_j \to \infty$ (the Maxwell–Boltzmann case) yields $\pi(\mathbf{n}) \propto \left(\prod_{j=1}^{g} n_j!\right)^{-1}$, which is uniform on all the allowed individual descriptions. In other words, what was previously written in the abstract unconstrained case here is valid, but *within the energy shell*. Further note that these results were obtained without having a detailed description of the energy shell.

In all the three cases described above, all the parameters α_j have the same value (all the cells are on a par, if allowed by the energy constraint); therefore, it is useful to introduce the new parameter $c = 1/\alpha_j$. The creation factor $\alpha_j + n_j$ becomes proportional to $1 + cn_j = c[(1/c) + n_j]$, and the three statistics relevant in physics are obtained for $c = 0, \pm 1$. For the Markov chain of macrostates $\mathbf{N}$, the creation factor becomes $g_i + cN_i = c[(g_i/c) + N_i]$. From Equation (7.115), the invariant weight (to be normalized) of a macrostate is given by

$$W(\mathbf{N}) = \prod_{i=1}^{d} \frac{c^{N_i} (g_i/c)^{[N_i]}}{N_i!}, \tag{7.116}$$

which becomes the usual weight for macrostates in statistical mechanics:

$$W_{\mathrm{BE}}(\mathbf{N}) = \prod_{i=1}^{d} \frac{g_i^{[N_i]}}{N_i!} = \prod_{i=1}^{d} \binom{g_i + N_i - 1}{N_i}, \tag{7.117}$$

for $c = 1$ (Bose–Einstein case),

$$W_{\mathrm{FD}}(\mathbf{N}) = \prod_{i=1}^{d} \frac{g_{i[N_i]}}{N_i!} = \prod_{i=1}^{d} \binom{g_i}{N_i}, \tag{7.118}$$

for $c = -1$ (Fermi–Dirac case) and

$$W_{\mathrm{MB}}(\mathbf{N}) = \prod_{i=1}^{d} \frac{g_i^{N_i}}{N_i!}, \tag{7.119}$$

for $c = 0$ (Maxwell–Boltzmann case).

In the unconstrained case, it is possible to eliminate the multivariate equilibrium distribution by marginalizing on a single category, and then obtain expected values for the multivariate distribution; on the contrary, in the constrained case, the initial symmetry of categories is broken by the energy function, and the marginal chain for a single category is very complicated. The only way to extract a univariate distribution from Equation (7.116) was envisaged by Boltzmann for the first time: it is the method of the most probable macrostate. In physics, the reason to maximize the probability of the macrostate given in Equation (7.116) instead of the microstate probability (7.115) is due to the assumed uniform distributions on the underlying microstates. Moreover, in physics, macrostates are obtained by merging many energy levels, in order to obtain large values of N_i and g_i so that Stirling's approximation for factorials can be used. However, it is possible to use a better procedure than applying Stirling's approximation as one knows that Equation (7.116) is exact. The maximum of $\ln W(\mathbf{N})$ with the constraints $\sum_{i=1}^{d} N_i = N = n$ and $\sum_{i=1}^{d} N_i\varepsilon_i = E$ is obtained by introducing Lagrange multipliers β, ν and by requiring that $\Delta(\ln W(\mathbf{N}) - \beta E + \nu N) = 0$ (see also Section 5.3.8). The variation is discrete. Considering the ith level, assume that N_i becomes $N_i \pm 1$, so that $\Delta E_i = \pm\varepsilon_i$ and $\Delta N_i = \pm 1$. Now define $W(N_i)$ as

$$W(N_i) = \frac{c^{N_i}(g_i/c)^{[N_i]}}{N_i!}; \tag{7.120}$$

considering that $\ln W(\mathbf{N}) = \sum_{i=1}^{d} \ln W(N_i)$ one can calculate the variation of the weight for a level if N_i becomes $N_i \pm 1$ using (7.116):

$$\Delta^+ \ln W(N_i) = \ln\frac{W(N_i+1)}{W(N_i)} = \ln\frac{c(g_i/c+N_i)}{N_i+1} = \ln\frac{g_i+cN_i}{N_i+1}, \tag{7.121}$$

and

$$\Delta^- \ln W(N_i) = \ln\frac{W(N_i-1)}{W(N_i)} = \ln\frac{N_i}{g_i+c(N_i-1)} = -\ln\frac{g_i+c(N_i-1)}{N_i}. \tag{7.122}$$

Note that the term $N_i + 1$ in Equation (7.121) allows an accommodation in an empty cell, whereas the term N_i in Equation (7.122) forbids a destruction in an empty cell. These terms are important when N_i is small. But all this procedure is intended for macrostates where $N_i \gg 1$. In this case $\Delta^+ \ln W(N_i)$ and $\Delta^- \ln W(N_i)$ become equal in absolute value, N_i can be treated as a continuous variable and one can write

$$\frac{\mathrm{d}}{\mathrm{d}N_i}\ln W(N_i) = \ln\frac{g_i+cN_i}{N_i} = \beta\varepsilon_i - \nu, \tag{7.123}$$

and deduce a simple analytical formula for the solution. If there exists a macrostate $\mathbf{N}^*$ such that

$$\ln \frac{(g_i + cN_i^*)}{N_i^*} = \beta\varepsilon_i - \nu, \tag{7.124}$$

that is

$$N_i^* = \frac{g_i}{\exp[\beta\varepsilon_i - \nu] - c}, \tag{7.125}$$

then it is the most probable macrostate, which for $c = 1, -1, 0$ represents the Bose–Einstein, the Fermi–Dirac and the Maxwell–Boltzmann case, respectively. The two new parameters β, ν must be chosen so that $\sum_{i=1}^{d} N_i^* = N$ and $\sum_{i=1}^{d} N_i^* \varepsilon_i = E$. Returning to the exact formulae (7.121) and (7.122), considering the transition $\mathbb{P}(\mathbf{N}_{ij}^{kl}|\mathbf{N})$, the variation of $\ln W(\mathbf{N})$ is just $\ln W(\mathbf{N}_{ij}^{kl}) - \ln W(\mathbf{N})$, which contains four terms if all levels are different, leading to

$$\begin{aligned}
&\ln W(\mathbf{N}_{ij}^{kl}) - \ln W(\mathbf{N}) \\
&= \Delta^+ \ln W(N_k) + \Delta^+ \ln W(N_l) + \Delta^- \ln W(N_i) + \Delta^- \ln W(N_j) \\
&= \ln \frac{N_i}{g_i + c(N_i - 1)} \frac{N_j}{g_j + c(N_j - 1)} \frac{g_k + cN_k}{N_k + 1} \frac{g_l + cN_l}{N_l + 1} \\
&= \ln \frac{\mathbb{P}(\mathbf{N}_{ij}^{kl}|\mathbf{N})}{\mathbb{P}(\mathbf{N}|\mathbf{N}_{ij}^{kl})} = \ln \frac{\pi(\mathbf{N}_{ij}^{kl})}{\pi(\mathbf{N})},
\end{aligned} \tag{7.126}$$

where the last equalities are a direct consequence of Equation (7.114) applied to macrostates. In the limit of large occupation numbers, if the starting $\mathbf{N}$ is $\mathbf{N}^*$, and applying Equation (7.123), one finds that

$$\Delta \ln W(\mathbf{N}^*) = \beta(\varepsilon_k + \varepsilon_l - \varepsilon_i - \varepsilon_j) = 0, \tag{7.127}$$

if $\mathbf{N}_{ij}^{kl}$ has the same energy as $\mathbf{N}^*$. This means that the probability is 'flat' around $\mathbf{N}^*$, as all its first neighbours have nearly the same probability. In words, starting from $\mathbf{N}^*$, and considering any transition which satisfies the energy constraint (that is such that $\varepsilon_k + \varepsilon_l = \varepsilon_i + \varepsilon_j$), the final vector is still in the region of maximum probability. For macroscopic physical systems, where n is roughly 10^{23}, the assumption of statistical mechanics is that the region of motion is overwhelmingly concentrated around $\mathbf{N}^*$, which then summarizes all the underlying dynamics and, at least, all equilibrium properties. Quite surprisingly, a formula similar to Equation (7.125) also holds true for small values of n. In fact, considering the *expected* flux

$$\phi_{ij}^{kl} = \sum_{\mathbf{N}} \mathbb{P}(\mathbf{N}_{ij}^{kl}|\mathbf{N})\mathbb{P}(\mathbf{N}), \tag{7.128}$$

which is the probability that a transition $i,j \to k,l$ occurs whatever the initial state $\mathbf{N}$ may be, and the reversed flux

$$\phi_{kl}^{ij} = \sum_{\mathbf{N}} \mathbb{P}(\mathbf{N}_{kl}^{ij}|\mathbf{N})\mathbb{P}(\mathbf{N}), \tag{7.129}$$

then assuming that the two fluxes are equal, one has that $\phi_{ij}^{kl}/\phi_{kl}^{ij} = 1$. Now, one can write

$$\phi_{ij}^{kl}/\phi_{kl}^{ij} = \frac{\mathbb{E}[N_i N_j (g_k + cN_k)(g_l + cN_l)]}{\mathbb{E}[N_k N_l (g_i + cN_i)(g_j + cN_j)]}. \tag{7.130}$$

If one can approximate the expectation of the product with the product of expectations, then from $\phi_{ij}^{kl}/\phi_{kl}^{ij} = 1$ one gets

$$\ln \frac{\mathbb{E}[N_i]}{g_i + c\mathbb{E}[N_i]} \frac{\mathbb{E}[N_j]}{g_j + c\mathbb{E}[N_j]} \frac{g_k + c\mathbb{E}[N_k]}{\mathbb{E}[N_k]} \frac{g_l + c\mathbb{E}[N_l]}{\mathbb{E}[N_l]} = 0; \tag{7.131}$$

in other words, if for any energy level $\ln[\mathbb{E}[N_i]/(g_i + c\mathbb{E}[N_i])] = \nu - \beta\varepsilon_i$ then for any conservative collision the two fluxes equalize. Then Equation (7.125) can be re-written for expected values also

$$\mathbb{E}[N_i] = \frac{g_i}{\exp[\beta\varepsilon_i - \nu] - c}. \tag{7.132}$$

Although both Equations (7.132) and (7.125) are approximate, in some applications (7.125) could be meaningless, whereas Equation (7.132) is always meaningful. For instance, consider a system with $g_i = 1$ for all the values of i. This is strange in physics, where g_i is an increasing function of i, but it can be the case in economics or in other fields. Then Equation (7.125) could still apply, because it was derived only for $N_i \gg 1$, while the usual Stirling's approximation also needs $g_i \gg 1$. In this particular case, macrostates and microstates do coincide. Further assume that $c = 1$. One knows from Equation (7.14) that the equilibrium distribution is flat, that is all the states are equiprobable. Hence there is no $\mathbf{N}^*$ at all! On the contrary, and this is confirmed by computer simulation, $\mathbb{E}[N_i]$ can be derived from the time average of occupation numbers for the ith level as time goes by, and an almost qualitative agreement with Equation (7.132) is easily obtained even for $n < 100$. Note that if there exists $\mathbf{N}^*$ such that $\pi(\mathbf{N}^*) \approx 1$, then $\mathbb{E}[N_i] = \sum N_i \pi(\mathbf{N}) \approx N_i^*$, but $\mathbb{E}[N_i]$ also exists when there is no predominant macrostate $\mathbf{N}^*$.

Finally, in order to justify the assumption leading from Equation (7.130) to Equation (7.131), note that if g random variables satisfy the constraint $\sum_1^g X_i = n$, then $\mathbb{V}(\sum_1^g X_i) = \sum_1^g \mathbb{V}(X_i) + \sum_{j \neq i} \mathbb{C}(X_i, X_j) = 0$ (see also Section 5.4). Even if all the variables X_i are not equidistributed, one finds that Equation (5.94) gives the order of magnitude for the Bravais–Pearson correlation which is

$$\rho(X_i, X_j) = -\frac{1}{g-1}. \tag{7.133}$$

Given that in any concrete case, the number g of levels involved in the dynamics is far larger than n, the correlation between the occupation numbers is negligible, and the corresponding approximation works fairly well.

7.8 Exercises

7.1 Consider the aperiodic Ehrenfest urn discussed at length in Chapter 6, set $n = 4$ and further consider a double selection, followed by a double independent creation. A transition path is D_1, D_2, C_1, C_2 where D_1, D_2 are conditioned on the starting state $k = 0, \ldots, 4$, whereas C_1, C_2 are independent. Therefore, one has $\mathbb{P}(C_1, C_2) = 1/4$ uniformly on all the four possibilities. Show that the transition matrix with $m = 2$ (binary moves) is different from the square of the matrix with $m = 1$ given by Equation (6.30).

7.2 Write a Monte Carlo program for the Ehrenfest–Brillouin model with unary moves. In the dichotomous case, compare the results for $\alpha_1 = \alpha_2 = 10$ and $\alpha_1 = \alpha_2 = 1/2$.

7.9 Summary

In the general case of m-ary moves, the Ehrenfest–Brillouin model is a Markov chain specified by the transition probability (7.53) and by the initial occupation vector $\mathbf{n}(0)$ (or, in a more abstract way, by the initial probability distribution $\mathbb{P}^{(0)}$). It describes the probabilistic dynamics of n objects jumping within g categories. The transition probability factorizes into 'random' destruction terms, as in the case of the Ehrenfest urn model, and creation terms following the generalized Pólya mechanism, as in the case of Brillouin's model. Each term is computed according to the current occupation state of the system with destructions preceding creations. For instance, in the case of binary moves, first an object is randomly selected from all the n objects and removed, then a second object is removed after being extracted from the remaining $n-1$ objects. Then the objects accommodate one after the other according to the Pólya mechanism. Before the first accommodation $n-2$ objects are left in the system and before the second accommodation there are $n-1$ objects.

If no exogenous constraint is imposed on the 'motion' of the chain, then it is possible to marginalize the multivariate Markov chain and follow the evolution of a single category as in Equation (7.54). Otherwise, in the presence of constraints, the time evolution of the Ehrenfest–Brillouin Markov chain is limited to the subset of all the possible occupation states satisfying the constraints. In the unconstrained cases, the invariant distribution of this model is a multivariate Pólya distribution, whereas in the constrained cases, it is still a Pólya but re-normalized over the subset of accessible states. In all the cases, this is also an equilibrium distribution.

The Ehrenfest–Brillouin model is remarkably general and flexible and it encompasses many popular models used in physics, economics and finance. In particular, as invariant and equilibrium distributions one recovers the popular Bose–Einstein, Fermi–Dirac and Maxwell–Boltzmann distributions of statistical physics as well as the equilibrium distribution for the celebrated Kirman's ant model. Moreover, all the limits of the Pólya distribution described in Chapter 5 apply.

Further reading

There are many introductory books on statistical mechanics. Jancel's book is a classical one with emphasis on foundational problems:

R. Jancel, *Foundations of Classical and Quantum Statistical Mechanics*, Pergamon Press, Oxford, UK (1969).

The seminal paper by Brillouin is listed in further reading for Chapter 4, whereas the encyclopaedia article written by P. and T. Ehrenfest is listed in Chapter 6. The Ehrenfest–Brillouin model was introduced in a series of papers written by D. Costantini and U. Garibaldi with applications to statistical physics in mind.

D. Costantini and U. Garibaldi, *A Probabilistic Foundation of Elementary Particle Statistics. Part I*, Studies in History and Philosophy of Modern Physics, **28**, 483–506 (1997).

D. Costantini and U. Garibaldi, *A Probabilistic Foundation of Elementary Particle Statistics. Part II*, Studies in History and Philosophy of Modern Physics, **29**, 37–59 (1998).

The most relevant paper, published by Synthese, is listed in further reading for Chapter 4.

A. Kirman's original paper including his 'ant model' was published in 1993:

A. Kirman, *Ants, Rationality, and Recruitment*, The Quarterly Journal of Economics, **108**, 137–156 (1993).

The data in Section 7.6 are taken from a paper by Gardini *et al.*

A. Gardini, G. Cavaliere and M. Costa, *A New Approach to Stock Price Modeling and Forecasting*, Journal of the Italian Statistical Society, **8**, 25–47 (1999).

whose results were later updated in

A. Gardini, G. Cavaliere and M. Costa, *Fundamentals and Asset Price Dynamics*, Statistical Methods & Applications, **12**, 211–226 (2003).

Note that market daily data are freely available from several websites. The model of Section 7.6 was introduced in

U. Garibaldi, M.A. Penco and P. Viarengo, *An Exact Physical Approach to Market Participation Models*, in R. Cowan and N. Jonard (eds.), *Heterogeneous Agents, Interactions and Economic Performance*, Lecture Notes in Economics and Mathematical Systems, **521**, 91–103, Springer, Berlin (2003).

Variations on the theme of the Ehrenfest–Brillouin model were used by G. Bottazzi *et al.*

G. Bottazzi, G. Fagiolo, G. Dosi and A. Secchi, *Modeling Industrial Evolution in Geographical Space*, Journal of Economic Geography, **7**, 651–672 (2007).

G. Bottazzi, G. Fagiolo, G. Dosi and A. Secchi, *Sectoral and Geographical Specificities in the Spatial Structure of Economic Activities*, Structural Change and Economic Dynamics, **19**, 189–202 (2008).

to analyze industrial evolution in geographical space both theoretically and empirically.

8

Applications to stylized models in economics

This chapter is devoted to three models where n objects move within g categories according to some prescribed transition probabilities. The natural way of presenting these models is in terms of transition probabilities over occupation states $\mathbf{Y} = \mathbf{n}$. Individual descriptions, occupation vectors as well as partition vectors can be given an interpretation in terms of economic variables. The three models are irreducible and aperiodic Markov chains; therefore they have a unique invariant distribution which is also the equilibrium distribution.

After reading this chapter you should be able to:

- intepret the random vectors discussed in the previous chapters of the book in terms of economic variables;
- understand the role of the methods discussed in Section 5.3.8 and in Chapter 6, in order to derive simple analytical results for comparison with Monte Carlo simulations and real data;
- understand three simple stylized economic models.

8.1 A model for random coin exchange

Consider g agents and n coins to be distributed among them. This problem was considered in the example in Section 2.3.2 as well as in the exercises of Chapter 6. In this problem, knowing that the ith coin belongs to the jth agent is a marginal individual description and the vector (random) variable $\mathbf{X} = (X_1 = x_1, \ldots, X_n = x_n)$, where $x_i \in \{1, \ldots, g\}$ gives full information on the position of the coins. The frequency (random) vector $\mathbf{Y} = (Y_1 = n_1, \ldots, Y_g = n_g)$ contains information on the number of coins belonging to each agent, with the constraint that $\sum_{i=1}^{g} n_i = n$. Finally, the numbers of agents with zero coin, one coin, etc. are included in the partition (random) vector $\mathbf{Z} = (Z_0 = z_0, \ldots, Z_n = z_n)$, with the two constraints that $\sum_{i=0}^{n} z_i = g$ and $\sum_{i=0}^{n} i z_i = n$. Usually, one considers a random interaction between two agents who randomly meet; one of them is a loser with probability $1/2$ and the other is a winner. The winner receives a coin from the loser. If the loser has no coins left, the move is not considered. After a little thought, one can avoid the problem of empty steps considering the following description of the game. Given an initial frequency vector $\mathbf{n}_0$:

1. at each step a *loser* is selected by chance from all the agents with at least one coin;
2. the loser gives one of his/her coins to a *winner* randomly selected from all the agents.

Note that, with this mechanism, the winner may well coincide with the loser. As further discussed in Exercises 8.1 and 8.2 below, this game can be described by a finite Markov chain with the following transition probability

$$\mathbb{P}(\mathbf{n}'|\mathbf{n}) = \frac{1-\delta_{n_i,0}}{g-z_0(\mathbf{n})}\frac{1}{g}, \tag{8.1}$$

where

1. after the move, the frequency vector $\mathbf{n}'$ differs from $\mathbf{n}$ at most by the change of position of one coin. In other words if agent i is selected as a loser and agent $j \neq i$ as a winner, one has that $n_i' = n_i - 1$ and $n_j' = n_j + 1$ and $\mathbf{n}' = \mathbf{n}_i^j$ with a destruction in category i followed by a creation in category j;
2. the numerator $1 - \delta_{n_i,0}$ means that the loser is selected from the agents with at least one coin;
3. the denominator $g - z_0(\mathbf{n})$ is the number of agents with at least one coin, as $z_0(\mathbf{n})$ is the number of agents with no coins;
4. the term $1/g$ represents the probability of selecting one of the agents at random;
5. the selections of the loser and of the winner are independent and the transition probability (8.1) is the product of the probability of selecting a loser and the probability of selecting a winner.

Exercises 8.1 and 8.2 below show how to determine the invariant probability distribution for this finite Markov chain using the eigenvector method as well as the Monte Carlo method. However, both methods become cumbersome for a large state space and in this particular case, the state space is a rapidly growing function of both g and n. For instance, for $n = 20$ coins and $g = 10$ agents, one already has more than ten million possible occupation states. Fortunately enough, for this Markov chain, the invariant distribution can be determined by imposing the detailed balance conditions.

8.1.1 The invariant distribution

Consider two subsequent states $(n_1,\ldots,n_i,\ldots,n_j,\ldots,n_g)$ and $(n_1,\ldots,n_i - 1,\ldots,n_j+1,\ldots,n_g)$ with $n_i > 0$ and $i \neq j$. If one tries to apply detailed balance, one has that the 'direct flux' is given by

$$\pi(\mathbf{n})\mathbb{P}(\mathbf{n}_i^j|\mathbf{n}) = \pi(\mathbf{n})\frac{1}{g - z_0(\mathbf{n})}\frac{1}{g}, \tag{8.2}$$

whereas the 'inverse flux' is

$$\pi(\mathbf{n}_i^j)\mathbb{P}(\mathbf{n}|\mathbf{n}_i^j) = \pi(\mathbf{n}_i^j)\frac{1}{g - z_0(\mathbf{n}_i^j)}\frac{1}{g}; \tag{8.3}$$

equating the two fluxes leads to

$$\pi(\mathbf{n})\frac{1}{g - z_0(\mathbf{n})}\frac{1}{g} = \pi(\mathbf{n}_i^j)\frac{1}{g - z_0(\mathbf{n}_i^j)}\frac{1}{g}, \tag{8.4}$$

which is always satisfied if

$$\pi(\mathbf{n}) = C(g - z_0(\mathbf{n})), \tag{8.5}$$

where C is a suitable normalization constant. In other words, the invariant probability distribution for this model is proportional to the number of agents with at least one coin in their pocket. If $n \gg g$, and starting from a uniform distribution of coins among agents, then the number of agents without coins is $z_0 = 0$, and it is very unlikely to see $z_0 > 0$ as time goes by. In other words, a very long time is needed to see a state in which some agent remains without coins. Setting $z_0 = 0$ in Equation (8.5) leads to a $\pi(\mathbf{n})$ uniform on all agent descriptions whose number is

$$W(g,n) = \binom{n+g-1}{n}. \tag{8.6}$$

Therefore, if $g \gg 1$, one finds that the marginal description of an agent is the geometric distribution discussed in Section 5.3.2. Given the assumed uniformity, the same result can be obtained using the Lagrange multiplier method for the partition vector. Note that the mode of the geometric distribution is just z_0, that is the most occupied state is expected to be z_0 contrary to the initial assumption! Indeed, it will be seen that for n and g finite, one can get coin distributions which significantly differ from the asymptotic geometric case.

Returning to the invariant distribution (8.5), in order to determine the normalization constant C, one notices that the total number of occupation states is given by Equation (8.6) as already discussed many times. The number of agent descriptions with k fixed agents possessing one coin at least (while all the other $n-k$ agents will remain with zero coins) is given by the ways of allocating $n-k$ coins to k agents:

$$\binom{n-k+k-1}{n-k} = \binom{n-1}{n-k} = \binom{n-1}{k-1}, \tag{8.7}$$

in fact, the first k coins are used as placeholders, and there are $n!/(k!(n-k)!)$ ways to choose these k agents. Therefore the number of agent descriptions with k agents possessing one coin at least is

$$W(k,g,n) = \binom{n}{k}\binom{n-1}{n-k}. \tag{8.8}$$

For instance, in the case of maximum coin concentration ($k = 1$), Equation (8.8) gives $W(1,g,n) = n$, as any agent can be the richest. Now, the total number of agent descriptions is just the sum of all the descriptions with k agents with at least one coin in their pockets (all the other agents being without coins); in other words, one has

$$W(g,n) = \binom{n+g-1}{n} = \sum_{k=1}^{g}\binom{n}{k}\binom{n-1}{n-k} = \sum_{k=1}^{g} W(k,g,n); \tag{8.9}$$

this equation represents the decomposition of the total number of states into all possible values of k. Coming back to Equation (8.5), one can write the following chain of equalities

$$\begin{aligned} 1 &= \sum_{\mathbf{n}} \pi(\mathbf{n}) = C\sum_{\mathbf{n}}(g - z_0(\mathbf{n})) = C\sum_{\mathbf{n}} k(\mathbf{n}) \\ &= C\sum_{k=1}^{g} kW(k,g,n) = C\sum_{k=1}^{g} k\binom{n}{k}\binom{n-1}{n-k}, \end{aligned} \tag{8.10}$$

so that one eventually finds

$$C = \left[\sum_{k=1}^{g} k\binom{n}{k}\binom{n-1}{n-k}\right]^{-1}. \tag{8.11}$$

The fourth equality in Equation (8.10) is justified by the fact that $W(k,g,n)$ is the number of terms with a given value of $k(\mathbf{n})$.

8.1.2 The expected wealth distribution

The marginal method presented in Section 5.3.8 can be used to obtain the expected wealth distribution (namely, the distributions of coins) in this case. The relevant random variable is the partition vector $\mathbf{Z} = (Z_0 = z_0, \ldots, Z_n = z_n)$ where z_0 is the number of agents without coins, z_1 is the number of agents with 1 coin, and so on, up to z_n which gives the number of agents with n coins. As mentioned many times, these numbers must satisfy the two following constraints:

$$\sum_{i=0}^{n} z_i = g, \tag{8.12}$$

and

$$\sum_{i=1}^{n} i z_i = n. \tag{8.13}$$

As in Equation (5.72) of Section 5.3.8, thanks to agent symmetry, one has that

$$\mathbb{P}(\mathbf{Z}=\mathbf{z}) = \frac{g!}{\prod_{i=0}^{n} z_i!}\mathbb{P}(\mathbf{Y}=\mathbf{n}) = \frac{g!}{\prod_{i=0}^{n} z_i!}C(g - z_0(\mathbf{n})). \tag{8.14}$$

Using the approximate method of Lagrange multipliers in order to find the most probable vector is not easy in this case, but the exact marginalization method works and, for each random variable Z_i, one can determine its average value $\mathbb{E}(Z_i)$.

As a consequence of Equations (5.74) and (5.75), the average number of agents whose occupation number is equal to i is

$$\mathbb{E}(Z_i) = g\mathbb{P}(n_1 = i), \tag{8.15}$$

as all the n_is are equidistributed. The first step for deriving $\mathbb{E}(Z_i)$ is studying the conditional average

$$\mathbb{E}(Z_i|k) = g\mathbb{P}(n_1 = i|k), \tag{8.16}$$

whose meaning is the marginal wealth distribution of an agent conditioned to $k = g - z_0$. From Equation (8.5), it is apparent that all agent descriptions with the same k have the same probability, and their number is $W(k,g,n)$ given in Equation (8.8). Therefore, $\mathbb{P}(n_1 = i|k)$ is given by the number of occupation vectors in which $g-1$ agents share $n-i$ coins divided by $W(k,g,n)$. This calculation can be divided into three parts. First, consider $\mathbb{P}(n_1 = 0|k)$; one has:

$$\begin{aligned}\mathbb{P}(n_1 = 0|k) &= \frac{W(k,g-1,n)}{W(k,g,n)} = \frac{\binom{g-1}{k}\binom{n-1}{k-1}}{\binom{g}{k}\binom{n-1}{k-1}} \\ &= \frac{\binom{g-1}{k}}{\binom{g}{k}} = \frac{(g-1)!}{(g-1-k)!}\frac{(g-k)!}{g!} = \frac{g-k}{g},\end{aligned} \tag{8.17}$$

then, consider $P(n_1 = i|k)$ with $k \geq 2$, and $i > 0$; as there are $k-1$ agents left with at least one coin, one has:

$$\mathbb{P}(n_1 = i|k) = \frac{W(k-1, g-1, n-i)}{W(k,g,n)}$$
$$= \frac{\binom{g-1}{k-1}\binom{n-i-1}{k-2}}{\binom{g}{k}\binom{n-1}{n-k}} = \frac{k}{g}\frac{\binom{n-i-1}{k-2}}{\binom{n-1}{k-1}}; \tag{8.18}$$

note that the above formula is valid for $n-i-1 \geq k-2$ and $i = 1, \ldots, n-1$. If these conditions are not satisfied, then $\mathbb{P}(n_1 = i|k) = 0$. Finally, for $k = 1$, one finds:

$$\mathbb{P}(n_1 = i|k = 1) = \frac{\delta_{i,n}}{g}, \tag{8.19}$$

for $i > 0$, as in this case all the coins are concentrated on a single agent. Eventually, from Equation (8.16) one finds

$$\begin{cases} \mathbb{E}(Z_0|k) = g - k \\ \begin{cases} \mathbb{E}(Z_i|k > 1) = k\frac{\binom{n-i-1}{k-2}}{\binom{n-1}{k-1}}, \ i = 1, \ldots, n-1 \\ \mathbb{E}(Z_i|k = 1) = \delta_{i,n}, \ i = 1, \ldots, n; \end{cases} \end{cases} \tag{8.20}$$

and further $\mathbb{E}(Z_i|k) = 0$ for $n-i-1 < k-2$ and $i = n$. The average marginal wealth distribution $\mathbb{E}(Z_i)$ is related to the average marginal conditional wealth distribution $\mathbb{E}(Z_i|k)$ by means of the following equation:

$$\mathbb{E}(Z_i) = \sum_{k=0}^{g} \mathbb{P}(k)\mathbb{E}(Z_i|k). \tag{8.21}$$

Therefore, $\mathbb{P}(k)$ is still needed. However, this is immediately given by Equation (8.5), recalling that there are $W(k,g,n)$ occupation vectors with the same value of k:

$$\mathbb{P}(k) = CkW(k,g,n) = Ck\binom{g}{k}\binom{n-1}{k-1}. \tag{8.22}$$

Note that one has

$$\sum_{i=0}^{n} \frac{\mathbb{E}(Z_i)}{g} = 1; \tag{8.23}$$

therefore, it is meaningful to compare $\mathbb{E}(Z_i)/g$ to the histograms for the relative frequency of agents with i coins in a Monte Carlo simulation of this model.

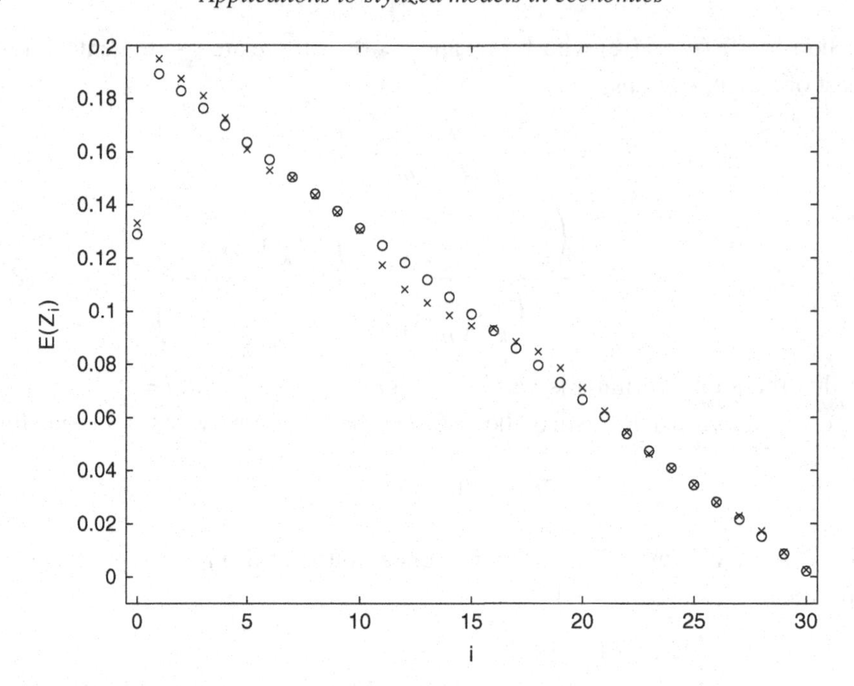

Figure 8.1. Comparison between the results of a Monte Carlo simulation of the random model for wealth distribution (crosses) and the analytical prediction of Equation (8.21) (open circles) in the case $n = 30$ and $g = 3$.

8.1.3 Behaviour of the expected wealth distribution

The expected wealth distribution is plotted in Fig. 8.1 for $n = 30$ coins and $g = 3$ agents. In the Monte Carlo simulation, based on the solution of Exercise 8.2, the values of the random variable Z_i were sampled and averaged over 10^5 Monte Carlo steps, after an equilibration run of 1000 steps. In the specific case plotted in Fig. 8.1, $\mathbb{E}(Z_i)$ is a (linearly) decreasing function of i, except for $\mathbb{E}(Z_0)$, which is much smaller than $\mathbb{E}(Z_1)$. This is not always the case, however. For $n = g = 30$, $\mathbb{E}(Z_i)$ is a strictly decreasing non-linear function of i. This is plotted in Fig. 8.2, without comparison with Monte Carlo simulations. In the thermodynamic limit, $(n, g, k \gg 1)$, Equation (8.20) can be approximated by a geometric distribution

$$\frac{\mathbb{E}(Z_{i+1}|k)}{k} \simeq \frac{k}{n}\left(1 - \frac{k}{n}\right)^i. \tag{8.24}$$

This means that the average fraction of agents with at least one coin follows a geometric distribution which becomes exponential in the continuous limit. In the latter limit, $\mathbb{E}(Z_i)$ becomes a mixture of exponential distributions with $\mathbb{P}(k)$ given by Equation (8.22) as a mixing measure. Indeed, from Equation (8.22), one can

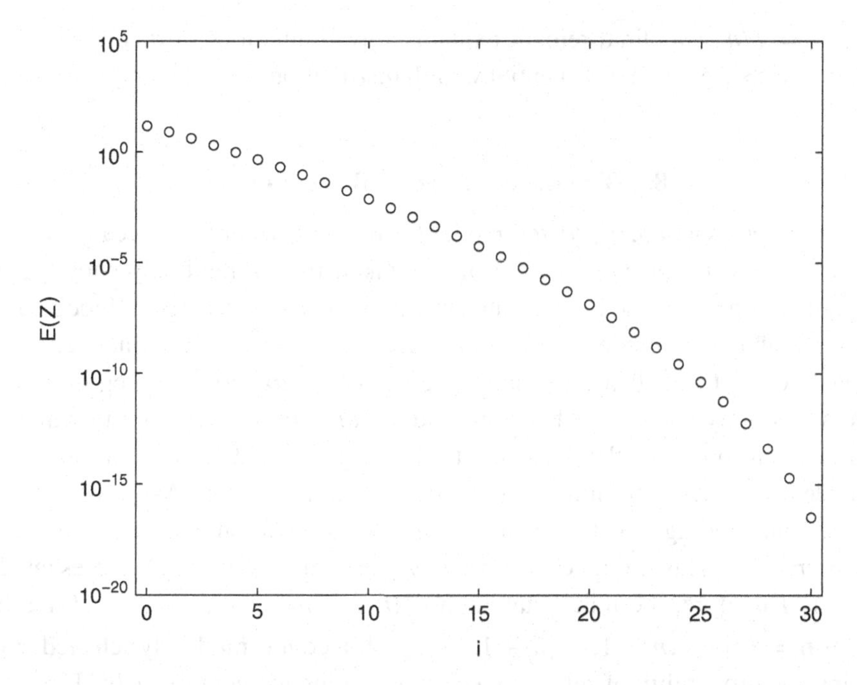

Figure 8.2. Behaviour of $\mathbb{E}(Z_i)$ for $n = g = 30$, in log-linear scale. Note that $\mathbb{E}(Z_i)$ is a strictly decreasing non-linear function of i.

show that

$$\frac{\mathbb{P}(k+1)}{\mathbb{P}(k)} = \frac{(g-k)(n-k)}{k^2}. \tag{8.25}$$

This ratio is greater than 1 for $k < k^* = ng/(n+g)$ and smaller than one for $k > k^*$. For $n = g$, $\mathbb{P}(k)$ is bell-shaped with maximum in k^* and $k^* + 1$ and $k^* = g/2$. For $n \gg g$, the distribution is left-skewed with a maximum very close to g as $k^* \simeq g(1 - g/n)$. Moreover, if $g(1 - g/n) > g - 1$, meaning that $g^2 < n$, the maximum value is $k^* = g$. In the case $n \gg g$, the mixing probability distribution $\mathbb{P}(k)$ is concentrated in a small number of values around k^*. In this large density case, if the condition $g \gg 1$ holds, one further finds that $\mathbb{E}(Z_i)/g$ is not different from a single geometric distribution

$$\frac{\mathbb{E}(Z_i)}{g} \simeq \frac{\mathbb{E}(Z_i|g)}{g} \simeq \frac{g}{n}\left(1 - \frac{g}{n}\right)^i. \tag{8.26}$$

In the continuous limit, Equation (8.26) becomes the exponential distribution

$$\frac{\mathbb{E}(Z_i)}{g} \simeq \lambda \exp(-\lambda i), \tag{8.27}$$

where $\lambda = g/n$. This final remark explains why simulations of this model in the case $n \gg g \gg 1$ give an exponential wealth distribution.

8.2 The taxation–redistribution model

In the previous section, a stylized model for the redistribution of wealth was discussed. In that model one coin out of n is taken from a randomly chosen agent with at least one coin and redistributed to a randomly selected agent. Indebtedness is not possible. The model can be modified along the following lines. There are always n coins to be distributed among g agents. A *taxation* is a step in which a coin is randomly taken out of n coins and a *redistribution* is a step in which the coin is given to one of the g agents. In the language of Chapter 6, a taxation is equivalent to a destruction and a redistribution to a creation. As in the previous model, coins and agents do not disappear and the taxation–redistribution model is *conservative*. This model considers only *unary* moves of the kind presented in Fig. 6.1. Therefore, the initial state is $\mathbf{n} = (n_1, \ldots, n_i, \ldots, n_j, \ldots, n_g)$ and the final state is $\mathbf{n}_i^j = (n_1, \ldots, n_i - 1, \ldots, n_j + 1, \ldots, n_g)$. If a coin is randomly selected out of n coins, the probability of selecting a coin belonging to agent i is n_i/n. Therefore, in this model, agents are taxed proportionally to their wealth measured in terms of the number of coins in their pockets. The redistribution step is crucial as it can favour agents with many coins (a rich gets richer mechanism) or agents with few coins (a taxation scheme leading to equality). This can be done by assuming that the probability of giving the coin taken from agent i to agent j is proportional to $w_j + n_j$ where n_j is the number of coins in the pocket of agent j and w_j is a suitable weight. Depending on the choice of w_j, one can obtain very different equilibrium situations. Based on the previous considerations, it is assumed that the transition probability is

$$\mathbb{P}(\mathbf{n}_i^j|\mathbf{n}) = \frac{n_i}{n} \frac{w_j + n_j - \delta_{i,j}}{w + n - 1}, \tag{8.28}$$

where $w = \sum_{i=1}^{g} w_i$ and the Kronecker symbol $\delta_{i,j}$ takes into account the case $i = j$. If the condition $w_j \neq 0$ is satisfied, then also agents without coins can receive them. If all the agents are equivalent, one has $w_j = \alpha$ uniformly, and $w = g\alpha = \theta$, so that Equation (8.28) becomes

$$\mathbb{P}(\mathbf{n}_i^j|\mathbf{n}) = \frac{n_i}{n} \frac{\alpha + n_j - \delta_{i,j}}{\theta + n - 1}. \tag{8.29}$$

This is the Ehrenfest–Brillouin model with unary moves discussed in Section 7.2, whose invariant and equilibrium distribution is the g-variate generalized Pólya distribution, given below for the convenience of the reader

$$\pi(\mathbf{n}) = \frac{n!}{\theta^{[n]}} \prod_{i=1}^{g} \frac{\alpha^{[n_i]}}{n_i!}. \tag{8.30}$$

In contrast to what is found in physics, here α is not limited to the values ± 1 and ∞. Indeed, the redistribution policy is characterized by the value of the parameter α. In parallel to physics, if α is small and positive, rich agents become richer, but for $\alpha \to \infty$ the redistribution policy becomes random: any agent has the same probability of receiving the coin. Finally, the case $\alpha < 0$ favours poor agents, but $|\alpha|$ is the maximum allowed wealth for each agent.

8.2.1 Marginal description and expected wealth distribution

As in the model described in the previous section, agents' equivalence leads to a simple relationship between the joint probability distribution of partitions and the probability of a given occupation vector. From Equations (5.72) and (8.30), one gets

$$\mathbb{P}(\mathbf{Z} = \mathbf{z}) = \frac{g!}{\prod_{i=0}^{n} z_i!} \mathbb{P}(\mathbf{Y} = \mathbf{n}) = \frac{g!}{\prod_{i=0}^{n} z_i!} \frac{n!}{\prod_{j=1}^{g} n_i!} \prod_{j=1}^{g} \frac{\alpha^{[n_i]}}{\theta^{[n]}}$$
$$= \frac{g!n!}{\prod_{i=0}^{n} z_i!(i!)^{z_i}} \prod_{j=1}^{g} \frac{\alpha^{[n_i]}}{\theta^{[n]}}, \tag{8.31}$$

where, as usual, z_i is the number of agents with i coins. In contrast to physics, where the constrained maximisation is necessary, the exact marginal method can be used here. All the agents are characterized by the same weight α. Given this equivalence, it is meaningful to focus on the behaviour of the random variable $Y = Y_1$ representing the number of coins of agent 1. If one begins from $Y_t = k$, one can define the following transition probabilities:

$$w(k, k+1) = \mathbb{P}(Y_{t+1} = k+1 | Y_t = k) = \frac{n-k}{n} \frac{\alpha + k}{\theta + n - 1}, \tag{8.32}$$

meaning that a coin is randomly removed from one of the other $n-k$ coins belonging to the other $g-1$ agents and given to agent 1 according to the weight α and the number of coins k;

$$w(k, k-1) = \mathbb{P}(Y_{t+1} = k-1 | Y_t = k) = \frac{k}{n} \frac{\theta - \alpha + n - k}{\theta + n - 1}, \tag{8.33}$$

meaning that a coin is randomly removed from agent 1 and redistributed to one of the other agents according to the weight $\theta - \alpha$ and the number of coins $n-k$; and

$$w(k,k) = \mathbb{P}(Y_{t+1} = k|Y_t = k) = 1 - w(k,k+1) - w(k,k-1), \tag{8.34}$$

meaning that agent 1 is not affected by the move. These equations define a birth–death Markov chain corresponding to a random walk with semi-reflecting barriers. This chain was studied in detail in Section 7.5.3. However, the invariant (and equilibrium) distribution can be directly obtained by marginalizing Equation (8.30). This leads to the dichotomous Pólya distribution, as discussed in Section 5.3:

$$\mathbb{P}(Y = k) = \frac{n!}{k!(n-k)!}\frac{\alpha^{[k]}(\theta-\alpha)^{[n-k]}}{\theta^{[n]}}. \tag{8.35}$$

As a consequence of the equivalence of all agents, from Equations (5.74) and (5.75), one obtains

$$\mathbb{E}(Z_k) = g\mathbb{P}(Y = k) = g\frac{n!}{k!(n-k)!}\frac{\alpha^{[k]}(\theta-\alpha)^{[n-k]}}{\theta^{[n]}}. \tag{8.36}$$

Equation (8.36) gives the expected wealth distribution for the taxation–redistribution model. Following Section 5.5, in the continuous thermodynamic limit, one can see that the equilibrium expected wealth distribution of the taxation–redistribution model is approximately described, for $\alpha > 0$, by the gamma distribution of parameters α and $n/\alpha g$.

8.2.2 Block taxation and the convergence to equilibrium

Consider the case in which taxation is done in a block: instead of extracting a single coin from an agent at each step, $m \leq n$ coins are randomly taken from various agents and then redistributed with the mechanism described above, that is with a probability proportional to the residual number of coins and to an a-priori weight. If $\mathbf{n} = (n_1,\ldots,n_g)$ is the initial occupation vector, $\mathbf{m} = (m_1,\ldots,m_g)$ (with $\sum_{i=1}^{g} m_i = m$) is the taxation vector and $\mathbf{m}' = (m'_1,\ldots,m'_g)$ (with $\sum_{i=1}^{g} m'_i = m$) is the redistribution vector, one can write

$$\mathbf{n}' = \mathbf{n} - \mathbf{m} + \mathbf{m}'. \tag{8.37}$$

This is essentially the Ehrenfest–Brillouin model with m-ary moves studied in Section 7.5.3. The main results of that section are summarized below interpreted in the framework of the model. The block taxation–redistribution model still has Equation (8.30) as its equilibrium distribution, as the block step is equivalent to m steps of the original taxation–redistribution model.

The marginal analysis for the block taxation–redistribution model in terms of a birth–death Markov chain is more cumbersome than for the original model given that, now, the difference $|\Delta Y|$ can vary from 0 to m. In any case, Equation (8.35) always gives the equilibrium distribution. In particular, this means that

$$\mathbb{E}(Y) = n\frac{\alpha}{\theta} = \frac{n}{g}, \tag{8.38}$$

and

$$\mathbb{V}(Y) = n\frac{\alpha}{\theta}\frac{\theta-\alpha}{\theta}\frac{\theta+n}{\theta+1} = \frac{n}{g}\frac{g-1}{g}\frac{\theta+n}{\theta+1}. \tag{8.39}$$

However, one can write

$$Y_{t+1} = Y_t - D_{t+1} + C_{t+1}, \tag{8.40}$$

where D_{t+1} is the random taxation for the given agent and C_{t+1} is the random redistribution to the given agent. The expected value of D_{t+1} under the condition $Y_t = k$ is

$$\mathbb{E}(D_{t+1}|Y_t = k) = m\frac{k}{n}; \tag{8.41}$$

this result holds true because m coins are taken at random out of the n coins and the probability of removing a coin from the first agent is k/n under the given condition. Moreover, under the conditions $Y_t = k$ and $D_{t+1} = d$, the probability of giving a coin back to agent 1 is $(\alpha + k - d)/(\theta + n - m)$, so that, after averaging over $D_{t+1}|Y_t = k$, one finds

$$\mathbb{E}(C_{t+1}|Y_t = k) = m\frac{\alpha + k - m\frac{k}{n}}{\theta + n - m}. \tag{8.42}$$

The expected value of $Y_{t+1} - Y_t$ conditioned on $Y_t = k$ can be found directly taking the expectation of (8.40) and using Equations (8.41) and (8.42). This leads to

$$\mathbb{E}(Y_{t+1} - Y_t|Y_t = k) = -\frac{m\theta}{n(\theta+n-m)}\left(k - n\frac{\alpha}{\theta}\right). \tag{8.43}$$

It is possible to make the following remarks on Equation (8.43):

1. Equation (8.43) is analogous to a mean reverting equation. If $Y_t = k$ is different from its expected value $n\alpha/\theta = n/g$, $\mathbb{E}(Y_{t+1}|Y_t = k)$ will then move back towards that value; this behaviour is represented in Fig. 8.3;
2. if $k = n\alpha/\theta$, then the chain is first-order stationary. This is shown in Fig. 8.3 when one begins with n/g, then one always gets $\mathbb{E}(Y_{t+1} - Y_t|Y_t = k) = 0$;

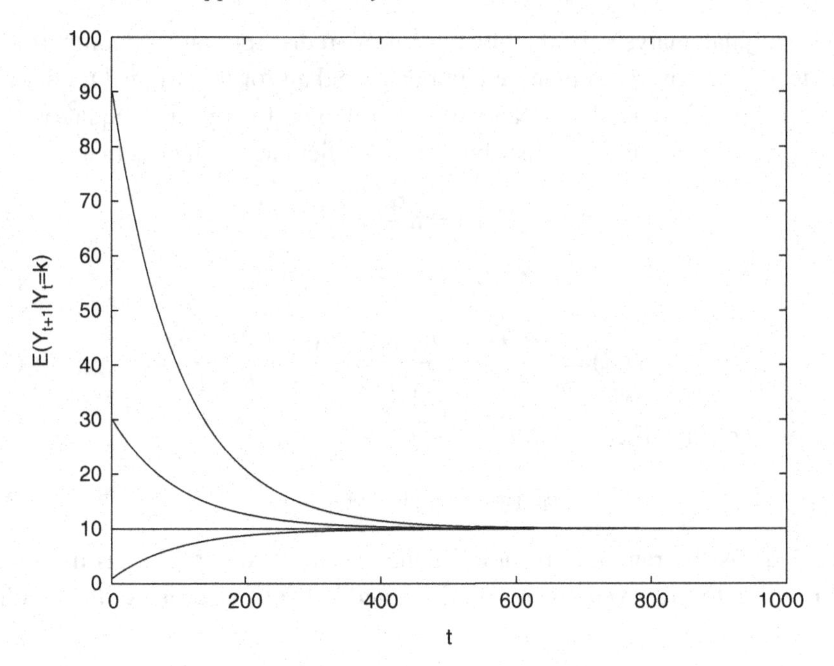

Figure 8.3. Average block dynamics for four different initial values of Y, namely $Y = 90$, $Y = 30$, $Y = 10$ and $Y = 1$ coins. There are $n = 100$ coins and $g = 10$ agents. The equilibrium expected value of Y is $n/g = 10$. The parameters of the block taxation–redistribution model are $m = 10$, $\alpha = 1$ and $\theta = 10$. Note that $r^{-1} = n(\theta + n - m)/(m\theta) = 100$ gives the order of magnitude of the number of steps necessary to reach equilibrium.

3. $r = m\theta/(n(\theta + n - m))$ is the intensity of the restoring force. The inverse of r gives the order of magnitude for the number of transitions needed to reach equilibrium. In the case plotted in Fig. 8.3, one has $r^{-1} = 100$. More exactly, after r^{-1} steps, one has $\mathbb{E}(Y_r - n/g) \approx e^{-1}(Y_0 - n/g)$;
4. if $m = n$, meaning that all the coins are taken and then redistributed, the new state has no memory of the previous one and statistical equilibrium is reached in a single step $r^{-1} = 1$.

Before concluding this section, it is interesting to discuss the case $\theta < 0$ in detail. In this case the marginal equilibrium distribution becomes the hypergeometric one as discussed in Section 5.1.2:

$$\mathbb{P}(Y = k) = \frac{\binom{|\alpha|}{k}\binom{|\theta - \alpha|}{n - k}}{\binom{|\theta|}{n}}, \tag{8.44}$$

with $\alpha = \theta/g$ and θ negative integers. The range of k is $(0,1,\ldots,\min(|\alpha|,n))$. If, for instance, $|\alpha| = 10n/g$, one has that $\theta = 10n$ and $r = 10m/(10n - n + m) \simeq (10m)/(9n)$. If $m \ll n$, this is not so far from the independent redistribution case. On the contrary, in the extreme case $|\alpha| = n/g$, then one finds that $r = m/(n-n+m) = 1$ and the occupation vector $\mathbf{n} = (n/g,\ldots,n/g)$ is obtained with probability 1. If an initial state containing individuals richer than $|\alpha|$ is considered, that is if one considers Equation (8.43) for $k > |\alpha|$, then $\mathbb{E}(D_{t+1}|Y_t = k)$ is still mk/n but $E(C_{t+1}|Y_t = k, D_{t+1} = d) = 0$ unless $k - d < |\alpha|$. More precisely, one has

$$\mathbb{E}(C_{t+1}|Y_t = k) = \begin{cases} m\dfrac{|\alpha| - k + m\dfrac{k}{n}}{|\theta| - n + m}, \text{ if } k - m\dfrac{k}{n} \leq |\alpha| \\ 0, \text{ if } k - m\dfrac{k}{n} > |\alpha|. \end{cases} \tag{8.45}$$

If the percent taxation is $f = m/n$, then one gets

$$\begin{aligned} &\mathbb{E}(Y_{t+1} - Y_t|Y_t = k) \\ &= \begin{cases} -\dfrac{f\theta}{\theta - n(1-f)}\left(k_t - \dfrac{n}{g}\right), \text{ if } k(1-f) \leq |\alpha| \\ -k(1-f), \text{ if } k(1-f) > |\alpha|. \end{cases} \end{aligned} \tag{8.46}$$

As $k(1-f)$ is the average value of Y after taxation, even if the agent is initially richer than $|\alpha|$, he/she can participate in redistribution if the percentage of taxation is high enough.

8.3 The Aoki–Yoshikawa model for sectoral productivity

In their recent book *Reconstructing Macroeconomics*, Masanao Aoki and Hiroshi Yoshikawa presented a stylized model for the economy of a country, where there are g economic sectors and the ith sector is characterized by the amount of production factor n_i, and by the level of productivity a_i. In the following, this model will be called the *Aoki–Yoshikawa Model* or AYM. For the sake of simplicity, n_i can be interpreted as the number of workers active in sector i, meaning that labour is the only production factor. In the AYM, the total endowment of production factor in the economy is exogenously given and set equal to n, so that:

$$\sum_{i=1}^{g} n_i = n. \tag{8.47}$$

The output of sector i is given by

$$Y_i = a_i n_i, \tag{8.48}$$

where, as above, a_i is the productivity of sector i. In general, productivities differ across sectors and can be ordered from the smallest to the largest:

$$a_1 < a_2 < \ldots < a_g. \tag{8.49}$$

The total output of the economy is

$$Y = \sum_{i=1}^{g} Y_i = \sum_{i=1}^{g} a_i n_i; \tag{8.50}$$

this quantity can be interpreted as the *Gross Domestic Product* or *GDP*. In the AYM, the GDP, Y, is assumed to be equal to an exogenously given aggregate demand D:

$$Y = D, \tag{8.51}$$

so that a second constraint is necessary

$$\sum_{i=1}^{g} a_i n_i = D. \tag{8.52}$$

Aoki and Yoshikawa are interested in finding the probability distribution of production factors (workers) across sectors. This is nothing other than the distribution of the occupation vector

$$\mathbf{n} = (n_1, n_2, \ldots, n_g), \tag{8.53}$$

when statistical equilibrium is reached.

The problem of the AYM coincides with a well-known problem in statistical physics already discussed in Chapter 7: the problem of determining the allocation of n particles into g energy levels ε_i so that the number of particles is conserved

$$\sum_i n_i = n, \tag{8.54}$$

and the total energy E is conserved

$$\sum_i \varepsilon_i n_i = E, \tag{8.55}$$

in conditions of statistical equilibrium. Even if this analogy is merely formal, it is very useful and one can immediately see that the levels of productivity a_i correspond to energy levels, whereas the demand D has a role analogous to the total energy E.

As mentioned before, after a first attempt in 1868, Ludwig Boltzmann solved this problem in 1877 using the *most probable occupation vector*, an approximate method also discussed in Section 5.3.8. One can introduce the individual description vectors $\mathbf{X}^{(n)} = (X_1 = x_1, \ldots, X_n = x_n)$, with $x_i \in \{1, \ldots, g\}$, where $x_i = j$ means that the ith worker is active in sector j; then, as discussed in Section 4.2.2, the number of distinct configurations belonging to a given occupation vector is $W(\mathbf{X} = \mathbf{x}|\mathbf{Y} = \mathbf{n})$:

$$W(\mathbf{X} = \mathbf{x}|\mathbf{Y} = \mathbf{n}) = \frac{n!}{\prod_{i=1}^{g} n_i!}. \tag{8.56}$$

When statistical equilibrium is reached, Boltzmann remarked that the probability $\pi(\mathbf{n})$ of an accessible occupation state is proportional to $W(\mathbf{X} = \mathbf{x}|\mathbf{Y} = \mathbf{n})$, in other words, this leads to

$$\pi(\mathbf{n}) = KW(\mathbf{x}|\mathbf{n}) = K\frac{n!}{\prod_{i=1}^{g} n_i!}, \tag{8.57}$$

where K is a suitable normalization constant. Occupation vectors maximizing $\pi(\mathbf{n})$ must minimize $\prod_{i=1}^{g} n_i!$ subject to the two constraints (8.54) and (8.55). For large systems, Stirling's approximation can be used for the factorial:

$$\log\left[\prod_{i=1}^{g} n_i!\right] \simeq \sum_{i=1}^{g} n_i(\log n_i - 1), \tag{8.58}$$

and the problem can be solved using Lagrange multipliers and finding the extremum of

$$L(\mathbf{n}) = -\sum_{i=1}^{g} n_i(\log n_i - 1) + \nu\left(\sum_{i=1}^{g} n_i - n\right) - \beta\left(\sum_{i=1}^{g} a_i n_i - D\right), \tag{8.59}$$

with respect to n_i. This leads to

$$0 = \frac{\partial L}{\partial n_i} = -\log n_i + \nu - \beta a_i, \tag{8.60}$$

or, equivalently, to

$$n_i^* = \mathrm{e}^{\nu}\mathrm{e}^{-\beta a_i}, \tag{8.61}$$

where, in principle, ν and β can be obtained from the constraints in Equations (8.47) and (8.52). An approximate evaluation of n_i^* is possible if $a_i = ia$ with $i = 1, \ldots, g$. This case is explicitly studied by Aoki and Yoshikawa. If $g \gg 1$, the sums in (8.47) and (8.52) can be replaced by infinite sums of the geometric series. In this case ν and β can be derived and replaced in (8.61) and one gets the most probable vector in terms of known parameters:

$$n_i^* = \frac{n}{r-1}\left(\frac{r-1}{r}\right)^i, \; i = 1, 2 \ldots \tag{8.62}$$

where $r = D/na$ is the aggregate demand per agent divided by the smallest productivity. In the limit $r \gg 1$, one gets

$$n_i^* \simeq \frac{n}{r}\mathrm{e}^{-i/r}. \tag{8.63}$$

Equation (8.63) gives the occupation vectors maximizing the probability given in Equation (8.57); these occupation vectors represent events and not random variables. However, if the economy is in the state $\mathbf{n}^*$, and if a worker is selected at random, the probability of finding her/him in sector i is given by

$$\mathbb{P}(Y = i|\mathbf{n}^*) = \frac{n_i^*}{n} \simeq \frac{na}{D}\exp\left(-\frac{na}{D}i\right). \tag{8.64}$$

Therefore, the marginal probability that a worker is in sector i, given the occupation vector $\mathbf{n}^*$, follows the exponential distribution.

All the results discussed above depend on the basic hypothesis for which Equation (8.57) holds true. This hypothesis is the equiprobability of all the individual descriptions $\mathbf{X} = \mathbf{x}$ compatible with the constraints (8.47) and (8.52). This is the typical assumption in classical statistical mechanics, where the uniform distribution on individual descriptions (i.e. on the accessible portion of phase space) is the only one compatible with the underlying deterministic dynamics. For Boltzmann, this assumption was not enough, and the dynamical part of his work (Boltzmann's equation, as well as the related H-Theorem) was introduced in order to prove that the most probable $\mathbf{n}^*$ summarizing the equilibrium distribution is actually reached as a consequence of atomic/molecular collisions. Indeed, the equilibrium distribution (if it exists) depends on the details of the dynamics according to which workers change sector. In physics, Brillouin's ideas and a generalized Ehrenfest urn model vindicate Boltzmann's attempt, which can also encompass quantum statistics. These results were discussed in Chapter 7. The relationship of the Ehrenfest–Brillouin model to the AYM is the subject of the next subsection.

8.3.1 Markovian dynamics

As discussed in Chapter 7, if $\mathbf{n}$ denotes the present state of the system:

$$\mathbf{n} = (n_1, \ldots, n_g), \tag{8.65}$$

where n_i is the number of workers in the ith sector with productivity a_i, then a unary move means that either n_i increases by one (creation) or n_i decreases by one (destruction or annihilation). One can write

$$\mathbf{n}^j = (n_1, \ldots, n_j + 1, \ldots, n_g) \tag{8.66}$$

for the creation of one unit and

$$\mathbf{n}_j = (n_1, \ldots, n_j - 1, \ldots, n_g) \tag{8.67}$$

for the annihilation of one unit. A unary move is an annihilation step followed by a creation step. It conserves the total number of workers, but does not fulfil the demand constraint, except for the trivial case $j = i$. In order to fix the ideas, assume $i < j$ so that one can write

$$\mathbf{n}_i^j = (n_1, \ldots, n_i - 1, \ldots, n_j + 1, \ldots, n_g), \tag{8.68}$$

to denote a unary move. Under the hypothesis of different sector productivities, in order to conserve demand, one should introduce binary moves at least, consisting of a sequence of two annihilations and two creations combined in a way that leaves the total production level unaffected. In a generic binary move, a worker moves from sector i to join sector l and another worker leaves sector j to enter sector m. The state following a binary move can be denoted by $\mathbf{n}_{ij}^{lm}$, where $n_i' = n_i - 1$, $n_j' = n_j - 1$, $n_l' = n_l + 1$ and $n_m' = n_m + 1$. After the move, the difference in the total product Y is $a_m + a_l - a_i - a_j$. When sector productivities are all different and incommensurable, this difference vanishes only if the two workers come back to their sectors ($l = i$ and $m = j$) or if they mutually exchange their sectors ($l = j$ and $m = i$). Indeed one has to take into account that $a_i \in \mathbb{R}$, $\forall i$ and that $n_i \in \mathbb{N}$, $\forall i$. In both cases, binary moves do not change the total number of workers per sector and any initial distribution of workers is conserved. The same applies to moves where r workers leave their sectors to join other sectors. If all the sectors have different real productivities, in order to fulfil the demand bound, workers have to rearrange so that the n_is do not vary.

In order to avoid this boring situation, one can assume that $a_i = ia$ where, as before, $i \in \{1, \ldots, g\}$. This means that all the productivities are multiples of the lowest productivity $a_1 = a$. In this case, binary transitions can conserve demand, but only a subset of occupation vectors can be reached from a given initial state fulfilling the demand constraint (8.52). As an example, consider a case in which there are three sectors with respective productivities $a_1 = a$, $a_2 = 2a$ and $a_3 = 3a$ and $n = 3$ workers. The initial demand is set equal to $D = 6a$. For instance, this situation is satisfied by an initial state in which all the three workers are in sector 2. Therefore, imagine that the initial occupation vector is $\mathbf{n} = (0, 3, 0)$. An allowed binary move leads to state $\mathbf{n}_{22}^{13} = (1, 1, 1)$ where two workers leave sector 2 to join sectors 1 and 3, respectively. This state also fulfils the demand bound as $a_1 n_1 + a_2 n_2 + a_3 n_3 = 6a$.

After defining binary moves and proper constraints on accessible states, it is possible to define a stochastic dynamics on the AYM using an appropriate transition probability. Following Chapter 7, a possible and suitable choice, for $i \neq j \neq l \neq m$, is:

$$\mathbb{P}(\mathbf{Y}_{t+1} = \mathbf{n}_{ij}^{lm} | \mathbf{Y}_t = \mathbf{n}) = A_{ij}^{lm}(\mathbf{n}) n_i n_j (1 + c n_l)(1 + c n_m) \delta(a_m + a_l - a_i - a_j), \tag{8.69}$$

where $A_{ij}^{lm}(\mathbf{n})$ is the normalization factor and c is a model parameter, whose meaning will be explained in the following. This equation can be justified by considering a binary move as a sequence of two annihilations and two creations. Forget the demand constraint for a while. When a worker leaves sector i, he/she does so with probability

$$\mathbb{P}(\mathbf{n}_i | \mathbf{n}) = \frac{n_i}{n}, \tag{8.70}$$

proportional to the number of workers in sector i before the move. In other words, the worker is selected at random among all the workers. When he/she decides to join sector l, this happens with probability

$$\mathbb{P}(\mathbf{n}^l | \mathbf{n}) = \frac{1 + c n_l}{g + c n}. \tag{8.71}$$

Note that the probability of any creation or destruction is a function of the actual occupation number, that is the occupation number seen by the moving agent. Therefore, in general, the worker will not choose the arrival sector independently from its occupation before the move, but he/she will be likely to join more populated sectors if $c > 0$ or he/she will prefer to stay away from full sectors if $c < 0$. Finally, he/she will be equally likely to join any sector if $c = 0$. Further notice that, if $c \geq 0$, there is no restriction in the number of workers who can occupy a sector, whereas for negative values of c, only situations in which $1/|c|$ is integer and no more than $1/|c|$ workers can be allocated in each sector are of interest.

This model is nothing other than the Ehrenfest–Brillouin model with constrained binary moves described in Section 7.7 with no degeneracy of the productivity levels, that is $g_i = 1$ for all the values of i. As discussed in Section 7.7, the general invariant and equilibrium distribution for the model is

$$\pi(\mathbf{n}) \propto \prod_{i=1}^{g} \frac{(1/c)^{[n_i]}}{n_i!}. \tag{8.72}$$

Equation (8.72) defines a generalized Pólya distribution (with positive, negative or zero prize) whose domain is just 'all the states $\mathbf{n}$ compatible with the constraints' or, equivalently, 'all the states $\mathbf{n}$ reachable from the initial state $\mathbf{n}_0$ by means of the transition probability (8.69)'. In Chapter 7, the reader has seen that the values $c = 0, \pm 1$ are the only ones appearing in the applications of (8.69) to physics.

Outside physics, there is no reason to be limited by these three possibilities. In the case $c = 0$, Equation (8.72) becomes

$$\pi(\mathbf{n}) \propto \frac{1}{\prod_{i=1}^{g} n_i!}, \quad (8.73)$$

meaning that workers' configurations are uniformly distributed. As mentioned above, this is the only case considered in the book by Aoki and Yoshikawa. Further note that for all the other values of c no equilibrium probability distribution is uniform either for sector occupations (this is indeed the case only for $c = \pm 1$) or for individual descriptions (as in the case $c = 0$).

Finally, applying the results of Chapter 7 to the present case, one can also show that the general solution of the conditional maximum problem for $\pi(\mathbf{n})$ is:

$$n_i^* = \frac{1}{\mathrm{e}^{-\nu}\mathrm{e}^{\beta a_i} - c}, \quad (8.74)$$

which coincides with (8.61) in the case $c = 0$. In the presence of degeneracy in the levels of productivity, Equation 8.74 would become equal to Equation 7.125. Then, according to the different laws for g_i as a function of i, one would have an even greater richness of distributional forms. The case with no degeneracy in productivity levels and $a_i = ia$ is studied in Exercise 8.3 by means of a Monte Carlo simulation.

8.4 General remarks on statistical equilibrium in economics

In all the simple models presented in this chapter, as well as in some of the examples discussed in Chapter 7, it is assumed that the distribution to be compared with empirical data is an equilibrium distribution in the sense defined in Chapter 6.

In economics, distributional problems appear at least in the following frameworks:

1. economic growth (and its dependence on firms' sizes);
2. allocation of resources (in particular the distribution of wealth).

Our presentation emphasizes the positive (rather than the normative) aspect of distributional problems. However, there is a general policy problem worth mentioning which was highlighted by Federico Caffè in his lectures [1]. When introducing the 1938 paper by A. Bergson [2], Caffè wrote

> ... when, in economic reasoning, the social wealth distribution is assumed 'given', this means that the existing distribution is accepted, without evaluating whether it is good or bad, acceptable or unacceptable ... this must be explicitly done further clarifying that the conclusions are conditioned on the acceptability of the distributional set-up.

Two main probabilistic methods were used to derive/justify observed empirical distributions:

1. the statistical equilibrium method (discussed in this book). According to this approach, the time evolution of an economic system is represented by an aperiodic, irreducible Markov chain and the distribution of relevant quantities is given by the invariant distribution of the Markov chain.
2. The diffusive (possibly non-equilibrium) method. According to this approach, the time evolution of an economic system is represented by a random walk (see also Sections 9.8 and 10.8).

The random walk method can be described in a nutshell as follows. Let $x(t) = \log(s(t))$, where $s(t)$ is a 'size' (wealth, firm size and so on), then one can write

$$x(t) = x(0) + \sum_{m=0}^{t-1} \xi(m), \tag{8.75}$$

where $\xi(m)$ are independent and identically distributed random variables with probability density with finite expected value $\mathbb{E}(\xi)$ and variance $\mathbb{V}(\xi)$. For large t, one has that $p(x,t)$ approaches the normal distribution

$$p(x,t) \simeq N(x; t\mathbb{E}(\xi), t\mathbb{V}(\xi)); \tag{8.76}$$

for fixed t, x is normal and s is log-normal, but $\lim_{t\to\infty} p(x,t)$ does not exist!

The statistical equilibrium method has been challenged from at least two points of view. Some scholars do not believe that economic systems can achieve any sort of equilibrium, including statistical equilibrium and, on the contrary, they claim that such systems are strongly out-of-equilibrium. Other scholars and, in particular, post-Keynesian economists, do not believe that the behaviour of macroeconomic systems can be explained in terms of the behaviour of interacting individuals. In other words, they challenge the micro-foundation of macroeconomics, possibly including a probabilistic microfoundation in terms of statistical equilibrium. Our provisional reply to these objections is that the concept of statistical equilibrium may prove useful even in economics and may be helpful to describe/justify empirical distributional properties. In other words, before rejecting the usefulness of this concept, it is worth studying its implications and understand what happens even in simple models. As argued by John Angle,[1] it may well be the case that the economic dynamics is fast enough to let relevant variables reach statistical equilibrium even in the presence of shocks moving the economic system out of this equilibrium.

Another common objection is that in economics, at odds with physics, there is no conservation of wealth or of the number of workers or of any other relevant

[1] Personal communication.

quantity. This might be true, but the example discussed in Section 7.6 shows how to study a market where the number of participants is not fixed, but yet statistical equilibrium is reached. In other words, the absence of conserved quantities in itself is not a major obstacle.

The last objection we consider does not deny the usefulness of statistical equilibrium, but of a probabilistic dynamical description. When considering large macroeconomic aggregates or a long time evolution, fluctuations may become irrelevant and only the deterministic dynamics of expected values is important. In other words, stochastic processes may be replaced by difference equations or even by differential equations for empirical averages. To answer this objection, in many parts of this book we have directed the attention of the reader to the phenomenon of lack of self-averaging, which is often there in the presence of correlations (see Section 5.2.1). In other words, when correlations are there, it is not always possible to neglect fluctuations and a description of economic systems in terms of random variables and stochastic processes becomes necessary.

8.5 Exercises

8.1 Consider a homogeneous Markov chain whose space state is given by occupation vectors of n objects in g categories with the following transition matrix

$$\mathbb{P}(\mathbf{n}_i^j|\mathbf{n}) = \frac{1-\delta_{n_i,0}}{g-z_0(\mathbf{n})}\frac{1}{g}; \tag{8.77}$$

this means that an object is selected at random from any category containing at least one object and randomly moves to any one of the other categories. Explicitly consider the case $n = g = 3$; write and diagonalize the transition matrix in order to find the stationary (or invariant) distribution.

8.2 Write a Monte Carlo simulation implementing the chain described by the transition probability (8.77) in the case $n = g = 3$.

8.3 Write a Monte Carlo program for the Aoki–Yoshikawa model described in Section 8.3 and based on the Ehrenfest–Brillouin model with constrained binary moves (see also Section 7.4).

8.6 Summary

In this chapter, three applications were presented for the methods introduced in previous chapters. There are always n objects to be allocated into g categories, but the objects change category according to a probabilistic dynamics. In all three cases, it is natural to study occupation vectors from the beginning of the analysis.

The first model is an example of random distribution of wealth. There are g agents with n coins. Two of the agents are selected at random, one is the loser and the other is the winner. If the loser has at least one coin, he/she gives a coin to the winner. Note that the loser and the winner may coincide. Otherwise, nothing happens: indebtedness is not possible. This is an aperiodic and irreducible Markov chain and the equilibrium wealth distribution can be obtained by means of the exact marginalization method.

In the second model, again, n coins are to be divided into g agents. A coin is randomly selected among all the coins and it is redistributed according to a probability depending on the number of coins in the pocket of each agent and on an a-priori weight. Also in this case, the model is an aperiodic and irreducible Markov chain, and the marginalization method leads to the equilibrium wealth distribution. This model can be interpreted as a stylized version of a taxation–redistribution policy.

The third model takes into account n workers active in g economic sectors. These sectors are characterized by different levels of productivity for input labour. A randomly selected worker moves from one sector to another based on a probability depending on the number of workers and an a-priori weight. This model is the Ehrenfest–Brillouin model of Chapter 7 reinterpreted in terms of macroeconomic variables. However, in macroeconomics there is no reason to limit oneself to the three cases sensible in physics (Bose–Einstein, Fermi–Dirac and Maxwell–Boltzmann).

Finally, it is interesting to briefly discuss the possibility of empirically validating or falsifying this class of models. With reference to the sectoral productivity model, in principle, both the empirical transition probabilities between sectors and the distribution of workers across sectors can be measured and compared with Equations (8.69) and (8.72), respectively. However, a priori, one can notice that these models have some unrealistic features. For instance, leaving one productive sector for another sector may take time because it is necessary to learn a new job (even if this is not always the case). Moreover, the version of the model presented in Section 8.3 does not explicitly take into account unemployment. Indeed, unemployment could be added by including a class with zero productivity. In any case, it is interesting to study what happens in stylized models leading to analytic solutions before considering more realistic models where only Monte Carlo simulations are possible.

Further reading

The random model for distribution of wealth was named the BDY game:

E. Scalas, U. Garibaldi and S. Donadio, *Statistical Equilibrium in Simple Exchange Games I: Methods of Solution and Application to the Bennati-Drăgulescu-Yakovenko (BDY) Game*, The European Physical Journal B, **53**, 267–272 (2006).

E. Scalas, U. Garibaldi and S. Donadio, *Erratum: Statistical Equilibrium in Simple Exchange Games I*, The European Physical Journal B, **60**, 271–272 (2007).

by the authors of the present book from the initials of Bennati,

E. Bennati, *Un metodo di simulazione statistica nell'analisi della distribuzione del reddito*, Rivista Internazionale di Scienze Economiche e Commerciali, **8**, 735–756 (1988).

E. Bennati, *La simulazione statistica nell'analisi della distribuzione del reddito. Modelli realistici e metodo di Montecarlo*, ETS Editrice, Pisa (1988).

Drăgulescu and Yakovenko.

A. Drăgulescu and V.M. Yakovenko, *Statistical Mechanics of Money*, The European Physical Journal B, **17**, 723–729 (2000).

These authors discussed versions of this model. Previous work on the BDY model was performed by means of Monte Carlo simulations in the limit $n \gg g \gg 1$ leading to the exponential distribution with probability density function given by Equation (8.27), sometimes incorrectly called the 'Gibbs–Boltzmann' distribution.

The taxation–redistribution model was introduced in the second paper of a series devoted to statistical equilibrium in simple exchange games

U. Garibaldi, E. Scalas and P. Viarengo, *Statistical Equilibrium in Simple Exchange Games II: The Redistribution Game*, The European Physical Journal B, **60**, 241–246 (2007).

The model on sectoral productivity of Aoki and Yoshikawa was discussed in detail in their book *Reconstructing Macroeconomics*:

M. Aoki and H. Yoshikawa, *Reconstructing Macroeconomics. A Perspective from Statistical Physics and Combinatorial Stochastic Processes*, Cambridge University Press, Cambridge UK (2007).

and its dynamic version was the subject of a paper written by the authors of this book:

E. Scalas and U. Garibaldi, *A Dynamic Probabilistic Version of the Aoki–Yoshikawa Sectoral Productivity Model*, Economics: The Open-Access, Open-Assessment E-Journal, **3**, 2009-15 (2009).

`http://www.economics-ejournal.org/economics/journalarticles/2009-15.`

Distributional properties of (macro)economic variables are discussed in the books by Aitchison and Brown,

J. Aitchison and J.A.C. Brown, *The Lognormal Distribution*, Cambridge University Press, Cambridge, UK (1957).

Steindl

J. Steindl, *Random Processes and the Growth of Firms - A Study of the Pareto Law*, Charles Griffin and Company, London (1965).

and Champernowne and Cowell,

D.G. Champernowne and F.A. Cowell, *Economic Inequality and Income Distribution*, Cambridge University Press, Cambridge, UK (1999).

as well as in the paper by Angle

J. Angle, *The Surplus Theory of Social Stratification and the Size Distribution of Personal Wealth*, Social Forces, **65**, 293–326 (1986).

Duncan K. Foley has discussed statistical equilibrium in economics in his papers. The reader is referred to:

D.K. Foley, *A Statistical Equilibrium Theory of Markets*, Journal of Economic Theory, **62**, 321–345 (1994).

References

[1] F. Caffè, *Lezioni di politica economica*, Bollati-Boringhieri (1978).
[2] A. Bergson, *A Reformulation of Certain Aspects of Welfare Economics*, Quarterly Journal of Economics, **52**, 310–334 (1938).

9

Finitary characterization of the Ewens sampling formula

This chapter is devoted to a finitary characterization of the so-called Ewens sampling formula. This formula emerges as a limiting case for the partition distribution coming out of the Pólya sampling distribution studied in Chapter 5 when the number of categories diverges ($g \to \infty$), but their weights vanish $\alpha_i \to 0$ leaving $\theta = \lim_{g\to\infty} \sum_{i=1}^{g} \alpha_i$ finite.

After reading this chapter you should be able to:

- understand the example of the Chinese restaurant;
- use an appropriate language to describe grouping of elements into clusters;
- derive the Ewens sampling formula using the methods described in Chapters 5 and 6, and in Section 5.3.8;
- interpret cluster dynamics in terms of economic variables.

9.1 Infinite number of categories

Up to now, the case of n objects to be allocated into d categories has been discussed. It must be stressed that an occupation vector $\mathbf{n} = (n_1, \ldots, n_d)$ is a partition of the n elements into k *clusters*, where k is the number of non-empty categories. These clusters are identified by the names of the corresponding categories. But, what happens if the number of categories is infinite? And what happens if the names of the categories are not known in advance? The situation with an infinite number of categories can naturally take 'innovations' into account. Consider for instance the case in which the n elements are agents who work in k firms. The occupation vector $\mathbf{n} = (n_1, \ldots, n_k)$ with $\sum_{i=1}^{k} n_i = n$ represents the distribution of agents into the k firms. Assume that, at a certain point, an agent decides to leave his/her firm (say the firm labelled by j) and create a new one (a one-person firm). The new situation is represented by the vector $\mathbf{n}' = (n_1, \ldots, n_j - 1, \ldots, n_k, n_{k+1} = 1)$. Now, there are $k+1$ firms (occupied categories) with a new distribution of agents among them. How can this new firm be named, before its birth? In other words, what does

$k+1$ mean in the symbol n_{k+1}? Another case might be when a one-person firm disappears as the worker leaves it and joins an existing firm.

It is essential to introduce the size distribution of these categories. If n is finite and $d \to \infty$, the partition vector is $\mathbf{z} = (z_0, z_1, \ldots, z_n)$, where $z_0 = d - k$ diverges and z_i is the number of clusters whose size is i. Now, one has $\sum_{i=1}^{n} z_i = k$ and $\sum_{i=1}^{n} iz_i = n$ with k not fixed and varying from 1 (all the elements together in the same cluster) to n (all the elements are singletons, that is all the elements belong to a different cluster). It turns out that the methods developed in the previous chapters can be used to quantitatively discuss these problems and to derive useful distributions that can be compared with empirical data.

9.1.1 Example: The Chinese restaurant process

In order to visualize the situation described above, it is useful to refer to the so-called *Chinese restaurant process*. There is a Chinese restaurant; usually, it is described as having an infinite number of tables each one with infinite capacity. A customer comes in and decides where to sit, all the tables being equivalent. This table is labelled 1. Then, a second customer arrives and decides where to sit. The second customer can sit at the same table as the previous one or choose a new table. In the latter case the new table is labelled 2. After the accommodation of the second customer, the restaurant may have either $k = 2$ tables occupied by 1 person ($a_1 = 1, a_2 = 1$) or $k = 1$ table occupied by 2 persons ($a_1 = 2$). Here, a_i denotes the occupation number of the ith table. In the former case, there are 2 clusters of size 1 ($z_1 = 2$), in the latter case, there is 1 cluster of size 2 ($z_1 = 0, z_2 = 1$). Again, when a third customer comes in, he/she can join a table occupied by the previous two customers or a new table (table 3). After the third customer sits down the possibilities are as follows:

1. all the three customers sit at the same table, then, $k = 1$, $a_1 = 3$ and $z_1 = 0, z_2 = 0, z_3 = 1$;
2. two customers sit at the same table and another one occupies a different table; then $k = 2$ and one has either $a_1 = 2, a_2 = 1$ or $a_1 = 1, a_2 = 2$ and $z_1 = 1, z_2 = 1$;
3. the three customers sit at separate tables; in this case $k = 3$ and $a_1 = 1, a_2 = 1, a_3 = 1$ with $z_1 = 3$.

The process is then iterated for the desired number of times. For instance, the following configuration is compatible with the allocation of $n = 10$ agents in the Chinese restaurant process: $k = 4$ tables are occupied, $a_1 = 3, a_2 = 5, a_3 = 1, a_4 = 1$ and $z_1 = 2, z_2 = 0, z_3 = 1, z_4 = 0, z_5 = 1, z_6 = 0, z_7 = 0, z_8 = 0, z_9 = 0, z_{10} = 0$. Note that in the scheme proposed above, tables are numbered (labelled) according to the order of occupation (*time-ordered description*) and there is no limit to the accommodation process.

If the restaurant owner knows that no more than n customers are going to arrive, there is an alternative way of describing the accommodation process, compatible with the fact that all empty tables are on a par for incoming customers. The owner can choose g tables and label them from 1 to g with $g \geq n$. In this case, the accommodation process can be described as usual. The customer decides either to join an occupied table, or to sit at a new table chosen equiprobably among the empty ones. Alternatively, tables are not labelled in advance. When a customer occupies a new table, she/he draws a label (without replacement) from 1 to g and tags the table at which she/he sits. After a little thought, the reader should recognize that these alternatives, respectively called *site process* and *label process*, are equivalent for the description of clustering. For instance, if $n = 3$ and $g = 4$, a possible situation is that two customers sit at the same table and another one occupies a different table, meaning that $z_1 = 1$ and $z_2 = 1$. Considering the sequences of accommodation, there are two possible time-ordered occupation vectors, denoted by $\mathbf{a} = (a_1 = 2, a_2 = 1)$ corresponding to the fact that the first table was occupied by two customers, and $\mathbf{a} = (a_1 = 1, a_2 = 2)$ meaning that the second table was occupied twice. The vector $\mathbf{a} = (a_1, \ldots, a_k)$ with $a_i > 0$ has as many components as the number k of distinct clusters, and $\sum_{i=1}^{k} a_i = n$. Note that k is not a fixed number, but it varies. Considering the site description, a possible occupation vector is $\mathbf{n} = (n_1 = 2, n_2 = 1, n_3 = 0, n_4 = 0)$ and every permutation of the site names is allowed for a total of 6 vectors. The vector $\mathbf{n} = (n_1, \ldots, n_g)$ has as many components as the number of sites, and $\sum_{i=1}^{g} n_i = n$. Note that the set of non-zero components of $\mathbf{n}$ coincides with the set of components of the vector $\mathbf{a}$, so that $\prod_{i=1}^{k} a_k! = \prod_{i=1}^{g} n_i!$. The site occupation vector behaves as a usual occupation vector of elements on categories; remember that, from now on, g represents the number of sites (or labels) introduced in order to describe the accommodation into clusters and not the number of nominal categories (called d in this chapter), which diverges.

9.1.2 Infinite number of categories and the Ehrenfest–Brillouin process

When the number of nominal categories is $d \gg n$ for fixed n, the frequency description $\mathbf{n} = (n_1, \ldots, n_d)$ has almost all terms equal to zero, and some caution is necessary.

In order to understand what happens, it is useful to consider the case in which d is large but finite, and assume that there are $k \leq n$ distinct clusters (categories initially occupied). Relabel them, and set $\mathbf{n} = (n_1, \ldots, n_k, n_{k+1})$, where the $(k+1)$th category collects all the $g - k$ categories empty at present. One can see that $n_1, \ldots, n_k$ are positive, whereas $n_{k+1} = 0$. In this setting, the transition probability for the symmetric Ehrenfest–Brillouin model (7.7) becomes

$$\mathbb{P}(\mathbf{n}_i^j|\mathbf{n}) = \begin{cases} \dfrac{n_i}{n} \dfrac{\alpha + n_j}{\theta + n - 1}, & \text{for } j \leq k \\ \dfrac{n_i}{n} \dfrac{\theta - k\alpha}{\theta + n - 1}, & \text{for } j = k+1 \end{cases} \tag{9.1}$$

where $\theta = d\alpha$, and $\theta - k\alpha$ is the total weight of the $d-k$ categories empty at present. If the weight α of each category is finite, for large d both θ and $\theta - k\alpha$ diverge, so that $\mathbb{P}(\mathbf{n}_i^j|\mathbf{n}) \to 0$ for $j \leq k$, and $P(\mathbf{n}_i^j|\mathbf{n}) \to 1$ for $j = k+1$. In other words, elements jump to a new category with probability one. The a-priori weights dominate on the empirical observations, so that no correlations are present. Pioneer behaviour dominates, new categories are always chosen and one expects to observe n singletons, that is n clusters occupied by a single object, that is $\mathbf{z} = (z_1 = n, z_2 = 0, \ldots, z_n = 0)$ with probability one.

An alternative and, now, significant situation is obtained if weight distribution is uniform as before, that is $\alpha_i = \theta/d$, for every i, and the total weight θ remains constant, that is $\lim_{d\to\infty}\sum_i^d \alpha_i = \theta < \infty$. Now, the initial weight θ and the number of elements are both finite, and interesting results are expected. In this limit $\lim_{d\to\infty}\alpha_i = \lim_{d\to\infty}\theta/d = 0$, and one can study the stochastic process characterized by the following transition probability

$$\mathbb{P}(\mathbf{n}_i^j|\mathbf{n}) = \begin{cases} \dfrac{n_i}{n}\dfrac{n_j}{\theta + n - 1}, & \text{for } j \leq k \\ \dfrac{n_i}{n}\dfrac{\theta}{\theta + n - 1}, & \text{for } j = k+1; \end{cases} \tag{9.2}$$

the process defined by Equation (9.2) is discussed below. Note that, due to the absence of the constant in the creation term, all the k non-empty categories will empty sooner or later, or better they will reach the equilibrium value 0 with a rate $\theta/(n(\theta + n - 1))$. All the mass will be transferred to new categories and the dead ones will never appear again. In other words, there is an incessant migration of objects into new categories and there is no equilibrium distribution. In order to define an irreducible and aperiodic Markov chain, it is necessary to look for new variables.

9.2 Finitary derivation of the Ewens sampling formula

9.2.1 Hoppe's urn and the auxiliary urn process for the label process

In the accommodation process, consider n individual random variables $U_1, \ldots, U_n$, whose range is a set of allowed possibilities (categories, strategies, and so on). The cluster description of the n elements is based on the elementary fact that two elements belong to the same cluster if and only if $U_i = U_j$. Given that in the cluster description labels are nothing else but tools for distinguishing clusters, a sequence of n random variables $X_1, \ldots, X_n$ is introduced, where $X_1 = 1$ (the first observed category, by definition) whatever U_1; then, one has $X_2 = 1$ if $U_2 = U_1$, whereas $X_2 = 2$ (the second observed category) if $U_2 \neq U_1$, and so on. Therefore, the range of X is the first, the second, ..., the kth category appearing in the sequence. The statement 'the fourth agent belongs to the first cluster' means that $U_4 = U_1$, while 'the fourth agent belongs to the second cluster' means that either $U_4 = U_2 \neq U_1$, or $U_4 = U_3 \neq U_2 = U_1$, in which cases the agent enters an already existing cluster (herding behaviour), or $U_4 \neq U_3 = U_2 = U_1$, in which case the agent is founding the second cluster (pioneer behaviour). In other words, the label j denotes the jth label that has been introduced. To be more specific, the current occupation vector with respect to time labels has components $a_{j,m} = \#\{X_i = j, i = 1,\ldots,m\}$, and k_m is the number of present labels, with $\sum_{j=1}^{k_m} a_{j,m} = m$. The conditional predictive distribution of X_{m+1} is taken from Equation (9.2), leading to (for $m = 0,1,\ldots,n-1$):

$$\mathbb{P}(X_{m+1} = j | a_{j,m}, m) = \begin{cases} \dfrac{a_{j,m}}{m+\theta}, & j \leq k_m \\ \dfrac{\theta}{m+\theta}, & j = k_m + 1 \end{cases} \tag{9.3}$$

with $\mathbb{P}(X_1 = 1) = 1$ by definition. This sampling process can be modelled by an urn process, Hoppe's urn, that can be traced back to A. De Moivre, according to Zabell. Initially, the urn contains a single black ball (the *mutator*) whose weight is θ. The following rules define the drawing process:

1. Whenever the black ball is drawn, a new colour is registered, the black ball is replaced in the urn together with a ball (of weight 1) of the newly registered colour. 'New' means not yet present in the urn.
2. If a coloured ball is drawn, it is replaced in the urn, together with a ball (of weight 1) of the same colour, as in the Pólya scheme.

The probability of a sequence is obtained from Equation (9.3) and is given by (understanding the subscript n in $\mathbf{a}_n$ and in k_n):

$$\mathbb{P}(X_1 = 1, X_2 = x_2, \ldots, X_n = x_n) = \frac{\theta^k}{\theta^{[n]}} \prod_{i=1}^{k} (a_i - 1)!; \tag{9.4}$$

the proof of Equation (9.4) is left to the reader as a useful exercise. Given an occupation vector $\mathbf{a}_n = (a_1, \ldots, a_k)$, with $\sum_{i=1}^{k} a_i = n$, all the possible corresponding sequences have the same probability. For instance, if $n = 4$, the sequences $X_1 = 1, X_2 = 2, X_3 = 3, X_4 = 2$ and $X_1 = 1, X_2 = 2, X_3 = 2, X_4 = 3$ are equiprobable. However, the sequences are not fully exchangeable according to the definition given in Chapter 4. For instance, the probability of the sequence $X_1 = 2$, $X_2 = 1$, $X_3 = 3$, $X_4 = 2$ is 0 due to the definition of time-ordered labels. Therefore, one cannot use Equation (4.59) to determine the probability of an occupation vector. In other words, Equation (4.59) cannot be used if the occupation number is given in a time-ordered way. In particular, all the sequences where 1 is not the first label are forbidden, as well as all the sequences where 3 appears before 2, 4 before 3 and 2, and so on. Taking this into account, if there are a_1 elements belonging to category 1, only $a_1 - 1$ can be freely chosen out of $n-1$ elements. Then, only $a_2 - 1$ elements in category 2 are free to be chosen out of the $n - a_1 - 1$ remaining elements, and so on. These considerations lead to the following equation for the weight $W(\mathbf{X}^{(n)}|\mathbf{a})$:

$$W(\mathbf{X}^{(n)}|\mathbf{a}) = \binom{n-1}{a_1 - 1}\binom{n-a_1-1}{a_2-1}\cdots\binom{a_{k-1}+a_k-1}{a_{k-1}-1}, \tag{9.5}$$

leading in its turn to

$$\begin{aligned}\mathbb{P}(\mathbf{a}) &= W(\mathbf{X}^{(n)}|\mathbf{a}) \cdot \frac{\theta^k}{\theta^{[n]}}\prod_{i=1}^{k}(a_i - 1)! \\ &= \frac{n!}{a_k(a_k + a_{k-1})\ldots(a_k + a_{k-1} + \ldots + a_1)}\frac{\theta^k}{\theta^{[n]}},\end{aligned} \tag{9.6}$$

called by Donnelly 'the size-biased permutation of the Ewens sampling formula' (see the discussion on this name below).

Next, the site-label process informally introduced in Section 9.1.1 is studied. At odds with the time-label process, facing the same individual random variables $U_1,\ldots,U_n$, a new set of n random variables $X_1^*,\ldots,X_n^*$ is introduced, whose range is a set of $g > n$ labels $L = \{1,\ldots,g\}$. Now, when U_1 is observed, a label is randomly chosen from the set L (without replacement) and the element is labelled by $l_1 \in L$, that is $X_1^* = l_1$. As for X_2^*, if $U_2 = U_1$, it will be labelled by l_1 otherwise a second draw is made from L and a new label, say l_2, is assigned to X_2^*, and so on. Continuing with this procedure, all the coincident values of U_i have the same label, whereas different values correspond to different labels. Comparing the site-label process to Hoppe's urn scheme, one can see that while $X_1 = 1$ with certainty, $X_1^* = i$ with probability $1/g$; and when the second label appears in Hoppe's scheme, a new label will be drawn from L. The predictive probability for the label process is

$$\mathbb{P}(X_{m+1}^* = j|m_j, m) = \begin{cases} \dfrac{m_j}{m+\theta}, & \text{for } m_j > 0 \\ \dfrac{1}{g-k_m}\dfrac{\theta}{m+\theta}, & \text{for } m_j = 0 \end{cases} \tag{9.7}$$

where k_m is the number of labels already drawn; indeed, as before, new labels are selected when the black ball of weight θ is drawn, but then, independently, the name of the category is randomly selected among all the remaining labels. The probability of a sequence can be directly obtained from (9.7). However, it is easier to note that every sequence $X_1 = 1, X_2 = x_2, \ldots, X_n = x_n$ corresponds to $g!/(g-k)!$ label sequences (see also the discussion in Section 9.1.1). Moreover, a little thought should convince the reader that the site-label process is exchangeable in the usual sense, and Equation (4.59) holds true. Therefore, one finds

$$\mathbb{P}(X_1^* = x_1^*, \ldots, X_n^* = x_n^*) = \frac{\mathbb{P}(X_1 = 1, X_2 = x_2 \ldots, X_n = x_n)}{g(g-1)\ldots(g-k+1)}. \quad (9.8)$$

In other words, the non-fully exchangeable sequence $X_1 = 1, X_2 = x_2 \ldots, X_n = x_n$ is partitioned into $g!/(g-k)!$ exchangeable sequences $X_1^*, \ldots, X_n^*$. Using Equations (4.59), (9.4) and (9.8), the distribution of the occupation vectors defined on the g states of the label urn turns out to be

$$\mathbb{P}(\mathbf{Y}_n^* = \mathbf{n}) = \frac{n!}{\prod_{i=1}^{g} n_i!} \mathbb{P}(\mathbf{X}^{*(n)} = \mathbf{x}^{*(n)}) = \frac{(g-k)!}{g!} \frac{n!}{\prod_{i \in A} n_i} \frac{\theta^k}{\theta^{[n]}}, \quad (9.9)$$

where A is the set of k labels representing the existing clusters and in Equation (9.4) one can set

$$\prod_{i=1}^{k} (a_i - 1)! = \prod_{i \in A} (n_i - 1)!. \quad (9.10)$$

9.2.2 The Ewens sampling formula

Equation (9.9) can be written in a way that directly leads to the Ewens sampling formula. In Section 5.3.8, the fact that $\prod_{i=1}^{g} n_i! = \prod_{i=0}^{n} (i!)^{z_i}$ was discussed (see Equation (5.71)). Similarly,

$$\prod_{i \in A} n_i = \prod_{i=1}^{n} i^{z_i}, \quad (9.11)$$

where, as usual, z_i is the number of clusters with i elements. Replacing Equation (9.11) into Equation (9.9) leads to

$$\mathbb{P}(\mathbf{Y}_n^* = \mathbf{n}) = \frac{(g-k)!}{g!} \frac{n!}{\theta^{[n]}} \prod_{i=1}^{n} \left(\frac{\theta}{i}\right)^{z_i}. \quad (9.12)$$

Note that, in this framework, $z_0 = g - k$ gives the number of empty sites (inactive labels). Again, thanks to the exchangeability of occupation vectors with respect to labels, using Equation (5.70), from Equation (9.12) one immediately obtains the celebrated Ewens sampling formula

$$\mathbb{P}(\mathbf{Z}_n = \mathbf{z}_n) = \frac{g!}{(g-k)! z_1! \cdots z_n!} \mathbb{P}(\mathbf{Y}_n^* = \mathbf{n}) = \frac{n!}{\theta^{[n]}} \prod_{i=1}^{n} \left(\frac{\theta}{i}\right)^{z_i} \frac{1}{z_i!}. \tag{9.13}$$

As an aside, note that combining Equations (5.72) and (9.8) leads to the following formula in terms of Hoppe's sequences:

$$\mathbb{P}(\mathbf{Z}_n = \mathbf{z}_n) = \frac{n!}{\prod_{i \in A} n_i!} \frac{1}{\prod_{i=1}^{n} z_i!} \mathbb{P}(\mathbf{X}^{(n)} = \mathbf{x})^{(n)} = \frac{n!}{\prod_{i=1}^{n} (i!)^{z_i} z_i!} \mathbb{P}(\mathbf{X}^{(n)} = \mathbf{x}^n), \tag{9.14}$$

so that one can directly obtain the joint cluster size distribution from a knowledge of the individual time-label vector (Hoppe's sequences) probability distribution, and the multiplicity factor is $W(\mathbf{X}^{(n)}|\mathbf{z})$. Note that $W(\mathbf{X}^{*(n)}|\mathbf{z})$ is given by $W(\mathbf{X}^{(n)}|\mathbf{z}) \cdot g!/(g-k)!$. In a sense, g is an auxiliary parameter which does not enter Equation (9.13). In the accommodation process, its minimum allowed value is indeed $g = n$, whereas for the birth–death processes we have in mind one must choose at least $g = n + 1$ as will become clearer in the following. In this framework, the Ewens sampling formula appears as the probability distribution of all partition vectors for the accommodation described by the site-label process (9.12) or by the time-label process (9.14) by the predictive probability given by Equation (9.7). This formula was introduced by W. Ewens in population genetics. If a random sample of n gametes is taken from a population and classified according to the gene at a particular locus, the probability that there are z_1 alleles represented once in the sample, z_2 alleles represented twice, and so on is given by (9.13) under the following conditions:

1. the sample size n is small if compared to the size of the whole population;
2. the population is in statistical equilibrium under mutation and genetic drift and the role of selection at the locus under scrutiny can be neglected;
3. every mutant allele is an 'innovation'.

As mentioned by Tavaré and Ewens (see further reading), Equation (9.13) provides a sort of null hypothesis for a non-Darwinian theory of evolution (a theory where selection plays no role). The case $\theta \to 0$ corresponds to the situation where all the objects occupy the initial category (all the genes are copies of the same allele, in the genetic interpretation). This can be immediately seen from the equations for the predictive probability: when $\theta \to 0$ no innovation is possible and all the elements stick together. The opposite case is the limit $\theta \to \infty$. In this case all the objects are singletons, that is they occupy a new category. The case $\theta = 1$ is also remarkable as it corresponds to the distribution of integer partitions induced by uniformly distributed random permutations meaning that each permutation has a probability equal to $(n!)^{-1}$. In this case, Equation (9.13) becomes

$$\mathbb{P}(\mathbf{Z}_n = \mathbf{z}_n) = \prod_{i=1}^{n} \left(\frac{1}{i}\right)^{z_i} \frac{1}{z_i!}, \tag{9.15}$$

which is quite old: it can be traced back at least to Cauchy.

9.2.3 The Ewens sampling formula as limit of Pólya partitions

Consider the symmetric Pólya distribution where α denotes the common weight of each category. We are interested in the limit in which this weight vanishes, $\alpha \to 0$, and the number of categories diverges, $g \to \infty$, while the total weight $\theta = g\alpha$ remains constant. For small α the rising factorial $\alpha^{[i]}$ can be approximated as follows:

$$\alpha^{[i]} = \alpha(\alpha+1)\cdots(\alpha+i-1) \simeq \alpha(i-1)!; \tag{9.16}$$

therefore, one has that the symmetric Pólya sampling distribution can be approximated as follows:

$$\mathbb{P}(\mathbf{n}) = \frac{n!}{(g\alpha)^{[n]}} \prod_{j=1}^{g} \frac{\alpha^{[n_j]}}{n_j!} = \frac{n!}{(g\alpha)^{[n]}} \prod_{i=1}^{n} \left(\frac{\alpha^{[i]}}{i!}\right)^{z_i} \simeq \frac{n!}{\theta^{[n]}} \prod_{i=1}^{n} \left(\frac{\alpha}{i}\right)^{z_i}, \tag{9.17}$$

where Equation (5.71) was used together with the fact that

$$\prod_{j=1}^{g} \alpha^{[n_j]} = \prod_{i=1}^{n} \left(\alpha^{[i]}\right)^{z_i}, \tag{9.18}$$

an equation that the reader can prove as a useful exercise (note that $\alpha^{[0]} = 1$). Now, recalling that $z_0 = g - k$ and that $\sum_{i=1}^{n} z_i = k$, for small α, as a consequence of Equation (5.70) and (9.17), one finds

$$\mathbb{P}(\mathbf{z}) = \frac{g!}{(g-k)!\prod_{i=1}^{n} z_i!}\mathbb{P}(\mathbf{n}) \simeq \frac{n!}{\theta^{[n]}} g^k \alpha^k \prod_{i=1}^{n} \left(\frac{1}{i}\right)^{z_i} \frac{1}{z_i!} = \frac{n!}{\theta^{[n]}} \prod_{i=1}^{n} \left(\frac{\theta}{i}\right)^{z_i} \frac{1}{z_i!}, \tag{9.19}$$

which is indeed Equation (9.13), that is the Ewens sampling formula.

9.3 Cluster number distribution

The number of clusters for the label process is a random variable K_n that can assume integer values $k \in \{1, \dots n\}$. The probability $\mathbb{P}(K_n = k)$ can be derived from Equation (9.9). It is sufficient to sum (9.9) over all the occupation vectors with exactly k clusters. Let $S(n,k)$ denote:

$$S(n,k) = \frac{(g-k)!n!}{g!} \sum_{\mathbf{n} \in B} \frac{1}{\prod_{i \in A} n_i}, \tag{9.20}$$

where B is the set of site occupation vectors compatible with k clusters. The dependence on g disappears because after choosing the k clusters out of g, the partial sum gives the same result for any possible choice leading to

$$S(n,k) = \frac{(g-k)!n!}{g!} \binom{g}{k} \sum_{\mathbf{a} \in C} \frac{1}{\prod_{i=1}^{k} a_i} = \frac{n!}{k!} \sum_{\mathbf{a} \in C} \frac{1}{\prod_{i=1}^{k} a_i}, \tag{9.21}$$

where C is the set of time occupation vectors with k clusters. Incidentally, Equation (9.21) is a definition of the so-called *unsigned (or 'signless') Stirling numbers of the first kind* appearing in combinatorics and number theory. No closed formula is available for them, but they satisfy the following recurrence equation

$$S(n+1,k) = S(n,k-1) + nS(n,k). \tag{9.22}$$

Eventually, one gets

$$\mathbb{P}(K_n = k) = \sum_{\mathbf{n} \in B} \mathbb{P}(\mathbf{Y}_n^* = \mathbf{n}) = S(n,k) \frac{\theta^k}{\theta^{[n]}}. \tag{9.23}$$

The moments of order q for the distribution defined by Equation (9.23) can be directly obtained by summing over the possible values of $k^q \mathbb{P}(K_n = k)$. However, it is easier to derive the expected value and the variance for the number of clusters from (9.3). Consider the indicator functions I_j with $j \in \{1, \dots, n\}$, such that $I_j = 0$ if there is no innovation and $I_j = 1$ if there is innovation. Then, one can write

$$K_n = \sum_{i=1}^{n} I_i, \tag{9.24}$$

and the expected value of K_n is the sum of the expected values of the indicator functions I_j:

$$\mathbb{E}(K_n) = \sum_{i=1}^{n} \mathbb{E}(I_i). \tag{9.25}$$

Note that, here, clusters are sequentially labelled. Therefore, from (9.3), one finds

$$\mathbb{P}(I_i = 1) = \frac{\theta}{\theta + i - 1},$$
$$\mathbb{P}(I_i = 0) = 1 - \mathbb{P}(I_i = 1), \tag{9.26}$$

leading to

$$\mathbb{E}(I_i) = 1 \cdot \mathbb{P}(I_i = 1) + 0 \cdot \mathbb{P}(I_i = 0) = \frac{\theta}{\theta + i - 1}. \tag{9.27}$$

Replacing Equation (9.27) into Equation (9.25) yields

$$\mathbb{E}(K_n) = \sum_{j=0}^{n-1} \frac{\theta}{\theta + j}. \tag{9.28}$$

The indicator functions I_j are independent dichotomous (Bernoullian) random variables, even if they are not equidistributed. The reader is invited to verify these statements. For this reason, the variance of K_n is just the sum of the variances of all the I_js. One has

$$\mathbb{V}(I_i) = \mathbb{E}(I_i^2) - \mathbb{E}^2(I_i), \tag{9.29}$$

and

$$\mathbb{E}(I_i^2) = 1^2 \cdot \mathbb{P}(I_i = 1) + 0^2 \cdot \mathbb{P}(I_i = 0) = \mathbb{E}(I_i) = \frac{\theta}{\theta + i - 1}, \tag{9.30}$$

yielding

$$\mathbb{V}(I_i) = \frac{\theta(i-1)}{(\theta + i - 1)^2}. \tag{9.31}$$

Thanks to the independence of the Bernoullian random variables, one immediately finds that

$$\mathbb{V}(K_n) = \sum_{j=0}^{n-1} \frac{\theta j}{(\theta + j)^2}. \tag{9.32}$$

9.4 Ewens' moments and site-label marginals

Returning to the site-label process, assume that n elements are classified into $g > n$ sites. As in Section 5.3.8, one can introduce the indicator function $\mathbb{I}_{Y_j^*=i} = I_{Y_j^*}^{(i)}$, which is one if and only if site j exactly contains i elements. As an immediate consequence of Equation (5.74), the number of clusters of size i is then given by

$$Z_i = \sum_{j=1}^{g} I_{Y_j^*}^{(i)}, \tag{9.33}$$

and

$$\mathbb{E}(Z_i) = \sum_{j=1}^{g} \mathbb{P}(Y_j^* = i) = g\mathbb{P}(Y_1^* = i), \tag{9.34}$$

due to the equidistribution of the g categories. Note that $\mathbb{P}(Y_1^* = i)$ is the marginal of (9.9) and can be obtained from the auxiliary urn model of the label process.

In the Chinese restaurant example, the site marginal description is simply the occupation number of a table fixed in advance. Indeed, after n accommodations, the occupation number of the table can be $1, \ldots, n$, but also 0.

Consider the sequence $\mathbf{X}_f^* = (X_1^* = 1, \ldots, X_i^* = 1, X_{i+1}^* \neq 1, \ldots X_n^* \neq 1)$; its probability is

$$\mathbb{P}(X_1^* = 1, \ldots, X_i^* = 1, X_{i+1}^* \neq 1, \ldots, X_n^* \neq 1) = \frac{\theta}{g} \frac{(i-1)!\theta^{[n-i]}}{\theta^{[n]}}. \tag{9.35}$$

This equation can be justified as follows. Based on (9.7), one finds that

$$\mathbb{P}(X_1^* = 1, X_2^* = 1, \ldots, X_i^* = 1) = \frac{1}{g} \frac{1}{\theta+1} \cdots \frac{(i-1)}{\theta+i-1} = \frac{\theta}{g} \frac{(i-1)!}{\theta^{[i]}}, \tag{9.36}$$

then, the next predictive step leads to

$$\begin{aligned}
&\mathbb{P}(X_1^* = 1, \ldots, X_i^* = 1, X_{i+1}^* \neq 1) \\
&= \mathbb{P}(X_{i+1}^* \neq 1 | X_1^* = 1, \ldots, X_i^* = 1)\mathbb{P}(X_1^* = 1, \ldots, X_i^* = 1) \\
&= (1 - \mathbb{P}(X_{i+1}^* = 1 | X_1^* = 1, \ldots, X_i^* = 1))\mathbb{P}(X_1^* = 1, \ldots, X_i^* = 1) \\
&= \left(1 - \frac{i}{\theta+i}\right) \frac{\theta}{g} \frac{(i-1)!}{\theta^{[i]}} = \frac{\theta}{\theta+i} \frac{\theta}{g} \frac{(i-1)!}{\theta^{[i]}},
\end{aligned} \tag{9.37}$$

and the second next predictive step to

$$\mathbb{P}(X_1^* = 1, \ldots, X_i^* = 1, X_{i+1}^* \neq 1, X_{i+2} \neq 1)$$
$$= \left(1 - \frac{i}{\theta + i + 1}\right) \frac{\theta}{\theta + i} \frac{\theta}{g} \frac{(i-1)!}{\theta^{[i]}} = \frac{\theta + 1}{\theta + i + 1} \frac{\theta}{\theta + i} \frac{\theta}{g} \frac{(i-1)!}{\theta^{[i]}}; \tag{9.38}$$

iterating this procedure leads to (9.35). Due to exchangeability, the probability of any sequence with the same number of 1s as $\mathbf{X}_f^*$ is the same. This yields

$$\mathbb{P}(Y_1^* = i) = \frac{n!}{i!(n-i)!} \mathbb{P}(\mathbf{X}_f^*) = \frac{\theta}{gi} \frac{\theta^{[n-i]}/(n-i)!}{\theta^{[n]}/n!}, \tag{9.39}$$

and, as a consequence of Equation (9.34), one can derive the expected cluster sizes

$$\mathbb{E}(Z_i) = \frac{\theta}{i} \frac{\theta^{[n-i]}/(n-i)!}{\theta^{[n]}/n!}. \tag{9.40}$$

From Equations (9.34) and (9.40), one can compute

$$\mathbb{P}(Y_i^* = 0) = \frac{g - \mathbb{E}(K_n)}{g}, \tag{9.41}$$

which is an increasing function of g; this is also the probability of finding an empty table after a random selection among them.

Regarding the second moments of the cluster size, the derivation is similar. For the sake of simplicity, denote by R a sequence of $n-i-j$ individual values different from both 1 and 2 (again, the choice of the category/cluster labels is arbitrary). Using the method outlined above, one can prove that

$$P(X_1^* = 1, \ldots, X_i^* = 1, X_{i+1}^* = 2, \ldots, X_{i+j}^* = 2, R)$$
$$= \frac{\theta^2}{g(g-1)} \frac{(i-1)!(j-1)!\theta^{[n-i-j]}}{\theta^{[n]}}. \tag{9.42}$$

As before, due to exchangeability, one finds

$$\mathbb{P}(Y_1^* = i, Y_2^* = j) = \frac{n!}{i!j!(n-i-j)!} \frac{\theta^2}{g(g-1)} \frac{(i-1)!(j-1)!\theta^{[n-i-j]}}{\theta^{[n]}}$$
$$= \frac{1}{g(g-1)} \frac{\theta^2}{ij} \frac{\theta^{[n-i-j]}/(n-i-j)!}{\theta^{[n]}/n!}, \tag{9.43}$$

where, thanks to equidistribution, 1 and 2 can be replaced by any couple of labels. Now, consider the random variable Z_i^2; from Equations (5.74) and (9.33), it is given by

$$Z_i^2 = \sum_{l=1}^{g} I_{Y_l^*}^{(i)} \sum_{k=1}^{g} I_{Y_k^*}^{(i)} = \sum_{l=1}^{g} I_{Y_l^*}^{(i)} + \sum_{l=1}^{g} \sum_{k=l+1}^{g} I_{Y_l^*}^{(i)} I_{Y_k^*}^{(i)}$$
$$= Z_i + \sum_{l=1}^{g} \sum_{k=l+1}^{g} I_{Y_l^*}^{(i)} I_{Y_k^*}^{(i)}, \tag{9.44}$$

as $(I_{Y_l^*}^{(i)})^2 = I_{Y_l^*}^{(i)}$. Using the result of Exercise 5.3, one finds

$$\mathbb{E}\left(Z_i^2\right) = \mathbb{E}(Z_i) + \sum_{l=1}^{g} \sum_{k=l+1}^{g} \mathbb{P}\left(Y_l^* = i, Y_k^* = i\right)$$
$$= \mathbb{E}(Z_i) + g(g-1)\,\mathbb{P}\left(Y_1^* = i, Y_2^* = i\right), \tag{9.45}$$

for integer $i \leq n/2$; otherwise, the second term vanishes. Replacing Equation (9.43) into Equation (9.45) with $i = j$ leads to

$$\mathbb{E}\left(Z_i^2\right) = \begin{cases} \mathbb{E}(Z_i) + \left(\dfrac{\theta}{i}\right)^2 \dfrac{n!}{(n-2i)!} \dfrac{\theta^{[n-2i]}}{\theta^{[n]}}, & \text{if } i \text{ integer} \leq \dfrac{n}{2} \\ \mathbb{E}(Z_i) & \text{otherwise.} \end{cases} \tag{9.46}$$

In a similar way, one can derive that

$$\mathbb{E}\left(Z_i Z_j\right) = \frac{\theta^2}{ij} \frac{n!}{(n-i-j)!} \frac{\theta^{[n-i-j]}}{\theta^{[n]}}, \quad i + j \leq n. \tag{9.47}$$

In the limit $n \gg \theta$, one has

$$\frac{\theta^{[n]}}{n!} \approx \frac{n^{\theta-1}}{(\theta-1)!}, \tag{9.48}$$

then the above-mentioned moments of Z_i can be approximated as follows:

$$\mathbb{E}(Z_i) \approx \frac{\theta}{i}\left(1 - \frac{i}{n}\right)^{\theta-1}, \tag{9.49}$$

$$\mathbb{E}\left(Z_i^2\right) \approx \mathbb{E}(Z_i) + \left(\frac{\theta}{i}\right)^2 \left(1 - \frac{2i}{n}\right)^{\theta-1}, \tag{9.50}$$

and

$$\mathbb{E}\left(Z_i Z_j\right) \approx \frac{\theta^2}{ij}\left(1 - \frac{i+j}{n}\right)^{\theta-1}, \tag{9.51}$$

and so on. In the continuous limit $n \to \infty$, if the fractional size $x = i/n$ is introduced, Equation (9.49) converges to the so-called *frequency spectrum* introduced by Ewens in his 1972 paper and defined by

$$f(x) = \frac{\theta}{x}(1-x)^{\theta-1}, \tag{9.52}$$

with $x \in (0,1]$. The interval open to the left means nothing else but that the minimum fractional size is $1/n$. The meaning of

$$f(x)\mathrm{d}x = \frac{\theta}{x}(1-x)^{\theta-1}\mathrm{d}x \tag{9.53}$$

is the average fraction of clusters whose fractional size is between x and $x+\mathrm{d}x$, whereas $\theta(1-x)^{\theta-1}\mathrm{d}x$ is the average mass fraction allocated to clusters whose fractional size is between x and $x+\mathrm{d}x$.

9.5 Alternative derivation of the expected number of clusters

The distribution of cluster sizes can be directly derived from Equation (9.3), without an extensive analysis of the label process. A little thought reveals that, given a cluster size frequency $\mathbf{Z}_n = \mathbf{z}_n$, if one samples an element, all elements being on a par, and one observes the size of the cluster to which this element belongs, the probability that the first observed cluster Y_1 contains i elements is just

$$\mathbb{P}(Y_1 = i|\mathbf{Z}_n = \mathbf{z}_n) = \frac{iz_i}{n}; \tag{9.54}$$

therefore, as a consequence of the total probability theorem, one finds

$$\mathbb{P}(Y_1 = i) = \sum_{\mathbf{z}_n}\mathbb{P}(Y_1 = i|\mathbf{Z}_n = \mathbf{z}_n)\mathbb{P}(\mathbf{Z}_n = \mathbf{z}_n) = \frac{i\mathbb{E}(Z_i)}{n}. \tag{9.55}$$

This leads to

$$\mathbb{E}(Z_i) = \frac{n}{i}\mathbb{P}(Y_1 = i); \tag{9.56}$$

in other words, the probabilistic study of the first cluster is sufficient to get the expected distribution of the cluster size. Consider the fundamental sequence $\mathbf{X}_f =$

$(X_1 = 1,\ldots,X_i = 1, X_{i+1} > 1,\ldots,X_n > 1)$. Considering that the marginal event $X_j > 1$ is the complement of $X_j = 1$, the very same method used to derive Equation (9.35) leads to

$$\mathbb{P}(X_1 = 1,\ldots,X_i = 1, X_{i+1} > 1,\ldots,X_n > 1) = \frac{\theta(i-1)!\theta^{[n-i]}}{\theta^{[n]}}; \tag{9.57}$$

then, one finds that

$$\mathbb{P}(Y_1 = i) = \binom{n-1}{i-1}\mathbb{P}(\mathbf{X}_f), \tag{9.58}$$

due to the fact that $X_1 = 1$ and the other $i-1$ variables corresponding to the first cluster can always be chosen among $n-1$ variables. Eventually, this leads to

$$\mathbb{P}(Y_1 = i) = \frac{\theta}{n}\frac{\theta^{[n-i]}/(n-i)!}{\theta^{[n]}/n!}, \tag{9.59}$$

in agreement with Equation (9.39). Combining Equation (9.56) with Equation (9.59), Equation (9.40) is recovered. Given that $(\theta+1)^{[n-1]} = \theta^{[n]}/\theta$, Equation (9.59) can be re-written as follows

$$\mathbb{P}(Y_1 = i) = \binom{n-1}{i-1}\frac{1^{[i-1]}\theta^{[n-i]}}{(\theta+1)^{[n-1]}}, \tag{9.60}$$

which is nothing other than the Polya$(i-1, n-1; 1, \theta)$. In the continuum limit, the probability density function of the fractional size of the first cluster (i/n) is then distributed according to Beta$(1,\theta)$.

9.6 Sampling and accommodation

Before introducing a Markovian probabilistic model leading to equilibrium cluster distributions given by the equations discussed in the previous sections of this chapter, it is useful to further discuss the meaning of the two processes discussed so far.

9.6.1 Sampling from a random distribution

Consider a system of n elements, and imagine that the only available information is that they are grouped in some number of clusters, say K_n, and there are $Z_{1,n}$ clusters with one element, $Z_{2,n}$ clusters with two elements, and so on, so that $\sum_{i=1}^{n} Z_{i,n} = K_n$.

The only certain piece of information is that $\sum_{i=1}^{n} iZ_{i,n} = n$. Further assume that $\mathbb{P}(\mathbf{Z}_n = \mathbf{z}_n)$ is given by Equation (9.13). Now, $\mathbf{Z}_n$ is no more completely unknown, it is still indeterminate, but its uncertainty is ruled by the sampling formula due to Ewens. While in statistics this uncertainty is based on Bayesian roots, in the case under scrutiny, it has a realistic root, as the system will be allowed to change in time, and $\mathbb{P}(\mathbf{Z}_n = \mathbf{z}_n)$ is proportional to the time fraction in which the cluster statistical distribution is given by $\mathbf{Z}_n = \mathbf{z}_n$. Sampling from this random distribution means taking a snapshot of the moving population, and then observing all the elements in a sequence. The sequence $X_1, \ldots, X_n$ describes the elements sampled without replacement. If the sampling hypothesis is that all sequences conditioned to the present partition value $\mathbf{z}_n$ are equiprobable, the probability of any sequence is given by Equation (9.4) and the predictive sampling probability is then (9.3).[1] This means that the first element observed belongs to the first cluster with probability 1, the second element either belongs to the first cluster with probability $1/(1+\theta)$ or it belongs to a new cluster (the second one) with probability $\theta/(1+\theta)$, and so on. After observing m agents, the $(m+1)$th belongs to a not yet observed (new) cluster with probability $\theta/(m+\theta)$, whereas the total probability of an already observed cluster is $m/(m+\theta)$.

Another type of sampling from the partition value $\mathbf{z}$ corresponding to k clusters is the so-called *species sampling*. It amounts to classifying the *size* of the k clusters in the order in which they appear if all the elements are equiprobably drawn without replacement. The distribution of the first pick was derived in the previous section. The cluster sizes are represented by the time-label frequency vector $\mathbf{a}$. A little thought leads to the following equation

$$\mathbb{P}(\mathbf{a}|\mathbf{z}) = \frac{a_1}{n}\frac{a_2}{n-a_1}\cdots\frac{a_{k-1}}{a_{k-1}+a_k}\frac{a_k}{a_k}\prod_{i=1}^{n} z_i!; \tag{9.61}$$

indeed, if all the clusters had different sizes and $\prod_{i=1}^{n} z_i! = 1$, these terms would describe the probability of drawing an element of the jth cluster from a population in which the elements belonging to the already observed clusters have been removed; if there were more than a single cluster of size i, the first time one selects a cluster of this size, the total probability mass is proportional to iz_i, the second time it is proportional to $i(z_i - 1)$, and so on. Equation (9.61) is called the size biased permutation of $\mathbf{z}$, as all the clusters belonging to $\mathbf{z}$ appear in (9.61), but their order of appearance is biased due to their size, meaning that different permutations have different probabilities. Given that $\mathbb{P}(\mathbf{a}) = \mathbb{P}(\mathbf{a}|\mathbf{z})\mathbb{P}(\mathbf{z})$, if $\mathbb{P}(\mathbf{a}|\mathbf{z})$ is given by (9.61)

[1] This duality between predictive inferences and sampling from a random distribution is well known in the realm of Bayesian statistics, and it is based on de Finetti's representation theorem for exchangeable sequences (see Section 4.3).

and $\mathbb{P}(\mathbf{z})$ is the Ewens sampling formula (9.13), it should be clear why Equation (9.6) was called the size-biased permutation of the Ewens sampling formula by Donnelly.

9.6.2 The accommodation process

Now change perspective and consider a population of n agents that can choose among a large number of possible strategies/choices. The names of these strategies are not interesting, and one wishes to study the way in which agents partition into groups, it being understood that two units belong to the same cluster if and only if $X_i = X_j$. The operational frame is the one of the Chinese restaurant described in Section 9.1.1. Agents choose sequentially, and each agent's choice is conditioned by the previous accommodation. Even if the possible strategies/choices are infinite, no more than n can be exploited by a finite population of n elements; therefore, we describe the possible choices by considering a system of $g > n$ labels or sites. Considering the conceptual difference between a cluster and a site, it is sufficient to say that when a cluster dies, it disappears forever, whereas a site can be occupied again. The only meaning of these labels is that of distinguishing between different clusters, and, *ceteris paribus*, the labels are set equiprobable. Then, the first agent chooses the jth site with uniform probability $1/g$. The second has two possibilities: if his/her choice is the same as the first ('herding'), he/she accommodates in the same site; if it is different ('innovation'), he/she accommodates in a site labelled, say, by i, with $i \neq j$, with probability $1/(g-1)$ uniform on the remaining empty sites. It is assumed that the choice is probabilistic, and influenced by the previous results with the weight of 'innovation' set equal to θ.

Hoppe's predictive probability (9.3) is a description of a sampling sequence in terms of time labels (the time of appearance of the clusters in sampling). On the contrary, the accommodation probability (9.7) is a description of a physical process, and it uses site labels that can be regarded as usual permanent categories.

The essential difference between the two types of label is given by the factor $1/(g - K_m)$ for the probability of innovations. In the case of time labels the name of a new cluster is already determined, whereas for sites it must be chosen. In this case one can pass from time labels to site labels by fixing the K_n cluster labels independently on anything. If g labels are available, each of the $g(g-1)\cdots(g-K_n+1) = g!/(g-K_n)!$ site descriptions is equally likely. These considerations lead to Equation (9.8).

As the clustering structure of the accommodation process is the same as (9.3), one can conclude that the size of the first cluster (also known as the size-biased pick) is ruled by Equation (9.59), the statistical cluster distribution has probability given by Equation (9.13) and the average number of clusters of size i is given by

(9.40). In this framework, all these distributions have an objective meaning: if one builds an ensemble of independent accommodation processes with the same initial conditions, the frequency of any event in the ensemble is roughly proportional to the corresponding probability. Note that $\mathbf{Z}_n$ is 'random', or, better, ruled by (9.13), because it is generated by a random mechanism (the accommodation process, agents' choices).

9.7 Markov chains for cluster and site dynamics

9.7.1 Ehrenfest–Brillouin with infinite categories

At the beginning of this chapter, we were looking for random variables suitable for building an irreducible and aperiodic Markov chain in the case of an infinite number of categories. This is why the site-label process was introduced. In the Chinese restaurant example, suppose that the n customers, once accommodated, are free to move from one table to another. If they join an already occupied table, their behaviour can be called 'herding'; if they move to an empty table, they are called 'pioneers'. Note that, in the usual accommodation process, whose probability is proportional to $\alpha_i + n_i$, if an element moves to a non-empty category, it cannot be identified either as a follower of the herd or as a follower of the initial distribution. On the contrary, when the number of categories becomes infinite and their weight vanishes, the two behaviours cannot give rise to the same choice. Indeed, the domain of the initial distribution is infinite and this means that the probability of choosing any finite number of fixed tables is zero. Consequently, with probability 1, a follower of the theoretical distribution is compelled to choose an empty table and, therefore, he/she is a pioneer, that is the founder of a new cluster.

The model outlined above has several possible interpretations in economics. If occupied tables represent firms, the elements may be agents who move from one firm to another, or create new firms. Elements may also be agents who choose different strategies (clusters). Another possible interpretation is that elements represent firms, and clusters economic sectors. It is possible to interpret elements as firms and clusters as regional sites as well, and so on. Considering the interpretation in terms of workers in firms, if the number of sites $g > n$ there is always room for opening a new firm. Among the g sites, there will always be $k \leq n < g$ living firms while the $g - k$ empty ones are ready to be occupied by pioneers.

In order to represent these features by means of a homogeneous Markov chain, the main route is introducing the 'label-site process', simplifying a method suggested by Kelly. This method is very similar to the Ehrenfest–Brillouin model, and is the subject of the next section. A second way is to abandon once and for all the labels of the categories and directly to study partition vectors. The Markov chain induced on partitions indeed has the Ewens sampling formula (9.13) as equilibrium distribution, but the proof is cumbersome (see Section 9.7.3 for further details).

9.7.2 The site-label Markov chain

Once again, if the population size is n, introduce $g \geq n$ fixed labels. Assume that at the beginning there are $k \leq n$ distinct clusters. Label them with the first k labels; the initial occupation vector is $\mathbf{n} = (n_1, \ldots, n_k, 0, \ldots 0)$. Define $A(\mathbf{n}) = \{i : n_i > 0\}$ as the set of all active labels (occupied sites), then $\#A(\mathbf{n}) = k$ is their number, and $g - k$ is the number of inactive labels (empty sites). The transition probability for the label process is:

$$\mathbb{P}(\mathbf{n}_i^j|\mathbf{n}) = \begin{cases} \dfrac{n_i}{n}\dfrac{n_j - \delta_{i,j}}{\theta + n - 1}, & \text{for } n_j > 0 \\ \dfrac{n_i}{n}\dfrac{1}{g-k}\dfrac{\theta}{\theta + n - 1}, & \text{for } n_j = 0. \end{cases} \tag{9.62}$$

The label process $\mathbf{Y}_0^*, \mathbf{Y}_1^*, \ldots, \mathbf{Y}_t^*, \ldots$, where $\mathbf{Y}_t^* = (n_1, \ldots, n_k, \ldots, n_g)$ with the constraint $\sum_1^g n_i = n$, describes the random occupation vector with respect to the g categories. This procedure produces an ergodic set of label states. In fact an inactive label j (i.e. $n_j = 0$) can be reactivated, as it can be chosen in the case of innovation. The transition probability (9.62) defines an irreducible aperiodic Markov chain, whose equilibrium distribution satisfies the detailed balance equations:

$$\mathbb{P}(\mathbf{n}) = \frac{(g-k)!}{g!}\frac{n!}{\prod_{j\in A} n_j}\frac{\theta^k}{\theta^{[n]}} \tag{9.63}$$

$$= \frac{(g-k)!}{g!}\frac{n!}{\theta^{[n]}}\prod_{i=1}^{n}\left(\frac{\theta}{i}\right)^{z_i}, \tag{9.64}$$

where in (9.64) the partition vector is used $\mathbf{z} = (z_1, \ldots, z_n)$, $z_i = \#\{n_j = i, j = 1, \ldots, g\}$, with $k = \sum_{i=1}^n z_i$ and $\sum_{i=1}^n iz_i = n$. Now z_i is the number of active labels (representing clusters) with $n_j = i$, and k is the number of active labels (that is the number of clusters in $\mathbf{n}$). As already discussed, note that $\prod_{j\in A(\mathbf{n})} n_j = \prod_{i=1}^n i^{z_i}$. Note that Equation (9.64) coincides with Equation (9.12). In order to verify that Equation (9.64) satisfies detailed balance, consider the following cases:

1. for $n_i > 1$ and $n_j > 0$ ($k(\mathbf{n}_i^j) = k(\mathbf{n})$), one has $\mathbb{P}(\mathbf{n}_i^j|\mathbf{n}) = Bn_in_j$ where $B = (n(\theta + n - 1))^{-1}$ and $\mathbb{P}(\mathbf{n}_i^j|\mathbf{n}) = B(n_i - 1)(n_j + 1)$, so that

$$\frac{\mathbb{P}(\mathbf{n}_i^j)}{\mathbb{P}(\mathbf{n})} = \frac{\mathbb{P}(\mathbf{n}_i^j|\mathbf{n})}{\mathbb{P}(\mathbf{n}|\mathbf{n}_i^j)} = \frac{n_i}{n_i - 1}\frac{n_j}{n_j + 1}; \tag{9.65}$$

2. for $n_i > 1$ and $n_j = 0$ ($k(\mathbf{n}_i^j) = k(\mathbf{n}) + 1$), one has $\mathbb{P}(\mathbf{n}_i^j|\mathbf{n}) = Bn_i\theta/(g-k)$ and $\mathbb{P}(\mathbf{n}_i^j|\mathbf{n}) = B(n_i - 1)$, so that

$$\frac{\mathbb{P}(\mathbf{n}_i^j)}{\mathbb{P}(\mathbf{n})} = \frac{\mathbb{P}(\mathbf{n}_i^j|\mathbf{n})}{\mathbb{P}(\mathbf{n}|\mathbf{n}_i^j)} = \frac{n_i}{n_i - 1}\frac{\theta}{g-k}; \tag{9.66}$$

3. for $n_i = 1$ and $n_j > 0$ ($k(\mathbf{n}_i^j) = k(\mathbf{n}) - 1$), one has $\mathbb{P}(\mathbf{n}_i^j|\mathbf{n}) = Bn_j$ and $\mathbb{P}(\mathbf{n}_i^j|\mathbf{n}) = B(n_j + 1)\theta/(g-k+1)$, so that

$$\frac{\mathbb{P}(\mathbf{n}_i^j)}{\mathbb{P}(\mathbf{n})} = \frac{\mathbb{P}(\mathbf{n}_i^j|\mathbf{n})}{\mathbb{P}(\mathbf{n}|\mathbf{n}_i^j)} = \frac{n_j}{n_j + 1}\frac{\theta}{g-k+1}; \tag{9.67}$$

4. for $n_i = 1$ and $n_j = 0$ ($k(\mathbf{n}_i^j) = k(\mathbf{n})$), one has $\mathbb{P}(\mathbf{n}_i^j|\mathbf{n}) = B\theta/(g-k)$ and $\mathbb{P}(\mathbf{n}_i^j|\mathbf{n}) = B\theta/(g-k)$, so that

$$\frac{\mathbb{P}(\mathbf{n}_i^j)}{\mathbb{P}(\mathbf{n})} = \frac{\mathbb{P}(\mathbf{n}_i^j|\mathbf{n})}{\mathbb{P}(\mathbf{n}|\mathbf{n}_i^j)} = 1. \tag{9.68}$$

It is straightforward to verify that the probability distribution defined by Equation (9.64) satisfies all the above cases. This means that it is the unique invariant probability and it coincides with the equilibrium probability.

The label is not connected to any physical characteristic of the cluster, it simply labels it. As time goes by, the same label will be used for different clusters, but after the extinction of a cluster a time interval will pass before its label is used again (due to the fact that $g > n$). Following the 'history' of a label, each passage from 1 to 0 indicates the death of a cluster, and each passage from 0 to 1 indicates the birth of a new cluster. This labelling process is analogous to that introduced by Hansen and Pitman (see further reading) in species sampling. They indeed write: 'it is just a device to encode species ... in a sequence of random variables'. As for partitions, one gets the Ewens sampling formula

$$\mathbb{P}(\mathbf{z}) = \frac{g!}{(g-k)!z_1!\cdots z_n!}P(\mathbf{n}) = \frac{n!}{\theta^{[n]}}\prod_{i=1}^{n}\left(\frac{\theta}{i}\right)^{z_i}\frac{1}{z_i!}. \tag{9.69}$$

Equation (9.69) coincides with Equation (9.13) and, in this framework, it appears as the equilibrium distribution on partitions generated by the label process $\mathbf{Y}^*$. Eventually, note that the minimum number of labels compatible with the description of n objects and the possibility of including innovations is simply $g = n+1$, so that one always has at least an empty category. This is so because in the extreme case of all singletons, one would like to distinguish for instance between $(1,1,\ldots,1,0)$ and $(0,1,\ldots,1,1)$.

9.7.3 Partition probabilistic dynamics

Consider the statistical description of clusters $\mathbf{z} = (z_1, \ldots, z_n)$ (the 'partition') as the state variable for a homogeneous Markov chain. Let $\mathbf{u}$ be the initial partition, $\mathbf{v}$ the partition after the destruction and $\mathbf{z}$ the partition after the creation. If the destruction affects a unit belonging to a cluster of size i, then it transforms a cluster of size i in a cluster of size $i-1$, therefore one has

$$v_i = u_i - 1, \quad v_{i-1} = u_{i-1} + 1. \tag{9.70}$$

Afterwards the created entity increases the number of clusters of size j if it joins to a cluster of size $j-1$, therefore one has

$$z_j = v_j + 1, \quad z_{j-1} = v_{j-1} - 1. \tag{9.71}$$

Hence, destructions in i are proportional to the number of entities belonging to an initial cluster of size i, that is iu_i, while creations in j are proportional to the number of entities belonging to a cluster of size $j-1$ after destruction, that is $(j-1)\,v_{j-1}$ leading to the following transition probability

$$P(\mathbf{z}|\mathbf{u}) = P\left(\mathbf{u}_i^j|\mathbf{u}\right) = \begin{cases} \dfrac{iu_i}{n}\dfrac{(j-1)\,v_{j-1}}{\theta+n-1}, & \text{for } j > 1 \\ \dfrac{iu_i}{n}\dfrac{\theta}{\theta+n-1}, & \text{for } j = 1 \end{cases} \tag{9.72}$$

as iu_i is the number of agents initially in some i-cluster, $(j-1)v_{j-1}$ is the number of agents in some $(j-1)$-cluster after destruction. If the agent joins some of them, the number of j-clusters increases by one. Once again, the transition probability (9.72) defines a homogeneous Markov chain irreducible and aperiodic, whose equilibrium distribution (the Ewens sampling formula (9.13)) can be derived from the detailed balance equations. Being statistical, the description in terms of partitions gives less information than the label description. If two agents exchange their places, the partition vector $\mathbf{z}$ does not change after this move, whereas the occupation vector $\mathbf{Y}^* = \mathbf{n}$ does vary.

9.8 Marginal cluster dynamics and site dynamics

Many studies on cluster growth introduce some (logarithmic) random walk, whose role is to mimic the behaviour of empirical systems. In other words, one considers a cluster in contact with some environment, whose model is never investigated. This is the case of the celebrated Gibrat's model. Let the random variable Y represent the size of a cluster, then its logarithmic size is given by $S = \log Y$ and one assumes that it follows a random walk:

$$S_t = S_0 + \sum_{i=1}^{t} U_i, \tag{9.73}$$

where U_i for $i = 1, \ldots, t$ is a set of independent and identically distributed random variables and S_0 is the initial value of the walk. Coming back to the size, one has

$$Y_t = V_0 \prod_{i=1}^{n} V_i, \tag{9.74}$$

where $V_i = \exp(U_i)$ are still independent and identically distributed random variables. In the case discussed here, attention can be concentrated on a fixed site, and the description of what happens in this site can be developed assuming that the behaviour of the whole population is ruled by some reasonable probabilistic mechanism.

9.8.1 The probabilistic dynamics for a cluster

It is remarkable that the probabilistic dynamics for a cluster can be exactly determined by marginalization. In fact, if a cluster has size i then the probability that its size increases or decreases is given by the following equations: for the increase,

$$w(i, i+1) = \frac{n-i}{n} \frac{i}{\theta + n - 1}, \quad i = 0, \ldots, n-1 \tag{9.75}$$

for the decrease,

$$w(i, i-1) = \frac{i}{n} \frac{n - i + \theta}{\theta + n - 1}, \quad i = 1, \ldots, n \tag{9.76}$$

and, if the size does not change,

$$w(i, i) = 1 - w(i, i+1) - w(i, i-1). \tag{9.77}$$

Note that Equation (9.75) holds also for $i = 0$, and one immediately gets

$$w(0, 1) = 0, \tag{9.78}$$

meaning that $i = 0$ is an absorbing state, as already discussed in Section 9.7.1. This behaviour is the Ewens limit of Equations (8.32), (8.33) and (8.34) in Chapter 8.

Hence, given that $\mathbb{E}(\Delta Y|Y=i) = w(i,i+1) - w(i,i-1)$, with $\Delta Y = Y_{t+1} - Y_t$, one finds

$$E(\Delta Y|Y=i) = -\frac{i}{n}\frac{\theta}{\theta+n-1} = -ri, \tag{9.79}$$

where r is the rate of approach of the conditional expected value to equilibrium for unary moves (see Section 8.2.2). Note that (9.79) is a particular case of (8.43) when the equilibrium mean is zero. In other words, each particular existing cluster is driven towards extinction with the rate defined by Equation (9.79). The matrix $w(i,j)$ drives the evolution of the represented cluster. All the states except for $i=0$ are transient, and the state $i=0$ is absorbing. In order to study the time behaviour of a cluster, one can use $w^{(s)}(i,j)$, that is the transition probability from i to j after s steps. Then $w^{(s)}(i,0)$ is the probability that a cluster of size i will die after s steps. If the random variable D_i denotes the number of time steps from size i to death, one can write

$$\mathbb{P}(D_i \leq s) = w^{(s)}(i,0). \tag{9.80}$$

Therefore, the difference $w^{(s)}(i,0) - w^{(s-1)}(i,0)$ gives the probability of dying exactly at the sth step for a cluster of size i. In symbols, one can write

$$\mathbb{P}(D_i = s) = \mathbb{P}(D_i \leq s) - \mathbb{P}(D_i \leq s-1) = w^{(s)}(i,0) - w^{(s-1)}(i,0), \tag{9.81}$$

as the event $\{D_i \leq s\}$ is the union of two disjoint events: $\{D_i \leq s\} = \{D_i = s\} \cup \{D_i \leq s-1\}$. The variable $\bar{\tau}_i$ given by

$$\bar{\tau}_i = \mathbb{E}(D_i) = \sum_{s=1}^{\infty} s(w^{(s)}(i,0) - w^{(s-1)}(i,0)) \tag{9.82}$$

is the expected duration time of a cluster of size i. Note that from (9.75), (9.76) and (9.77), one has $w^{(s)}(i,0) - w^{(s-1)}(i,0) = w^{(s-1)}(i,1)/n$, as dying at the sth step is the same as reaching size 1 at the $(s-1)$th step and then dying with transition probability $w(1,0) = 1/n$ from (9.75). Hence, one finds

$$\bar{\tau}_i = \sum_{s=1}^{\infty} s\frac{w^{(s-1)}(i,1)}{n}, \tag{9.83}$$

where $w^{(0)}(i,1) = \delta_{i,1}$. The expected duration time of a new cluster (of size 1, $\bar{\tau}_1$) is the expected life of a newborn cluster. It can be exactly computed from Equation (9.83), but an alternative derivation is possible. In fact considering that

$r = \theta/(\theta + n - 1)$ is the innovation rate, that is the expected number of new clusters at each step, and using $\mathbb{E}(K_n)$, the stationary expected number of clusters, the average life of a new cluster is just

$$\bar{\tau}_1 = \frac{\mathbb{E}(K_n)}{r}, \tag{9.84}$$

where $\mathbb{E}(K_n)$ is given by Equation (9.28). A recurrence relation exists for $\bar{\tau}_i$, so that it can be found avoiding the sum in Equation (9.83). In fact if a cluster has size i (and its expected duration time is $\bar{\tau}_i$), at the following step there are three possibilities, and three possible expected duration times, all increased by the duration of the step $(= 1)$. Therefore

$$\bar{\tau}_i = w(i, i+1)\{\bar{\tau}_{i+1} + 1\} + w(i, i)\{\bar{\tau}_i + 1\} + w(i, i+1)\{\bar{\tau}_{i-1} + 1\}, \tag{9.85}$$

that is

$$\bar{\tau}_i = 1 + w(i, i+1)\bar{\tau}_{i+1} + w(i, i)\bar{\tau}_i + w(i, i-1)\bar{\tau}_{i-1}, \tag{9.86}$$

and replacing Equations (9.75), (9.76) and (9.77), reordering and defining $\Delta\bar{\tau}_i = \bar{\tau}_i - \bar{\tau}_{i-1}$, $i \geq 2$ with $\bar{\tau}_0 = 0$ leads to

$$\Delta\bar{\tau}_i = \Delta\bar{\tau}_{i-1}\frac{n - i + 1 + \theta}{n - i + 1} - \frac{n(n - 1 + \theta)}{(i - 1)(n - i + 1)}, \quad \Delta\bar{\tau}_1 = \frac{\mathbb{E}(K_n)}{u}. \tag{9.87}$$

9.8.2 The probabilistic dynamics for a site

If a site is considered instead of a cluster, there is the possibility of rebirth. When a site is empty, an agent occupying another site can decide to move to the empty site. If the site contains i agents, the probabilistic dynamics is given by the same equations as before, namely Equations (9.75), (9.76) and (9.77). All the other occupied sites are merged in a single one (the thermostat), whose weight is $n - i$, plus θ, that is the weight of the empty sites. Note that the cluster dynamics described in the previous section does not depend on the number of clusters K_n and thus it can be exactly discussed. Considering the site dynamics, if the site is empty, $i = 0$, one can introduce the rebirth term and its complement

$$\begin{aligned} w(0,1) &= \frac{1}{g - K_n}\frac{\theta}{\theta + n - 1}, \\ w(0,0) &= 1 - w(0,1); \end{aligned} \tag{9.88}$$

in this way, the state $Y^* = 0$ is no longer absorbing. Note that $w(0,1)$ depends on K_n and g, so that this term contains an exogenous random variable. This is not essential for the equilibrium distribution of the cluster size. Starting from a cluster whose size is i, the history of the cluster (that ends when the size reaches $i = 0$) must be distinguished from that of the site, that sooner or later reactivates. Then, the birth–death chain defined by Equations (9.75), (9.76), (9.77) and (9.88) is irreducible and aperiodic, and its stationary and equilibrium distribution satisfies the detailed balance condition

$$\mathbb{P}(Y^* = i+1)\frac{i+1}{n}\frac{n-i-1+\theta+\alpha}{\theta+n-1} = \mathbb{P}(Y^* = i)\frac{n-i}{n}\frac{i-\alpha}{\theta+n-1}, \tag{9.89}$$

for $i = 1,\dots,n-1$, and

$$\mathbb{P}(Y^* = 1)w(1,0) = \mathbb{P}(Y^* = 0)w(0,1), \tag{9.90}$$

thus leading to

$$\frac{\mathbb{P}(Y^* = i+1)}{\mathbb{P}(Y^* = i)} = \frac{n-i}{i+1}\frac{i}{\theta+n-i-1} = \frac{\mathbb{E}(Z_{i+1})}{\mathbb{E}(Z_i)}, \tag{9.91}$$

where $\mathbb{E}(Z_i)$ is just the expected cluster size given by (9.40) in the Ewens population, whereas $\mathbb{P}(Y^* = i)$ is the marginal distribution of the Ewens population (Equation (9.39)). The term $\mathbb{P}(Y^* = 0)$ depends on g and K_n. If, in Equation (9.88), K_n is replaced by $\mathbb{E}(K_n)$, at best, the chain mimics what happens at the fixed site if its history is drawn from the exact joint dynamics of the whole population. But what is essential from the cluster point of view is that $w(0,1)$ is positive. If for instance $g \gg n$, each fixed site is going to be empty for a long period of time. If one chooses a site at random, all sites being equiprobable, the site will be empty with probability $(g - \mathbb{E}(K_n))/g$ and $\mathbb{E}(K_n)/g$ is the probability of finding a living cluster. Then, randomly choosing a cluster, all clusters being equiprobable, and observing, its size is described by

$$\mathbb{P}(Y^* = i | i > 0) = \frac{\mathbb{E}(Z_i)}{\mathbb{E}(K_n)},\ i = 1,\dots,n, \tag{9.92}$$

it being understood that any observation concerns living clusters (non-empty sites).

9.8.3 A simple example on workers and firms

Consider a population of n workers following an aggregation dynamics driven by Equations (9.2), and therefore, at equilibrium, described by the Ewens sampling formula parameters n and θ. Given that $\mathbb{E}_n(Z_{i,n}) = (n/i)\cdot \text{Polya}(i-1,n-1;1,\theta)$, in order to simplify calculations suppose that $\theta = 1$. Then one gets $\text{Polya}(i-1,n-$

$1;1,1) = 1/n$, and $\mathbb{E}(Z_{i,n}) = 1/i$. Now Polya$(i-1,n-1;1,1)$ is the probability distribution of the first pick, whose meaning is: if one samples a worker at random, that is all workers are on a par, and one asks the size of the firm to which the worker belongs, then the answer will be uniformly distributed on $\{1,\ldots,n\}$. One can assume that the worker spends his time in all possible firm sizes uniformly. For instance, there is the same probability of finding him alone (as a singleton) as of finding him employed in a monopolistic firm of size n. The expected number of firms for size $1,2,\ldots,n$ is $1,1/2,1/3,\ldots,1/n$, that is one expects a singleton in every snapshot of the population, a couple every two snapshots, ..:, a monopolistic size n every n snapshots. Hence the expected mass of workers in firms of size i is $i\mathbb{E}(Z_{i,n}) = 1$, which is uniform, and explains why the first pick (sampling a size biased by the number of workers) is uniform. On the contrary, if one chooses a firm, equiprobably drawn among the existing ones, and observes its size, one finds that this random variable is distributed as $\mathbb{E}(Z_{i,n})/\mathbb{E}(K_n) = i^{-1}/\mathbb{E}(K_n)$, with $\mathbb{E}(K_n) = \sum_{i=1}^{n} i^{-1}$, and the odds of a singleton against the monopoly are $n:1$. If one samples the site-labels, the denominator is g, and one finds that $1-\mathbb{E}(K_n)/g$ is the probability of finding an empty site, while i^{-1}/g is the absolute probability for the size i. Following the link given by Equation (9.55), $i^{-1}/\mathbb{E}(K_n)$ is the equilibrium probability for the marginal chain of a firm as long as it is alive. It can be interpreted as the visiting time relative frequency for the size i during the life of the cluster.

Suppose that $n = 10$: the expected number of firms is $\mathbb{E}(K_n) = 2.93$. Assume that the first pick gives $a_1 = 4$, whose probability is $\mathbb{P}(a_1 = 4) = 1/10$. What can we say about a_2? It is not difficult to show that in general $\mathbb{P}(a_2|a_1) = \text{Polya}(a_2-1,n-a_1-1;1,\theta)$, that is: the second pick has the same distribution as the first, within the restricted domain $n-a_1$; and this holds true for all k clusters. Hence the sequence $a_1 = 4, a_2 = 6$ has probability $1/10 \cdot 1/6 = 1/60$. If one considers the permutation $a_1 = 6, a_2 = 4$, its probability is $1/10 \cdot 1/4 = 1/40$, which is greater than $1/60$, due to the intuitive fact that the order of appearance is positively correlated with the size of the cluster. Both permutations are the only possibilities belonging to $(z_4 = z_6 = 1)$, whose probability is given by Equation (9.13), and is just equal to $(1/4)\cdot(1/6) = 1/24 = 1/40 + 1/60$.

Considering the continuum limit $n \to \infty$, the probability of a size-biased sequence $x_1,x_2,\ldots$,with $x_i = a_i/n$, is obtained considering that residual fractions $x_1, x_2/(1-x_1),\ldots,x_{i+1}/(1-x_1-\ldots-x_i)$ are all distributed according to the distribution Beta$(1,\theta)$. This way of assigning probability to the sequence is called the Residual Allocation Model (RAM), as all residuals are independent; moreover, being also identically distributed as Beta$(1,\theta)$, the distribution is called GEM by Tavaré and Ewens (see further reading), after Griffiths, Engen and McCloskey.

9.9 The two-parameter Ewens process

9.9.1 Definition of the process

Hansen and Pitman generalized the Ewens sampling formula described in this chapter. Their generalization can be discussed in a rather elementary way with the tools developed above. Equation (9.2) is modified as follows

$$\mathbb{P}(X_{n+1}=j|X_1=x_1,\ldots,X_n=x_n)=\begin{cases}\dfrac{n_j-\alpha}{\theta+n}, & \text{for } j\leq k\\ \dfrac{\theta+\alpha k}{\theta+n}, & \text{for } j=k+1\end{cases} \tag{9.93}$$

with $0\leq\alpha<1$ and $\theta+\alpha>0$. In Equation (9.93), if the parameter α is positive, the probability of choosing a new cluster increases proportionally to the number of existing clusters $K_n=k$, whereas for each existing cluster, the probability of being occupied decreases with respect to the Ewens case. This process can be modelled by means of a two-parameter Hoppe's urn, as described by Zabell in his 1992 paper *Predicting the Unpredictable* (see further reading):

> Imagine an urn containing both colored and black balls, from which balls are both drawn and then replaced. Each time a colored ball is drawn, it is put back into the urn, together with a new ball with the same color having a unit weight, each time a black ball is selected, it is put back into the urn, together with two new balls, one black and having weight α, and one of a new color, having weight $1-\alpha$. Initially the urn contains a single black ball (the mutator).

In other words, if an already observed colour is drawn, the updating mechanism is Pólya (the weight of the colour increases by one), while a new colour receives an initial weight equal to $1-\alpha$. If $\alpha=0$ the model reduces to a conventional Hoppe's urn. The case $\alpha>0$ implies an increasing weight α for the mutator every time it is drawn (a new colour appears). However, the model can be extended to $\alpha<0$, in which case the weight of the mutator decreases every time it is drawn. If $\theta=k|\alpha|$, after k colours have appeared no new colour can be drawn. Hence, sampling coincides with a k-dimensional Pólya sampling with uniform initial weights equal to $|\alpha|$. The only difference is that in the k-dimensional Pólya sampling the names of the categories are known in advance, and the initial weights are not forced to be uniform.

The method described in Section 9.5 applied to the predictive probability (9.93) immediately leads to the distribution of the first cluster picked:

$$\mathbb{P}(Y_1=i)=\frac{\theta(1-\alpha)^{[i-1]}}{n(i-1)!}\frac{(\theta+\alpha)^{[n-i]}/(n-i)!}{\theta^{[n]}/n!}, \tag{9.94}$$

that can also be written similarly to Equations (9.58) and (9.59) as

$$\mathbb{P}(Y_1 = i) = \binom{n-1}{i-1} \frac{(1-\alpha)^{[i-1]}(\theta+\alpha)^{[n-i]}}{(1+\theta)^{[n-1]}}. \tag{9.95}$$

Equation (9.95) is a generalization of Equation 9.60 and is simply the dichotomous Pólya distribution of initial weights $1-\alpha$ and $\theta+\alpha$ for $i-1$ hits over $n-1$ trials. In the continuum limit, the density of the fractional size for the first cluster (i/n) is then distributed according to Beta$(1-\alpha, \theta+\alpha)$. This result is not surprising at all. Indeed, the two-parameter Pólya urn described above coincides with the usual dichotomous Pólya urn with initial weights $1-\alpha$ and $\theta+\alpha$ and unitary prize, when the first drawn colour is taken into account. If one finds $i-1$ times the first drawn colour in the successive $n-1$ draws, then the final size of the first colour drawn from the urn is i. As before, one can find $\mathbb{E}(Z_i)$ multiplying (9.95) by n/i. Regarding Markovian dynamics, referring to the methods developed in Section 9.8, the first moment of the equilibrium cluster size distribution is obtained from the equilibrium marginal distribution for a site. The only difference is that the first two Equations (9.75) and (9.76) are replaced by

$$w(i,i+1) = \frac{n-i}{n} \frac{i-\alpha}{\theta+n-1}, \tag{9.96}$$

and by

$$w(i,i-1) = \frac{i}{n} \frac{n-i+\theta+\alpha}{\theta+n-1}, \tag{9.97}$$

respectively.

9.9.2 Expected number of clusters

A typical example of the usefulness of the finite approach to the two-parameter Ewens distribution is the exact calculation of the equilibrium average number of clusters as a function of α, θ, n. Let us consider the balance between death and birth. If the system is described by $\{Z_{i,n}\}$, considering unary changes, a cluster dies if and only if the extracted unit belongs to a singleton, hence the probability of a death is $z_{1,n}/n$. A new cluster is born if the extracted unit is a pioneer, and, given that the pioneering probability depends on the actual number of present clusters, we must distinguish the case in which the accommodating unit comes from a singleton, where the actual number of clusters is $K_{n-1} = K_n - 1$, from the alternative one, where the actual number of clusters is $K_{n-1} = K_n$. Hence

$$\begin{cases} \mathbb{P}(\text{death}|\mathbf{z}_n) = \dfrac{z_{1,n}}{n}, \\ \mathbb{P}(\text{birth}|\mathbf{z}_n) = \dfrac{z_{1,n}}{n}\dfrac{\theta + (k_n - 1)\alpha}{n-1+\theta} + \left(1 - \dfrac{z_{1,n}}{n}\right)\dfrac{\theta + k_n\alpha}{n-1+\theta}. \end{cases} \tag{9.98}$$

Now

$$\begin{aligned} &\frac{z_{1,n}}{n}\frac{\theta + (k_n - 1)\alpha}{n-1+\theta} + \left(1 - \frac{z_{1,n}}{n}\right)\frac{\theta + k_n\alpha}{n-1+\theta} \\ &\quad = \frac{1}{n(n-1+\theta)}\big(z_{1,n}(\theta + (k_n - 1)\alpha - \theta - k_n\alpha) + n(\theta + k_n\alpha)\big) \\ &\quad = \frac{1}{n(n-1+\theta)}\big(-z_{1,n}\alpha + n(\theta + k_n\alpha)\big). \end{aligned} \tag{9.99}$$

The unconditional death (and birth) probabilities are:

$$\begin{cases} \mathbb{P}(\text{death}) = \sum_{\mathbf{z}_n} \mathbb{P}(\text{death}|\mathbf{z}_n)\mathbb{P}(\mathbf{z}_n) = \dfrac{\mathbb{E}(Z_{1,n})}{n}, \\ \mathbb{P}(\text{birth}) = \dfrac{1}{n(n-1+\theta)}\big(-\mathbb{E}(Z_{1,n})\alpha + n(\theta + \mathbb{E}(K_n)\alpha)\big), \end{cases} \tag{9.100}$$

and the equilibrium balance implies

$$\mathbb{E}(Z_{1,n}) = -\mathbb{E}(Z_{1,n})\frac{\alpha}{n-1+\theta} + \frac{\theta + \mathbb{E}(K_n)\alpha}{n-1+\theta}, \tag{9.101}$$

that gives

$$\mathbb{E}(K_n) = \mathbb{E}(Z_{1,n})\frac{n-1+\theta+\alpha}{n\alpha} - \frac{\theta}{\alpha}. \tag{9.102}$$

Introducing in (9.102) the value

$$\mathbb{E}(Z_{1,n}) = \theta n\frac{(\theta+\alpha)^{[n-1]}}{\theta^{[n]}} = n\frac{(\theta+\alpha)^{[n-1]}}{(\theta+1)^{[n-1]}}, \tag{9.103}$$

from (9.95), we find

$$\mathbb{E}(K_n) = \frac{\theta}{\alpha}\left(\frac{(\theta+\alpha)^{[n]}}{\theta^{[n]}} - 1\right). \tag{9.104}$$

The fact that $\mathbb{E}(K_n)$ in (9.104) is exactly equal to $\sum_{i=1}^{n}\mathbb{E}(Z_{i,n})$ from (9.95) is easy to check numerically. Further it is easy to show that for small α,

$$\begin{aligned}\frac{(\theta+\alpha)^{[n]}}{\theta^{[n]}} &= \frac{(\theta+\alpha)}{\theta}\frac{(\theta+\alpha+1)}{\theta+1}\cdots\frac{(\theta+\alpha+n-1)}{\theta+n-1}\\ &= \left(1+\frac{\alpha}{\theta}\right)\left(1+\frac{\alpha}{\theta+1}\right)\cdots\left(1+\frac{\alpha}{\theta+n-1}\right)\\ &= 1+\sum_{i=1}^{n}\frac{\alpha}{\theta+i-1}+o(\alpha).\end{aligned} \tag{9.105}$$

Hence

$$\mathbb{E}(K_n) = \frac{\theta}{\alpha}\sum_{i=1}^{n}\frac{\alpha}{\theta+i-1}+o(\alpha) = \sum_{i=1}^{n}\frac{\theta}{\theta+i-1}+o(\alpha), \tag{9.106}$$

where

$$\sum_{i=1}^{n}\frac{\theta}{\theta+i-1} = \mathbb{E}_{0,\theta}(K_n) \tag{9.107}$$

is the well-known expected number of clusters in the one-parameter Ewens case.

In the opposite limit $\alpha \to 1$, we have

$$\mathbb{E}(K_n) \to \theta\left(\frac{(\theta+1)^{[n]}}{\theta^{[n]}}-1\right) = \theta\left(\frac{\theta+n}{\theta}-1\right) = n, \tag{9.108}$$

as the population is composed of all singletons with probability one.

For large n, (9.104) can be approximated by

$$\mathbb{E}(K_n) \simeq \frac{\theta}{\alpha}\left(\frac{\Gamma(\theta)}{\Gamma(\theta+\alpha)}n^{\alpha}-1\right) \simeq \frac{\Gamma(\theta+1)}{\Gamma(\theta+\alpha)}\frac{n^{\alpha}}{\alpha}. \tag{9.109}$$

9.10 Summary

In this chapter the problem of allocating n elements into $d \gg n$ categories/clusters was discussed. This problem allows us to take into account *innovations*, namely when an element joins an empty category: in practice, this can be interpreted as a novelty not seen before.

It turns out that the joint distribution of cluster sizes is described by the so-called Ewens sampling formula. Two derivations of this formula were presented. The first makes use of the so-called site label process, in which, every time a new category/cluster appears, a label is assigned randomly drawn from a predefined set. The second directly uses the order of appearance in order to time-label a new cluster.

It is possible to modify the Ehrenfest–Brillouin Markov chain discussed in Chapter 7 in order to take innovations into account. When the resulting model is properly interpreted it turns out that a cluster sooner or later disappears. It is then interesting to study the distribution of lifetimes for a cluster of given size as well as its expected lifetime. On the contrary, a site can resuscitate leading to an aperiodic and irreducible Markov chain, with the Ewens sampling formula being the joint probability distribution for the size of the sites. Suitable birth–death chains describe the marginal probabilistic dynamics evolution for a cluster and for a site.

Finally, a generalization of the Ewens distribution to two parameters is possible. Both parameters rule the probability of innovations which, in this case, also depends on the total number of clusters.

Further reading

Ewens introduced his sampling formula in a 1972 paper

W.J. Ewens, *The Sampling Theory of Selectively Neutral Alleles*, Theoretical Population Biology, **3**, 87–112 (1972).

There is a tutorial paper written by Tavaré and Ewens that summarizes many relevant results including some of the equations derived in this chapter

S. Tavaré and W.J. Ewens, *Multivariate Ewens Distribution* in N.L. Johnson, S. Kotz and N. Balakrishnan (eds), *Discrete Multivariate Distributions*, Wiley, New York (1997).

Donnelly's paper introducing the size-biased permutation of the Ewens sampling formula appeared in 1986

P. Donnelly, *Partition Structures, Pólya Urns, the Ewens Sampling Formula and the Ages of Alleles*, Theoretical Population Biology, **30**, 271–288 (1986).

One year later, Hoppe published his own paper on the sampling theory of neutral alleles

F.M. Hoppe, *The Sampling Theory of Neutral Alleles and an Urn Model in Population Genetics*, Journal of Mathematical Biology, **25**, 123–159 (1987).

where his urn model was discussed.

All the above papers concern the one-parameter Ewens sampling formula and are devoted to its applications in biology. The two-parameter generalization discussed in Section 9.9 was introduced by Hansen and Pitman

B. Hansen and J. Pitman, *Prediction Rules for Exchangeable Sequences Related to Species Sampling*, Statistics and Probability Letters, **46**, 251–256 (2000).

but Zabell anticipated it in his 1992 work on predictive inferences

S. Zabell, *Predicting the Unpredictable*, Synthese, **90**, 205–232 (1992).

Aoki applied these methods to the formation of clusters in economics

M. Aoki, *New Approaches to Macroeconomic Modeling: Evolutionary Stochastics Dynamics, Multiple Equilibria and Externalities as Field Effects*, Cambridge University Press, Cambridge, UK (1996).

M. Aoki, *Cluster Size Distribution of Economic Agents of Many Types in a Market*, Journal of Mathematical Analysis and Applications, **249**, 32–52 (2000).

M. Aoki, *Modeling Aggregate Behavior and Fluctuations in Economics*, Cambridge University Press, Cambridge, UK (2002).

The finitary version discussed in this chapter was studied in a series of papers co-authored by Garibaldi

U. Garibaldi, D. Costantini and P. Viarengo, *A Finite Characterization of Ewens Sampling Formula*, Advances in Complex Systems, **7**, 265–284 (2003).

U. Garibaldi, D. Costantini and P. Viarengo, *A Finitary Characterization of Ewens Sampling Formula* in T. Lux, S. Reitz and E. Samanidou (eds), *Nonlinear Dynamics and Heterogeneous Interacting Agents*, Lecture Notes in Economics and Mathematical Systems, **550**, Springer, Berlin (2005).

U. Garibaldi, D. Costantini and P. Viarengo, *The Two-parameter Ewens Distribution: a Finitary Approach*, Journal of Economic Interaction and Coordination, **2**, 147–161 (2007).

Readers interested in Gibrat's model may consult his book

R. Gibrat, *Les Inégalités Économiques*, Librairie du Recueil Sirey, Paris (1931).

A recent account of his ideas is also available

M. Armatte, *Robert Gibrat and the Law of Proportional Effect* in W.J. Samuels (ed.), *European Economists of the Early 20th Century* vol. I, Edward Elgar Publishing, Northampton, UK (1998).

Random walks and multiplicative processes are dealt with in many textbooks already quoted in previous chapters.

10

The Zipf–Simon–Yule process

This chapter is devoted to a modification of the Chinese restaurant example in which the probability of choosing a new table is a fixed parameter u and no longer depends on the number of previous customers. This leads to a non-exchangeable accommodation process. The final sections of this chapter deal with the continuous limit of finite birth-and-death chains.

After reading this chapter, you should be able to:

- understand why Zipf's process is not exchangeable;
- understand the meaning of Simon's regression;
- derive the Yule distribution in various ways;
- write a Monte Carlo program for cluster dynamics leading to cluster-size distributions with power-law tails;
- discuss the continuous limit of discrete birth-and-death Markov chains.

10.1 The Ewens sampling formula and firm sizes

To understand the rationale for this chapter, it is useful to compare the Ewens sampling formula with empirical data on firm sizes. In Section 9.7.1, various economic interpretations of the model under scrutiny were given. Indeed, one can consider the elements as workers and the clusters as firms. For the dynamics discussed in Section 9.8.2, and for $n \gg \theta$, one finds that $\mathbb{E}(Z_i)$ is given by Equation (9.49). One can further note that, by definition,

$$\sum_{i=1}^{n} \mathbb{E}(Z_i) = \mathbb{E}(K_n), \tag{10.1}$$

and that, in the limit of large n, due to Equation (9.28), one finds

$$\mathbb{E}(K_n) = \sum_{i=0}^{n-1} \frac{\theta}{\theta + i} \approx \theta \ln \frac{n+\theta}{\theta} + \gamma, \tag{10.2}$$

where γ is Euler's constant. Note that $\mathbb{E}(Z_i)$ is the expected number of firms of size i when the size of the population is n. The meaning of $\mathbb{E}(K_n)$ in the firm size problem is the equilibrium expected number of firms, which increases logarithmically with total number of individual agents n. For a comparison with empirical data, $\mathbb{E}(Z_i)$ and $\mathbb{E}(K_n)$ are the theoretical quantities to be compared with empirical values. In particular, $\mathbb{E}(Z_i)/\mathbb{E}(K_n)$ is the expected fraction of firms of size i. From the marginal point of view, this ratio gives the time fraction spent by a firm in the state corresponding to size i. A remarkable case of Equation (9.49) is found when $\theta = 1$. If $\theta = 1$, then $\mathbb{E}(Z_i) = 1/i$. In other words, the expected firm size follows a power law. This looks promising, but it is deceiving. In the case of USA firms, for instance, the rough number of firms is $k = 5.5 \cdot 10^6$ and the number of workers is about $n = 105 \cdot 10^6$. In the Ewens model, given n, the only parameter to be fixed is θ, whose best estimate can be found from the empirical value of k. Inverting (10.2), one gets $\widehat{\theta} = 1.24 \cdot 10^6$, and in that region the normalized Equation (9.49) cannot be distinguished from

$$\mathbb{E}(Z_i) \approx \frac{\widehat{\theta}}{i} \exp(-i\widehat{\theta}/n). \tag{10.3}$$

This can be represented by the log-series distribution

$$L(i) = -\frac{1}{\log(1-y)} \frac{y^i}{i}, \; y = 1, 2, \ldots \tag{10.4}$$

where the parameter is $y = \exp(-\widehat{\theta}/n)$. In this limit, the normalizing constant given in Equation (10.2) becomes

$$\mathbb{E}(K_n) = \widehat{\theta} \ln \frac{n}{\widehat{\theta}}. \tag{10.5}$$

It follows that the size distribution is approximately proportional to $\widehat{\theta}/i$ for small sizes, but has an exponential tail for large sizes. This is sufficient to exclude a Ewens-like dynamics producing equilibrium probabilities whose tail follows a power law. Therefore, it is necessary to define a new cluster dynamics leading to a power-law distribution of cluster sizes.

10.2 Hoppe's vs. Zipf's urn

Consider the label process for Hoppe's urn. Using the indicator functions $\mathbb{I}_{n_j=0}$ which is 1 when $Y_j^* = n_j = 0$ and 0 for $Y_j^* = n_j > 0$ and its complement

$\mathbb{I}_{n_j>0} = 1 - \mathbb{I}_{n_j=0}$, Equation (9.7) can be written in a single line

$$\mathbb{P}(X^*_{n+1} = j|n_j, n) = \frac{n}{n+\theta}\frac{n_j}{n}\mathbb{I}_{n_j>0} + \frac{\theta}{n+\theta}\frac{1}{g-K_n}\mathbb{I}_{n_j=0}, \quad (10.6)$$

where all the parameters and variables have the same meaning as in Chapter 9. Once more, note that at the beginning of this process, the probability of selecting one of the g categories/sites is given by $1/g$ and this is automatically included in Equations (9.7) and (10.6). In the right-hand side of Equation (10.6), the first term describes herding as it gives the probability of joining an already existing cluster of size j, whereas the second term describes innovation because it is the probability of founding a new cluster in an empty site as a pioneer. In Chapter 9, it was shown that the label process is exchangeable. For example, the probability of the sequence $X^*_1 = i, X^*_2 = j, X^*_3 = i$ (with $i \neq j$) is

$$\mathbb{P}(X^*_1 = i, X^*_2 = j, X^*_3 = i) = \frac{1}{g(g-1)}\frac{\theta^2}{\theta(\theta+1)(\theta+2)}, \quad (10.7)$$

and it coincides with the probability of any permutation of the values, that is with the probability of the sequences $X^*_1 = i, X^*_2 = i, X^*_3 = j$ and $X^*_1 = j, X^*_2 = i, X^*_3 = i$. The label process for Zipf's urn can be defined in a similar way. One can write for $n > 0$

$$\mathbb{P}(X^*_{n+1} = j|n_j, n) = (1-u)\frac{n_j}{n}\mathbb{I}_{n_j>0} + u\frac{1}{g-K_n}\mathbb{I}_{n_j=0}, \quad (10.8)$$

where $0 < u < 1$ is the probability of innovation which is now independent of n and its complement $1-u$ is the probability of herding, whereas for $n = 0$, one has

$$\mathbb{P}(X^*_1 = j) = \frac{1}{g}. \quad (10.9)$$

A direct application of Equation (10.8) on the sequences $X^*_1 = i, X^*_2 = j, X^*_3 = i$ and $X^*_1 = i, X^*_2 = i, X^*_3 = j$ immediately shows that this process is no longer exchangeable. Indeed, one has

$$\mathbb{P}(X^*_1 = i, X^*_2 = j, X^*_3 = i) = \frac{1}{2}\frac{u(1-u)}{g(g-1)}, \quad (10.10)$$

whereas

$$\mathbb{P}(X^*_1 = i, X^*_2 = i, X^*_3 = j) = \frac{u(1-u)}{g(g-1)}. \quad (10.11)$$

Given that the process of individual sequences is not exchangeable, the methods introduced in Section 5.3.8 and used in Chapters 7, 8 and 9 cannot be applied in order to derive the distribution of cluster sizes. In 1955, Simon, aware or not of the problem, studied the expected cluster dynamics as a way to circumvent this problem and obtain quantitative results. His method is described in the next section using the terminology developed in the previous chapters.

10.3 Expected cluster size dynamics and the Yule distribution

Simon refers to the sequential construction of a text adding a new word at each step. This is equivalent to considering a new customer entering the Chinese restaurant and choosing an existing table (a word already written in the text) or a new table (a new word not yet used in the text). Here, instead of using site labels as in the previous section, it is useful to refer to time labels. Equation (10.8) modifies to

$$\mathbb{P}(X_{n+1} = j|n_j, n) = (1-u)\frac{n_j}{n}\mathbb{I}_{j\le K_n} + u\mathbb{I}_{j=K_n+1}. \tag{10.12}$$

Simon assumes that you are writing a text using the following rules. You have already written n words. If $z_{i,n}$ is the number of words appearing i times in the text, the probability of using one of these words at the next step is proportional to $iz_{i,n}$; the probability of using a new word not yet present in the text is constant and equal to u with $0 < u < 1$. Note that the first word in the text is new with probability 1. Moreover, if $u = 0$, no innovation is possible and $\mathbb{P}(Z_{n,n} = 1) = 1$, in other words, there is only a cluster containing n repetitions of the same word. On the contrary, if $u = 1$, every word included in the text is a new word not yet used; therefore, in this case, the text is made up of n distinct words meaning that there are n clusters each containing 1 element, so that one has $\mathbb{P}(Z_{1,n} = n) = 1$. If $Y_{n+1} = i$ denotes the event that the $(n+1)$th word is among the words that occurred i times, and $Y_{n+1} = 0$ means that the $(n+1)$th word is a new one, Equation (10.12) leads to the following probabilities conditioned on the partition vector $\mathbf{Z}_n = \mathbf{z}_n$:

$$\begin{cases} \mathbb{P}(Y_{n+1} = i|\mathbf{Z}_n = \mathbf{z}_n) = (1-u)\dfrac{iz_{i,n}}{n}, \\ \mathbb{P}(Y_{n+1} = 0|\mathbf{Z}_n = \mathbf{z}_n) = u, \quad n > 0 \\ \mathbb{P}(Y_1 = 0) = 1. \end{cases} \tag{10.13}$$

The evolution of the partition vector is as follows. If $Y_{n+1} = i$, it implies that a cluster of size i is destroyed and a cluster of size $i+1$ is created meaning that $z_{i,n+1} = z_{i,n} - 1$ and $z_{i+1,n+1} = z_{i+1,n} + 1$. If $Y_{n+1} = 0$, it means that a cluster of size 1 is created, that is $z_{1,n+1} = z_{1,n} + 1$. Consider now the random variable difference

$Z_{i,n+1} - Z_{i,n}$; it can assume two values, one has $Z_{i,n+1} - Z_{i,n} = 1$ if a cluster of size $(i-1)$ is destroyed and a cluster of size i is created, and $Z_{i,n+1} - Z_{i,n} = -1$ if a cluster of size i is destroyed and a cluster of size $i+1$ is created. As a consequence of Equation (10.13), one has for $i = 2, \ldots, n$

$$\mathbb{P}(Z_{i,n+1} - Z_{i,n} = 1 | \mathbf{Z}_n = \mathbf{z}_n) = (1-u)\frac{(i-1)z_{i-1,n}}{n}, \tag{10.14}$$

and

$$\mathbb{P}(Z_{i,n+1} - Z_{i,n} = -1 | \mathbf{Z}_n = \mathbf{z}_n) = (1-u)\frac{iz_{i,n}}{n}. \tag{10.15}$$

Therefore, considering that $\mathbb{E}(Z_{i,n} | \mathbf{Z}_n = \mathbf{z}_n) = z_{i,n}$ for any i, one finds that

$$\mathbb{E}(Z_{i,n+1} | \mathbf{Z}_n = \mathbf{z}_n) - z_{i,n} = (1-u)\left(\frac{(i-1)z_{i-1,n}}{n} - \frac{iz_{i,n}}{n}\right), \tag{10.16}$$

an equation valid for $i = 2, \ldots, n$. For $i = 1$, as a consequence of Equation (10.13), one finds that

$$\mathbb{E}(Z_{1,n+1} | \mathbf{Z}_n = \mathbf{z}_n) - z_{1,n} = u - (1-u)\frac{z_{1,n}}{n}. \tag{10.17}$$

Defining the unconditional expected value of $Z_{i+1,n}$

$$\bar{z}_{i+1,n} = \mathbb{E}(Z_{i+1,n}) = \sum_{\mathbf{z}_{n-1}} \mathbb{E}(Z_{i+1,n} | \mathbf{Z}_n = \mathbf{z}_n)\mathbb{P}(\mathbf{Z}_n = \mathbf{z}_n), \tag{10.18}$$

one can derive the following equations by taking the expectations of (10.16) and (10.17):

$$\bar{z}_{i,n+1} - \bar{z}_{i,n} = (1-u)\left(\frac{(i-1)\bar{z}_{i-1,n}}{n} - \frac{i\bar{z}_{i,n}}{n}\right), \tag{10.19}$$

and

$$\bar{z}_{1,n+1} - \bar{z}_{1,n} = u - (1-u)\frac{\bar{z}_{1,n}}{n}. \tag{10.20}$$

In principle, the recurrence Equations (10.19) and (10.20) can be directly solved. However, Simon suggests looking for solutions such that $\bar{z}_{i,n} \propto n$ corresponding to a steady growth of the clusters, that is, the fraction of clusters of size i tends to a constant value as n grows. With Simon's *Ansatz*, one has that

$$\frac{\bar{z}_{i,n+1}}{\bar{z}_{i,n}} = \frac{n+1}{n}, \tag{10.21}$$

equivalent to

$$\bar{z}_{i,n+1} - \bar{z}_{i,n} = \frac{\bar{z}_{i,n}}{n}, \tag{10.22}$$

for $i = 1, \ldots, n$. Replacing Equation (10.22) for $i = 1$ into Equation (10.20) gives

$$\bar{z}^*_{1,n} = \frac{nu}{2-u} = \frac{\rho}{1+\rho} nu, \tag{10.23}$$

where $\rho > 1$ is a parameter defined as

$$\rho = \frac{1}{1-u}. \tag{10.24}$$

If Equation (10.22) is replaced into Equation (10.19), the recurrence equation simplifies to

$$\bar{z}^*_{i,n} = \frac{(1-u)(i-1)}{1+(1-u)i} \bar{z}^*_{i-1,n} = \frac{i-1}{\rho+i} \bar{z}^*_{i-1,n}. \tag{10.25}$$

The iteration of (10.25) leads to a closed-form solution for $\bar{z}^*_{i,n}$:

$$\begin{aligned} \bar{z}^*_{i,n} &= \frac{i-1}{\rho+i} \frac{i-2}{\rho+i-1} \cdots \frac{1}{\rho+2} \bar{z}^*_{1,n} = \frac{\Gamma(i)\Gamma(\rho+2)}{\Gamma(\rho+i+1)} \bar{z}^*_{1,n} \\ &= (\rho+1) \frac{\Gamma(i)\Gamma(\rho+1)}{\Gamma(\rho+i+1)} \bar{z}^*_{1,n} = \rho B(i,\rho+1) nu, \end{aligned} \tag{10.26}$$

where Equation (10.23) and the definition of Euler's beta function (4.101) were used. Direct replacement shows that Equation (10.26) is a solution of Equation (10.19). Note that the ratio $\bar{z}_{i,n}/n$ on the left-hand side of Equation (10.22) is not a frequency. The expected number of clusters (of words in the text) is given by

$$\mathbb{E}(K_n) = \sum_{i=1}^{n} \mathbb{E}(Z_{i,n}) = \sum_{i=1}^{n} \bar{z}^*_{i,n} = nu \sum_{i=1}^{n} \rho B(i,\rho+1), \tag{10.27}$$

and, summing up to infinity, it is possible to prove that

$$\sum_{i=1}^{\infty} \rho B(i,\rho+1) = 1. \tag{10.28}$$

Therefore, from Equation (10.13) one has that $\mathbb{E}(K_n) = 1 + (n-1)u \approx \nu$ for large n, meaning that there is a constant flow of new words. In other words, Simon's

Ansatz means that the relative frequency of clusters of size i (the number of words that appeared i times in the text) given by $\mathbb{E}(Z_{i,n})/\mathbb{E}(K_n) \approx \bar{z}^*_{i,n}/nu$ is invariant once steady growth is reached. In order to complete this justification of Simon's method, it is necessary to convince oneself that Equation (10.28) holds true. Using

$$f_i = \rho B(i, \rho + 1), \tag{10.29}$$

one has that $\forall i, f_i > 0$, moreover $\forall i, f_i < 1$. Regarding the limiting properties, one has that $\lim_{i\to\infty} f_i = 0$ and that $\lim_{i\to\infty} if_i = 0$. From

$$f_i = \frac{i-1}{\rho + i} f_{i-1}, \tag{10.30}$$

one can derive the recurrence relation

$$\left(1 + \frac{1}{\rho}\right) f_i = (i-1)\frac{1}{\rho}(f_{i-1} - f_i), \tag{10.31}$$

valid for any $i \geq 2$, whereas, for $i = 1$, one has

$$\left(1 + \frac{1}{\rho}\right) f_1 = 1. \tag{10.32}$$

Now define $S_n = \sum_{i=1}^{n} f_i$. Using the previous two Equations (10.31) and (10.32), one arrives at

$$\left(1 + \frac{1}{\rho}\right) S_n = 1 + \frac{1}{\rho} S_n - \frac{n}{\rho} f_n. \tag{10.33}$$

In order to derive (10.33), one must write Equation (10.31) for $i = 2, \ldots, n$ and add all the $(n-1)$ equations and finally add also Equation (10.32). Solving Equation (10.33) for S_n leads to

$$\sum_{i=1}^{n} f_i = 1 - \frac{n}{\rho} f_n, \tag{10.34}$$

so that one finds

$$\sum_{i=1}^{\infty} f_i = \lim_{n\to\infty}\left(1 - \frac{n}{\rho} f_n\right) = 1. \tag{10.35}$$

In summary, Equation (10.29) defines a legitimate probability distribution of the integers $i \geq 1$ called a *Yule distribution*. Let Y be a random variable distributed according to the Yule distribution, then one has

$$\mathbb{P}(Y = i) = f_i = \text{Yule}(i; \rho) = \rho B(i, \rho + 1); \tag{10.36}$$

in particular, from Equation (10.26), one can see that the ratio $\bar{z}^*_{i,n}/nu$ tends to the Yule distribution. For the sake of completeness, and without proof, note that, if Y follows the Yule distribution, its expected value is

$$\mathbb{E}(Y) = \sum_{i=1}^{\infty} i f_i = \frac{\rho}{\rho - 1}, \tag{10.37}$$

defined for $\rho > 1$, and its variance is

$$\mathbb{V}(Y) = \frac{\rho^2}{(\rho - 1)^2(\rho - 2)}, \tag{10.38}$$

defined for $\rho > 2$.

It is now time to summarize the previous results. If at each step a unit is added to the system (a word is added to the text) according to Equation (10.13), the expected number of clusters of size i (the expected number of words with i occurrences) converges to $\bar{z}^*_{i,n}/nu$. Note that $\bar{z}_{i,n} = 0$ for $i > n$ and the convergence holds true only for $n \gg i$. In other words, considering a fixed size i, if the population grows, it is possible to find a size n_0 such that for $n > n_0$, one has $\bar{z}_{i,n} \propto n$ reaching a steady growth. Eventually, one can see that for $i \gg \rho$

$$f_i = \rho B(i, \rho + 1) = \rho \frac{\Gamma(i)\Gamma(\rho + 1)}{\Gamma(i + \rho + 1)} = \rho \frac{\Gamma(\rho + 1)}{i(i+1)\cdots(i+\rho)} \approx \frac{\rho\Gamma(\rho + 1)}{i^{\rho+1}}, \tag{10.39}$$

so that one gets the prescribed power-law tail of the distribution with $f_i \approx i^{-(\rho+1)}$ whose continuous limit is Pareto-like.

10.4 Birth and death Simon–Zipf's process

Simon himself realized that a complete solution of his problem needs a birth–death process, in which the length of the text (the size of the population) is fixed, and clusters are created and destroyed according to a suitable probabilistic model. He tackled this question in Section III of his paper. This discussion can be traced back at least to Steindl's book (see further reading, Chapter 8). In Simon's paper, Equation (10.19) was interpreted in this way: the term

$$\frac{1-u}{n}\left((i-1)\bar{z}_{i-1,n} - i\bar{z}_{i,n}\right) \tag{10.40}$$

is the increment of $\bar{z}_{i,n}$ due to a new arrival, whereas

$$-\frac{\bar{z}_{i,n}}{n} \tag{10.41}$$

is the decrease due to the probability of a destruction. If the two contributions balance, it means that $\bar{z}^*_{i,n}$ is the invariant distribution of a Markov chain. While in the original model (10.19), the parameter n represents the size of the text, in the new interpretation it represents time. This is quite unclear.

More than 50 years after Simon's seminal work, it is possible to present clear statements. The first statement deals with the lack of exchangeability for the creation process; this forbids closed formulae for the sampling distribution. This problem was discussed in Section 10.2. Further, assume that the system is 'built up' to size n following (10.13), so that the cluster-size expected values exactly follow Simon's equations. Then growth is stopped, the system is left to evolve following unary moves, where a word is randomly cancelled (all words being equivalent) and a new word is produced following (10.13): then the marginal transition probability is just given by Equations (9.75), (9.76) and (9.77) with

$$\frac{\theta}{\theta+n-1} = u. \tag{10.42}$$

Therefore, the equilibrium distribution is the Ewens sampling formula[1] given by Equation (9.13) with

$$\theta = \frac{u(n-1)}{1-u}. \tag{10.43}$$

Alternatively, consider n-ary moves: if at each step the system is razed to the ground and rebuilt following Zipf's scheme, the equilibrium distribution coincides with that of Simon's method. Therefore, the equilibrium distribution strongly depends on the number of destructions-per-step. If one wishes to use a creation probability such as (10.13), in order to escape the Ewens basin of attraction, one has to change the death model in a substantial way.

Indeed, Simon's suggestion was to remove a whole cluster, and then to reassign the corresponding number of items to the population based on Zipf's scheme. Suppose that the destruction step consists in eliminating all the occurrences of a word, while creation consists in adding a number of words equal to the size of

[1] One might ask what is the rate of approach to equilibrium. As in the Ehrenfest–Brillouin case, one has that $r = \theta/[n(\theta+n-1)]$ is the rate of approach of the expected value to its equilibrium value, and r^{-1} is the related number of steps. Using the values of Section 10.1, one finds that $r^{-1} \approx [n(\theta+n)]/\theta \approx 9 \cdot 10^9$, that is the number of unary moves needed to reach equilibrium. There are nearly 86 moves per worker.

the destroyed cluster. First, one eliminates a cluster, all clusters being equivalent, and then one puts back the corresponding number of items into the population following Equation (10.13). At each step, the size of the population first diminishes by a random size m equal to the size of the destroyed cluster, and then returns to its natural size n via m accommodations. Using the interpretation in terms of firms, at each step a firm dies, and all its workers either join existing firms or found new ones with individual probability $1-u$ or u, respectively. Newborn firms accommodate in some empty site, and they are labelled by their site. The Markov chain of the site process is ergodic, but an analytic solution is cumbersome because the transition matrix is very complex, due to the lack of exchangeability for the creation process. In Section II of his paper, Simon suggests that this approach is reasonable. Consider Equation (10.16) and compare it with (10.17). If all $z_{i,n}$ are equal to their equilibrium expected value $\bar{z}_{i,n}$ the increment of $z_{i,n}$ is given by (10.22). On the contrary, if $z_{i-1,n} = \bar{z}_{i-1,n}$ but $z_{i,n} = \bar{z}_{i,n} + \varepsilon_{i,n}$, then, its expected increment $\mathbb{E}(z_{i,n+1} - z_{i,n})$ differs from the expected increment by a restoring term

$$-(1-u)\frac{i\varepsilon_{i,n}}{n}. \tag{10.44}$$

In words, on average, a fluctuation away from the expected value is smoothed by new arrivals.

10.5 Marginal description of the Simon–Zipf process

If a cluster of size i is considered, belonging to a population whose dynamics is described by the mechanism described in the previous section, one can see that the time evolution cannot be a simple function of i, $n-i$ and u. In fact, the death mechanism deeply differs from the Ewens case; it amounts to 'killing' a whole cluster, with equal probability for all the k active clusters (with $1 \le k \le n$). Two cases are possible: the first is to assume that a destruction occurs at each step; in the second case, one sets the probability of removing each existing cluster equal to $1/n$, so that the probability of removals is $k/n \le 1$, with the proviso that if no killing occurs the system is left in the initial state. The second version can be interpreted supposing that each cluster is represented by its elder worker. If at each step a unit is chosen, killing occurs when the elder worker is chosen. With this assumption the probability of being killed, that is moving from size i to size 0, is

$$w(i,0) = \frac{1}{n}. \tag{10.45}$$

If another cluster dies, the cluster of size i can increase up to the (random) size of the destroyed cluster. When $n \gg 1$, as the probability of joining the cluster is proportional to i/n, the mechanism can be simplified assuming that no more than one unit is allowed to eventually join the cluster, with probability given by $(1-u)i/n$, that is

$$w(i,i+1) = (1-u)\frac{i}{n}. \tag{10.46}$$

The size of the cluster does not change with probability

$$w(i,i) = 1 - w(i,0) - w(i,i+1). \tag{10.47}$$

A rebirth term $w(0,1)$ is necessary in order to avoid that the state $i=0$ is absorbing. The transition probabilities are summarized in the following table giving $\{w(i,j) : i,j = 0,\ldots,n\}$, with $w(0,0) = 1 - w(1,0)$:

$$\mathbb{W} =$$

$$\begin{pmatrix} w(0,0) & w(0,1) & 0 & 0 & \ldots & 0 & 0 \\ \frac{1}{n} & w(1,1) & \frac{1-u}{n} & 0 & \ldots & 0 & 0 \\ \frac{1}{n} & 0 & w(2,2) & \frac{2(1-u)}{n} & \ldots & \ldots & 0 \\ \ldots & \ldots & \ldots & \ldots & \ldots & \ldots & \ldots \\ \frac{1}{n} & 0 & 0 & 0 & \ldots & w(n-1,n-1) & \frac{(n-1)(1-u)}{n} \\ \frac{1}{n} & 0 & 0 & 0 & \ldots & 0 & w(n,n) \end{pmatrix}.$$

It is apparent that the above Markov chain is not reversible as the dynamics consists in a step-by-step growth followed by a sudden breakdown. Therefore, the invariant distribution cannot be found using detailed balance. As discussed in Chapter 6, the invariant distribution $p(i) = \mathbb{P}(Y = i), i = 0,\ldots,n$, must satisfy the Markov equation

$$p(i) = \sum_{j=1}^{n} p(j)w(j,i),\ i = 0,1,\ldots n. \tag{10.48}$$

Then, for $i = 0$, one has

$$\begin{aligned} p(0) &= p(0)w(0,0) + p(1)w(1,0) + \ldots + p(n)w(n,0) \\ &= p(0)w(0,0) + (1-p(0))\frac{1}{n}. \end{aligned} \tag{10.49}$$

For $i = 1$, Equation (10.48) becomes

$$p(1) = p(0)w(0,1) + p(1)w(1,1). \tag{10.50}$$

For $1 \leq i < n$ the generic term is

$$p(i) = p(i-1)w(i-1,i) + p(i)w(i,i); \tag{10.51}$$

it can be re-written as

$$p(i)(1 - w(i,i)) = p(i-1)w(i-1,i). \tag{10.52}$$

Using Equations (10.45), (10.46) and (10.47) leads to

$$p(i)\left(\frac{1}{n} + (1-u)\frac{i}{n}\right) = p(i-1)\frac{(1-u)(i-1)}{n}. \tag{10.53}$$

This equation coincides with (10.25). The explicit form of $p(i)$ is

$$p(i) = \frac{\theta_{i-1}}{\varphi + \theta_i} \cdots \frac{\theta_1}{\varphi + \theta_2} \frac{\theta_0}{\varphi + \theta_1} \frac{\varphi}{\varphi + \theta_0}, \tag{10.54}$$

where $\theta_i = w(i,i+1)$ and $\varphi = 1/n$. The equation

$$p(0) = \frac{\varphi}{\varphi + \theta_0} \tag{10.55}$$

can be interpreted as the probability (the fraction of time) that the label is idle. Conditioning on the time fraction in which the cluster is alive, that is introducing

$$\pi_i = \mathbb{P}(Y = i | i > 0) = \frac{p(i)}{1 - p(0)} = p(i)\frac{\varphi + \theta_0}{\theta_0}, \tag{10.56}$$

one obtains

$$\pi_i = \frac{\theta_{i-1}}{\varphi + \theta_i} \cdots \frac{\theta_1}{\varphi + \theta_2} \frac{\varphi}{\varphi + \theta_1}, \tag{10.57}$$

a result that does not depend on the rebirth term θ_0. Note that

$$\frac{\theta_{i-1}}{\varphi + \theta_i} = \frac{\dfrac{i-1}{n\rho}}{\dfrac{1}{n} + \dfrac{i}{n\rho}} = \frac{i-1}{\rho + i}, \tag{10.58}$$

so that

$$\pi_1 = \frac{1}{1+(1-u)} = \frac{\rho}{\rho+1},$$
$$\pi_2 = \frac{\rho}{\rho+1}\frac{1}{\rho+2},$$
$$\pi_3 = \frac{\rho}{\rho+1}\frac{1}{\rho+2}\frac{2}{\rho+3},\ldots \qquad (10.59)$$

Therefore, it turns out that

$$\pi_i = \rho\frac{\Gamma(i)\Gamma(\rho+1)}{\Gamma(\rho+i+1)} = \rho B(i,\rho+1)\ i = 1,\ldots,n-1; \qquad (10.60)$$

this is exactly the value of the Yule distribution (10.29). The term π_n is different, as $\theta_n = 0$; therefore, one finds

$$\pi_n = \frac{\theta_{n-1}}{\varphi+\theta_n}\pi_{n-1} = \frac{n-1}{\rho}\pi_{n-1} \neq \frac{n-1}{\rho+n}\pi_{n-1}. \qquad (10.61)$$

This is not surprising as the support of π_i is finite, while for the Yule distribution f_i given by Equation (10.29) it is unlimited. The equilibrium distribution of the marginal chain is then the right-censored Yule distribution

$$\pi_i = f_i = \rho B(i,\rho+1),\ \text{for}\ i = 1,\ldots,n-1$$
$$\pi_n = \sum_{i=n}^{\infty} f_i = \frac{\Gamma(n)\Gamma(\rho+1)}{\Gamma(\rho+n)}, \qquad (10.62)$$

where the last term contains all the mass in the tail of the Yule distribution. The chain described above has the same structure as the so-called *problem of runs*, because at each step the state, if it moves, either increases its size by one or it jumps down to zero. The cluster suffers a 'sudden death' which is not the consequence of subsequent departures of all its units.

10.6 A formal reversible birth-and-death marginal chain

In Section 10.3 the following recurrence equation was derived:

$$\bar{z}_i(1+(1-u)i) = \bar{z}_{i-1}(1-u)(i-1), \qquad (10.63)$$

for the equilibrium expected number of clusters of size i. If $\bar{z}_i/(nu)$ is interpreted as the equilibrium distribution of the marginal process for a cluster (firm) P_i, then

$$\frac{P_i}{P_{i-1}} = \frac{\bar{z}_i}{\bar{z}_{i-1}} \tag{10.64}$$

implies that

$$P_i(1+(1-u)i) = P_{i-1}(1-u)(i-1). \tag{10.65}$$

A probabilistic dynamics compatible with (10.65) is the following: at each step a cluster (firm) can lose or gain a unit (worker) at most (unary moves). The increment–decrement probability is

$$\begin{aligned} w(i,i+1) &= \psi_i = A\frac{(1-u)i}{n}, \\ w(i,i-1) &= \phi_i = A\frac{1+(1-u)i}{n}. \end{aligned} \tag{10.66}$$

Then (10.65) can be interpreted as the balance equation of the chain, with the additional normalization constant A. This chain produces the right-truncated Yule distribution

$$\pi_i = \mathbb{P}(Y=i|i \leq n) = \frac{f_i}{F_n}, i=1,\ldots,n \tag{10.67}$$

as equilibrium probability, where $F_n = \sum_{i=1}^{n} f_i$. Whereas in the previous case, the marginal chain is obtained from a microscopic dynamics, in this case, the destruction–creation mechanism does not look 'agent-based' in a clear way. The factor $1/n$ in $w(i,i-1)$ reflects the possibility that one of the z_i firms may be destroyed, while the two factors $(1-u)i/n$ account for the addition of a moving unit. In fact, in the model in which a firm is destroyed (closed) with probability $1/n$, the corresponding workers open new firms with probability u, and with probability $1-u$ they join existing firms proportionally to the size of these firms; it is reasonable to set

$$\begin{aligned} \mathbb{P}(\Delta Z_i = 1) &= (1-u)\frac{(i-1)z_{i-1}}{n}, \\ \mathbb{P}(\Delta Z_i = -1) &= \frac{z_i}{n} + (1-u)\frac{iz_i}{n}, \end{aligned} \tag{10.68}$$

if, for $n \to \infty$, one assumes that at most one free worker joins a cluster of size i or a cluster of size $i-1$. If he/she joins a cluster of size $i-1$, one has that $z_i \to z_i+1$, while if he/she joins a cluster of size i, one has that $z_i \to z_i - 1$. In the limit

$n \to \infty$, it is reasonable to assume that these three events are separate. Then one has $\mathbb{E}(\Delta Z_i) = 0$ if (10.63) holds. In (10.66), the increasing term ψ_i is proportional to the size of the cluster and to the probability of a herding choice, whereas the decreasing term ϕ_i is still proportional to the size of the cluster and to the probability of a herding choice, plus a term taking the probability of death into account.

10.7 Monte Carlo simulations

In order to simulate the dynamics of clusters, one can consider a system of n elements which can be found in $g = n+1$ sites, assuming that the initial state is $\mathbf{Y}^* = (1,1,\ldots,1,0)$. At each step a cluster is removed. If k is the number of active clusters, for each cluster, the probability of removal is simply given by $1/k$. If m is the size of the removed cluster, these items are returned to the system following Zipf's scheme. Accommodations are such that elements either join existing clusters (proportionally to their size) or move to free sites with innovation probability u. However, they cannot join the cluster which was just destroyed. The herding probabilities of joining already existing clusters sum up to $1-u$. A Monte Carlo program written in *R* is listed below.

```
# Program clusters.R
# Number of objects
n<-50
# Number of sites
g<-n+1
# Number of steps
T<-2000
# Probability of innovation
u<-0.5
# Initial state
Y<-rep(c(1), times=n)
Y<-c(Y,0)
# Array of results
A<-Y
# Main cycle
for (t in 1:T) {
# destruction
# sites with at least one item
indexp<-which(Y>0)
# number of clusters
Kp<-length(indexp)
# a cluster is selected and removed
```

```
R<-sample(1:Kp,1)
irem<-indexp[R]
# size of the removed cluster
m<-Y[irem]
# the cluster is removed
Y[irem]<-0
# empty sites
index0<-which(Y==0)
# number of empty sites
K0<-length(index0)
# update of active sites
indexp<-which(Y>0)
# update of cluster number
Kp<-length(indexp)
#number of items in the system
N<-n-m
# creation
for (i in 1:m) {
# when all the sites are empty, a new site is filled
# with uniform probability. It never coincides with the
# previously destroyed site!
if (N==0) {
Y[irem]<-1
index0<-which(Y==0)
K0<-length(index0)
F<-sample(1:K0,1)
icre<-index0[F]
Y[icre]<-1
Y[irem]<-0
} # end if on N==0
if (N>0) {
rand<-runif(1)
if (rand<=u) { # innovation
# a new cluster is created
# update of empty clusters
Y[irem]<-1
index0<-which(Y==0)
K0<-length(index0)
# an empty site is selected and filled
F<-sample(1:K0,1)
```

```
ifill<-index0[F]
Y[ifill]<-1
Y[irem]<-0
} # end if on innovation
if (rand >u) { # herding
# number of active clusters
indexp<-which(Y>0)
# frequency vector
prob<-Y[indexp]/N
# cumulative probability
cumprob<-cumsum(prob)
# pointer to selected site
indexsite<-min(which((cumprob-runif(1))>0))
indexsite<-indexp[indexsite]
Y[indexsite]=Y[indexsite]+1
} # end if on herding
} # end if on N>0
N<-N+1
} # end for on i
A<-c(A,Y)
} # end for on t
A<-matrix(A,nrow=T+1,ncol=g,byrow=TRUE)
# Frequencies
norm<-0
f<-rep(c(0),times=n)
for (j in 1:n) {
f[j]<-length(which(A==j))
norm<-norm + f[j]
} # end for on j
f<-f/norm
# Yule distribution
rho<-1/(1-u)
fth1<-rep(c(0),times=n)
for (j in 1:n) {
fth1[j] = rho*beta(j,rho+1);
} # end for on j
k<-c(1:n)
# Plot of results
plot(log(k),log(f),xlab="i",ylab="f(i)")
lines(log(k),log(fth1))
```

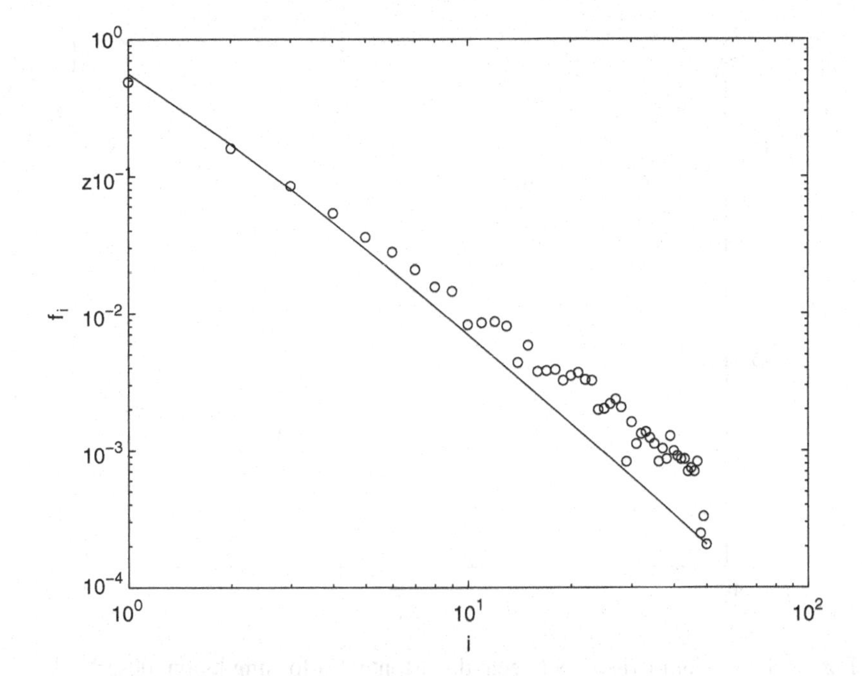

Figure 10.1. Comparison between the Monte Carlo simulation described in Section 10.7 and the Yule distribution. In the case represented in this figure, the innovation parameter is $u = 0.2$ and the number of elements is $n = 50$; the circles represent the relative frequency of clusters with i elements over $T = 2000$ steps, the continuous line is the Yule distribution of parameter $\rho = 1/(1-u)$.

Some results of the simulations described above are presented in Fig. 10.1 and 10.2. In both cases $n = 50$, but in Fig. 10.1, u is set equal to 0.2, whereas $u = 0.5$ in Fig. 10.2. The agreement with the Yule distribution is better for the larger innovation parameter. Note that small values of u yield a more rapid growth of clusters, larger sizes and shorter mean life, while large values of u imply smaller growth rate, greater dispersion as innovations are more frequent and larger mean life.

10.8 Continuous limit and diffusions

It is now timely to further analyze the continuous limit of the discrete models that have been discussed so far. It turns out that birth-and-death chains have *diffusion processes* (also known as *diffusions*) as their continuous limit. A detailed discussion of diffusions is outside the scope of the present book. However, a heuristic treatment of the limit is possible. Below, we limit ourselves to unidimensional motions such as the marginal chains of a category or of a site. Moreover, only unary moves are dealt with.

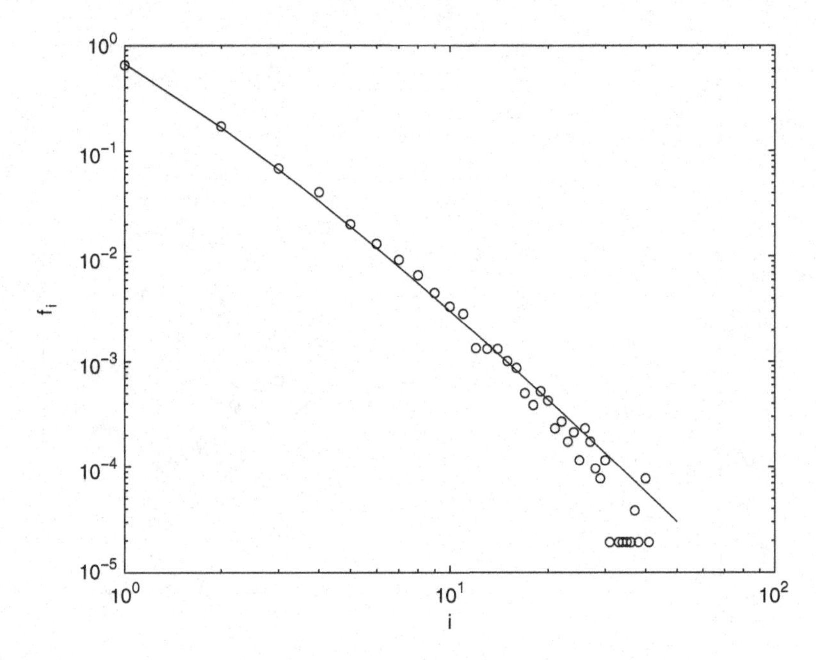

Figure 10.2. Comparison between the Monte Carlo simulation described in Section 10.7 and the Yule distribution. In the case represented in this figure, the innovation parameter is $u = 0.5$ and the number of elements is $n = 50$; the circles represent the relative frequency of clusters with i elements over $T = 2000$ steps, the continuous line is the Yule distribution of parameter $\rho = 1/(1-u)$.

10.8.1 From birth-and-death chains to diffusions

It is possible to prove that the continuous limit of birth-and-death chains is a homogeneous diffusion process. Even if a detailed discussion and the proof of this limit are outside the scope of this book, it is possible to give a simple method for evaluating the continuous limit of the stationary solution for the birth-and-death marginal Markov chain. Here, the chain is homogeneous, irreducible and aperiodic, so that it admits a unique invariant distribution which is also the equilibrium distribution. Hopefully, these convenient properties are conserved in the continuous limit.

A birth-and-death Markov chain is characterized by the three transition probabilities $\psi_k = w(k, k+1)$, $\phi_k = w(k, k-1)$ and $w(k,k) = 1 - \psi_k - \phi_k$, and by suitable boundary conditions. The integer variable $Y = k$ denotes the state of the system and can only vary by one unit. All the one-step transitions by more than one unit are forbidden. Thanks to the total probability theorem, the t-step transition matrix of the chain can be written in the following way:

$$w^{(t)}(j,k) = w^{(t-1)}(j,k-1)\psi_{k-1} + w^{(t-1)}(j,k)(1 - \psi_k - \phi_k) + w^{(t-1)}(j,k+1)\phi_{k+1}. \quad (10.69)$$

In order to connect this chain to diffusion processes, a physical analogy is useful. Assume that $Y(t-1) = k$ represents the position of a diffusing particle on a one-dimensional lattice. At the next time step t, the particle can jump only to the two nearest-neighbour sites denoted by $Y = k-1$ and $Y = k+1$ or stay at its current position. If Δx represents the distance between lattice sites and Δt is the time interval between jumps, introducing the probability density $p(x,t|x_0,0)$ such that $w^{(t)}(x_0,x) = p(x,t|x_0,0)\Delta x$, Equation (10.69) can be re-written as

$$\begin{aligned} p(x,t|x_0,0)\Delta x = & \, p(x-\Delta x, t-\Delta t|x_0,0)\Delta x \cdot \psi(x-\Delta x) \\ & + p(x,t-\Delta t|x_0,0)\Delta x \cdot (1-\psi(x)-\phi(x)) \\ & + p(x+\Delta x, t-\Delta t|x_0,0)\Delta x \cdot \phi(x+\Delta x), \end{aligned} \tag{10.70}$$

where $p(x,t|x_0,0)\Delta x$ replaces $w^{(t)}(j,k)$. Note that the factor Δx can be simplified leading to

$$\begin{aligned} p(x,t|x_0,0) = & \, p(x-\Delta x, t-\Delta t|x_0,0)\psi(x-\Delta x) \\ & + p(x,t-\Delta t|x_0,0)(1-\psi(x)-\phi(x)) \\ & + p(x+\Delta x, t-\Delta t|x_0,0)\phi(x+\Delta x). \end{aligned} \tag{10.71}$$

It is now interesting to see what happens if Δx and Δt vanish. This cannot be arbitrarily done. First of all, it is necessary to assume that the infinitesimal mean $\mu(x)$ and variance $\sigma^2(x)$ of the process exist, so that

$$\mu(x) = \lim_{\Delta t \to 0} \frac{1}{\Delta t}[\mathbb{E}((X(t+\Delta t) - X(t)|X(t) = x)], \tag{10.72}$$

and

$$\sigma^2(x) = \lim_{\Delta t \to 0} \frac{1}{\Delta t}[\mathbb{V}((X(t+\Delta t) - X(t)|X(t) = x)]. \tag{10.73}$$

Then, one observes that, if the diffusing particle is in position $X(t) = x$ at time t, the difference $X(t+\Delta t) - X(t)$ is equal to Δx with probability $\psi(x)$, equal to 0 with probability $1-\psi(x)-\phi(x)$ and equal to $-\Delta x$ with probability $\phi(x)$, so that

$$\mathbb{E}(X(t+\Delta t) - X(t)|X(t) = x) = (\psi(x) - \phi(x))\Delta x, \tag{10.74}$$

and

$$\mathbb{V}(X(t+\Delta t) - X(t)|X(t) = x) = [\psi(x) + \phi(x) - (\psi(x) - \phi(x))^2](\Delta x)^2. \tag{10.75}$$

Comparing Equations (10.74) and (10.75) to (10.72) and (10.73) for finite, but small, Δt, one gets

$$\mu(x)\Delta t \approx (\psi(x) - \phi(x))\Delta x, \tag{10.76}$$

and

$$\sigma^2(x)\Delta t \approx [\psi(x) + \phi(x) - (\psi(x) - \phi(x))^2](\Delta x)^2. \tag{10.77}$$

These equations become exact in the so-called *diffusive limit*, with $\Delta x \to 0$ and $\Delta t \to 0$, but with the ratio $(\Delta x)^2/\Delta t = A$ kept constant (or converging to a constant):

$$\mu(x) = \lim_{\Delta x, \Delta t \to 0} (\psi(x) - \phi(x))\frac{\Delta x}{\Delta t}, \tag{10.78}$$

and

$$\sigma^2(x) = \lim_{\Delta x, \Delta t \to 0} [\psi(x) + \phi(x) - (\psi(x) - \phi(x))^2]\frac{(\Delta x)^2}{\Delta t}. \tag{10.79}$$

Now, replacing

$$(\Delta x)^2 = A\Delta t, \tag{10.80}$$

$$\psi(x) \approx \frac{1}{2A}[\sigma^2(x) + \mu(x)\Delta x], \tag{10.81}$$

$$\phi(x) \approx \frac{1}{2A}[\sigma^2(x) - \mu(x)\Delta x], \tag{10.82}$$

into Equation (10.71), expanding $p(x, t - \Delta t|x_0, 0)$ and $p(x \pm \Delta x, t - \Delta t|x_0, 0)$ in a Taylor series around (x, t) as well as $\mu(x \pm \Delta x)$ and $\sigma^2(x \pm \Delta x)$ in Taylor series around x, and keeping the terms up to order Δt (including $(\Delta x)^2$), straightforward but tedious calculations lead to the *Fokker–Planck equation*, also known as the *Kolmogorov forward equation*

$$\frac{\partial}{\partial t}p(x,t|x_0,0) = -\frac{\partial}{\partial x}[\mu(x)p(x,t|x_0,0)] + \frac{1}{2}\frac{\partial^2}{\partial x^2}[\sigma^2(x)p(x,t|x_0,0)]. \tag{10.83}$$

In particular, in Section 10.9, the stationary solution of Equation (10.83) is derived, coinciding with the continuous limit of the stationary probability of the birth-and-death Markov chain. When the stationary condition is reached, $p(x) = \lim_{t\to\infty} p(x,t|x_0,0)$, one has

$$\lim_{t\to\infty} \frac{\partial}{\partial t}p(x,t|x_0,0) = 0, \tag{10.84}$$

so that Equation (10.83) reduces to

$$-\frac{\mathrm{d}}{\mathrm{d}x}[\mu(x)p(x)] + \frac{1}{2}\frac{\mathrm{d}^2}{\mathrm{d}x^2}[\sigma^2(x)p(x)] = 0. \tag{10.85}$$

Example: the marginal Ehrenfest–Brillouin chain

In order to concretely illustrate the method described above, one can start from the marginal birth–death Markov chain of Section 8.2.1, leading to the dichotomous Pólya distribution as an invariant measure whose continuous limit is the beta distribution. In this case, for $n \gg 1$, one has

$$\begin{aligned}
\psi_i &= \frac{n-i}{n}\frac{\alpha_1 + i}{\theta + n - 1} = \left(1 - \frac{i}{n}\right)\frac{\alpha_1 + n\frac{i}{n}}{\theta + n - 1} \\
&\simeq \left(1 - \frac{i}{n}\right)\left(\frac{\alpha_1}{n} + \frac{i}{n}\right) = (1-x)\left(\frac{\alpha_1}{n} + x\right) \\
&= x(1-x) + \frac{\alpha_1}{n}(1-x),
\end{aligned} \tag{10.86}$$

where the position $x = i/n$ was assumed; analogously, one finds

$$\phi_i = \frac{i}{n}\frac{\alpha_2 + n - i}{\theta + n - 1} = \frac{i}{n}\frac{\alpha_2 + n\left(1 - \frac{i}{n}\right)}{\theta + n - 1} \simeq x(1-x) + \frac{\alpha_2}{n}x. \tag{10.87}$$

Note that $\theta = \alpha_1 + \alpha_2$. Now, given that

$$\mathbb{E}(\Delta Y^* | Y^* = i) = \psi_i - \phi_i, \tag{10.88}$$

and

$$\mathbb{V}(\Delta Y^* | Y^* = i) = \psi_i + \phi_i - (\psi_i - \phi_i)^2, \tag{10.89}$$

using the above approximations in the large n limit, and recalling that $\Delta x = 1/n$, one can write

$$\mathbb{E}(\Delta X | X = x) \approx \frac{\alpha_1}{n^2}(1-x) - \frac{\alpha_2}{n^2}x = \frac{1}{n^2}(\alpha_1 - \theta x), \tag{10.90}$$

and

$$\mathbb{V}(\Delta X | X = x) \approx \frac{1}{n^2}2x(1-x). \tag{10.91}$$

Rescaling time according to $\Delta t = 1/n^2$, one finds that the discrete Markov chain converges to the diffusion process $X(t)$ with the following infinitesimal parameters

$$\begin{aligned}\mu(x) &= (\alpha_1 - \theta x),\\ \sigma^2(x) &= 2x(1-x).\end{aligned} \tag{10.92}$$

Solving Equation (10.85) with infinitesimal parameters (10.92), one can find the stationary distribution (see Section 10.9). One can start from

$$\frac{2\mu(x)}{\sigma^2(x)} = \frac{2(\alpha_1 - \theta x)}{2x(1-x)} = \frac{\alpha_1 - \theta x}{x(1-x)}, \tag{10.93}$$

and then derive

$$\begin{aligned}s(x) &= \exp\left[-\int \frac{2\mu(u)}{\sigma^2(u)} du\right] = \exp\left[-\log(x^{\alpha_1}(1-x)^{\theta-\alpha_1})\right]\\ &= \frac{1}{x^{\alpha_1}(1-x)^{\theta-\alpha_1}}.\end{aligned} \tag{10.94}$$

The stationary solution is given by (10.119); setting $C_1 = 0$, one has

$$p(x) = \frac{C_2}{2} x^{\alpha_1 - 1}(1-x)^{\theta-\alpha_1-1} = A x^{\alpha_1-1}(1-x)^{\theta-\alpha_1-1}, \tag{10.95}$$

that is the Beta$(\alpha_1, \theta - \alpha_1)$, the continuous limit of the Pólya distribution.

Example: the continuous limit for the Ewens marginal chain

In the case of the Ewens marginal chain discussed in Chapter 9, one has

$$\begin{aligned}\psi_i &= \frac{n-i}{n}\frac{i}{n-1+\theta} \approx \frac{i(n-i)}{n^2},\\ \phi_i &= \frac{i}{n}\frac{n-i+\theta}{n-1+\theta} \approx \frac{i(n-i)}{n^2} + \frac{i\theta}{n^2},\end{aligned} \tag{10.96}$$

so that, one finds

$$\mathbb{E}(\Delta Y^* | Y^* = i) = \psi_i - \phi_i \approx -\frac{-i\theta}{n^2}, \tag{10.97}$$

and

$$\mathbb{V}(\Delta Y^* | Y^* = i) = \psi_i + \phi_i - (\psi_i - \phi_i)^2 \approx 2\frac{i(n-i)}{n^2} + \frac{i\theta}{n^2} \approx 2\frac{i(n-i)}{n^2}. \tag{10.98}$$

Setting $x = i/n$ and $\Delta x = 1/n$ leads to

$$\mathbb{E}(\Delta X|X = x) \approx -\frac{\theta}{n^2}x, \tag{10.99}$$

and

$$\mathbb{V}(\Delta X|X = x) \approx \frac{2x(1-x)}{n^2}. \tag{10.100}$$

Rescaling time intervals according to $\Delta t = 1/n^2$, the following infinitesimal parameter can be obtained:

$$\begin{aligned} \mu(x) &= -\theta x, \\ \sigma^2(x) &= 2x(1-x). \end{aligned} \tag{10.101}$$

Now, as a consequence of (10.101) one has

$$\frac{2\mu(x)}{\sigma^2(x)} = -\frac{2\theta x}{2x(1-x)} = -\frac{\theta}{1-x}, \tag{10.102}$$

and

$$s(x) = \exp\left[-\int \frac{2\mu(u)}{\sigma^2(u)} du\right] = \exp[-\theta \log(1-x)] = (1-x)^{-\theta}. \tag{10.103}$$

The term

$$\frac{1}{s(x)\sigma^2(x)} \propto x^{-1}(1-x)^{\theta-1} \tag{10.104}$$

is proportional to the frequency spectrum, the continuous limit of the expected fractional cluster size given in Equation (9.52).

Example: the formal chain of Section 10.6

The approximate marginal chain for the Zipf–Simon–Yule process discussed in Section 10.5 is not reversible and a continuous limit does not exist. On the contrary, consider the birth-and-death chain discussed in Section 10.6. The expectation and the variance for the increment of the cluster size driven by (10.66), with $A = 1$, are given by

$$\mathbb{E}(\Delta Y^*|Y^* = i) = \psi_i - \phi_i = -\frac{1}{n}, \tag{10.105}$$

and

$$\mathbb{V}(\Delta Y^*|Y^* = i) = \psi_i + \phi_i - (\psi_i - \phi_i)^2 = \frac{1}{n} + \frac{2(1-u)i}{n} - \left(\frac{1}{n}\right)^2. \tag{10.106}$$

Introducing $x = i/n$ for large n, one obtains

$$\mathbb{E}(\Delta X|X=x) \approx -\frac{1}{n^2}, \qquad (10.107)$$

and

$$\mathbb{V}(\Delta X|X=x) \approx \frac{1}{n^2}2(1-u)x. \qquad (10.108)$$

Then, rescaling time to $\Delta t = 1/n^2$, the discrete Markov chain (10.66) converges to the diffusion process $X(t)$ whose infinitesimal parameters are

$$\mu(x) = -1,$$
$$\sigma^2(x) = 2(1-u)x. \qquad (10.109)$$

Solving Equation (10.85) with infinitesimal parameters (10.109), one can find the stationary distribution (see Section 10.9). Now,

$$\frac{2\mu(x)}{\sigma^2(x)} = -\frac{1}{(1-u)x} = -\frac{\rho}{x}. \qquad (10.110)$$

Setting

$$s(x) = \exp\left[-\int \frac{2\mu(u)}{\sigma^2(u)}\mathrm{d}u\right] = \exp\left[\int \frac{\rho}{u}\mathrm{d}u\right] = \exp[\rho \ln x] = x^{\rho}, \qquad (10.111)$$

the stationary solution has the form (10.119); setting $C_1 = 0$ leads to

$$p(x) = \frac{C_2\rho}{2(1-u)xx^{\rho}} = \frac{A}{x^{\rho+1}}, \qquad (10.112)$$

that is the Pareto distribution, which is the continuous limit of the Yule distribution.

10.9 Appendix 1: invariant measure for homogeneous diffusions

If a stationary probability density $p(z)$ exists, it must satisfy

$$p(z) = \int \mathrm{d}x\, p(z,t|x,0)p(x). \qquad (10.113)$$

From Equation (10.83), one can see that $p(x)$ is a solution of the following differential equation

$$0 = -\frac{\mathrm{d}}{\mathrm{d}z}[\mu(z)p(z)] + \frac{1}{2}\frac{\mathrm{d}^2}{\mathrm{d}z^2}[\sigma^2(z)p(z)]. \qquad (10.114)$$

Moreover, under appropriate conditions, it is possible to prove that

$$\lim_{t\to\infty} p(z,t|x,0) = p(z), \tag{10.115}$$

irrespective of the initial state. A first integration of Equation (10.114) leads to

$$-\mu(z)p(z) + \frac{1}{2}\frac{\mathrm{d}}{\mathrm{d}z}[\sigma^2(z)p(z)] = \frac{C_1}{2}, \tag{10.116}$$

with C_1 constant. It is useful to introduce the integrating factor

$$s(z) = \exp\left[-\int^z \mathrm{d}u \frac{2\mu(u)}{\sigma^2(u)}\right], \tag{10.117}$$

by which, Equation (10.116) can be multiplied. This leads to

$$\frac{\mathrm{d}}{\mathrm{d}z}[s(z)\sigma^2(z)p(z)] = C_1 s(z). \tag{10.118}$$

Setting $S(z) = \int^z \mathrm{d}u\, s(u)$, a last integration gives

$$p(z) = C_1 \frac{S(z)}{s(z)\sigma^2(z)} + C_2 \frac{1}{s(z)\sigma^2(z)}; \tag{10.119}$$

the constants C_1 and C_2 must be determined in order to satisfy the constraints $p(z) \geq 0$ and $\int \mathrm{d}z\, p(z) = 1$. If this can be done, there exists a stationary solution.

10.10 Summary

Consider the Chinese restaurant process described at the beginning of Chapter 9. If the innovation process is *à la* Zipf, that is, it is such that the probability u of choosing a new table is independent of the number n of previous customers, both time- and site-label accommodation processes are no longer exchangeable. Therefore, many nice analytical results developed starting from Section 5.3.8 cannot be used any more. For this reason, it is necessary to apply either approximate methods or Monte Carlo simulations. For instance, one can consider n agents moving within $g = n+1$ sites. At each step an active site is selected and all the agents leave it. Then, the removed agents re-enter the system according to Zipf's rule (10.8).

Simon derived an exact regression equation for the expected cluster growth. Once a stationary regime is reached, it turns out that the relative frequency of the expected number for clusters of size i is given by the Yule distribution which has a power-law tail.

It is possible to describe this growth process with a suitable approximate marginal birth-and-death chain, whose equilibrium distribution is the right-censored Yule distribution. A formal birth- and death-chain can also be written whose equilibrium distribution is the right-truncated Yule distribution. In both cases, the dynamics is on the size of a cluster.

The final part of this chapter is devoted to the relationship between birth-and-death chain and diffusion processes for the Ehrenfest–Brillouin and Ewens marginal chains. It is also shown that the continuous limit of the formal Zipf-Simon-Yule chain leads to the continuous Pareto distribution.

Further reading

In 1925, G. Udny Yule published his paper on a mathematical theory of evolution based on the conclusions of Dr. J.C. Willis

G.U. Yule, *A Mathematical Theory of Evolution, Based on the Conclusions of Dr. J.C. Willis*, Philosophical Transactions of the Royal Society of London. Series B, Containing Papers of a Biological Character, **213**, 21–87 (1925).

It contains a detailed discussion of what later became known as the Yule distribution. Thirty years later, Herbert Simon published his paper discussing a class of skew distribution functions, where he recovered the Yule distribution in the solution of his regression equations

H. Simon, *On a Class of Skew Distribution Functions*, Biometrika, **42**, 425–440 (1955).

These results are critically discussed in two recent papers on which this chapter is based:

D. Costantini, S. Donadio, U. Garibaldi and P. Viarengo, *Herding and Clustering: Ewens vs. Simon–Yule Models*, Physica A, **355**, 224–231 (2005).

U. Garibaldi, D. Costantini, S. Donadio and P. Viarengo, *Herding and Clustering in Economics: The Yule-Zipf-Simon Model*, Computational Economics, **27**, 115–134 (2006).

The continuous limit of birth-and-death Markov chains is discussed both in the book by Cox and Miller:

D.R. Cox and H.D. Miller, *The Theory of Stochastic Processes*, Wiley, New York (1965).

and in the treatise by Karlin and Taylor:

S. Karlin and H.M. Taylor, *A Second Course in Stochastic Processes*, Academic Press, New York (1981).

This limit leads to diffusion processes (also known as diffusions). A rigorous discussion of diffusion can be found in the two volumes written by Rogers and Williams:

L.C.G. Rogers and D. Williams, *Diffusions, Markov Processes and Martingales: Volume 1, Foundations*, Cambridge University Press, Cambridge, UK (2000).

L.C.G. Rogers and D. Williams, *Diffusions, Markov Processes and Martingales: Volume 2, Itô Calculus*, Cambridge University Press, Cambridge, UK (2000).

Appendix

Solutions to exercises

Chapter 2

2.1 There are $n!$ possible different orders of draw for the n objects (permutations). But it is not their name that is observed, rather their X value. Therefore, exchanging two objects which belong to the same class does not produce a different s-description. As a trivial example consider the following: if all the objects belong to the same class i, there is only a single s-description, $\varsigma = (i, i, \ldots, i)$. The permutations within classes must be eliminated and this means dividing $n!$ by the product $n_1! \ldots n_g!$.

2.2 Consider the set $U = \{u_1, u_2, \ldots, u_n\}$, and the sequence $(1, 0, \ldots, 0)$. We can put this sequence in correspondence with the set $\{u_1\}$, which is a subset of U. Therefore, each sequence $(i_1, i_2, \ldots, i_n)$, with $i_j = 0, 1$ can be put in one-to-one correspondence with a particular subset of U containing those and only those elements u_j corresponding to $i_j = 1$. The sequence $(0, 0, \ldots, 0)$ corresponds to the empty set, that is $S = \varnothing$, whereas $(1, 1, \ldots, 1)$ corresponds to $S = U$. Therefore, the total number of subsets (proper and improper) of U is 2^n. Further notice that, from Newton's binomial formula, one has

$$(a+b)^n = \sum_{m=0}^{n} \binom{n}{m} a^m b^{n-m}, \tag{A2.1}$$

so that

$$2^n = \sum_{m=0}^{n} \binom{n}{m}. \tag{A2.2}$$

One can show that this is the decomposition of all samples into samples of the same size m.

2.3 Taking the parts in order:

1. The system is composed of $n = 5$ objects and $g = 3$ categories. The constraints for the partition vector are $\sum_{i=0}^{n} z_i = 3$, and $\sum_{i=0}^{n} iz_i = 5$, and they are satisfied by the choice. Therefore, the assumption is consistent.
2. There is a molecule with energy equal to 3ε, another one with energy equal to 2ε, and the third one is energyless.

2.4 This problem is equivalent to the one of finding the number of ways of allocating n objects into g boxes. This problem has already been solved in Chapter 1, Section 1.2.1.

Chapter 3

3.1 The rule is true for $n = 3$. The three sets are denoted by A, B and C. For $\mathbb{P}(A) > 0$, $\mathbb{P}(A \cap B) > 0$, given the identity:

$$\mathbb{P}(A \cap B \cap C) = \frac{\mathbb{P}(A \cap B \cap C)}{\mathbb{P}(A \cap B)} \frac{\mathbb{P}(A \cap B)}{\mathbb{P}(A)} \mathbb{P}(A), \tag{A3.1}$$

replacing the definition of conditional probability, one gets

$$\mathbb{P}(A \cap B \cap C) = \mathbb{P}(A)\mathbb{P}(B|A)\mathbb{P}(C|A \cap B). \tag{A3.2}$$

This holds for any permutation of events, that is, for instance,

$$\mathbb{P}(A \cap B \cap C) = \mathbb{P}(C)\mathbb{P}(A|C)\mathbb{P}(B|A \cap C). \tag{A3.3}$$

(Bayes' theorem is based on this symmetry.)

Suppose the rule is true for $n - 1$, can one prove that it is true also for n? The inductive hypothesis is

$$\mathbb{P}(A_1 \cap \ldots \cap A_{n-1}) = \mathbb{P}(A_1)\mathbb{P}(A_2|A_1)\ldots\mathbb{P}(A_{n-1}|A_{n-2} \cap \ldots \cap A_1). \tag{A3.4}$$

Now, it is enough to apply Bayes' rule

$$\mathbb{P}(A_1 \cap \ldots \cap A_{n-1} \cap A_n) = \mathbb{P}(A_n|A_1 \cap \ldots \cap A_{n-1})\mathbb{P}(A_1 \cap \ldots \cap A_{n-1}), \tag{A3.5}$$

and this immediately leads to the thesis by replacing $\mathbb{P}(A_1 \cap \ldots \cap A_{n-1})$ into the inductive hypothesis

$$\mathbb{P}(A_1 \cap \ldots \cap A_n) = \mathbb{P}(A_1)\mathbb{P}(A_2|A_1)\ldots\mathbb{P}(A_n|A_{n-1} \cap \ldots \cap A_1). \tag{A3.6}$$

3.2 First, complete the algebra:

1) You are in front of three closed doors, numbered from 1 to 3. Denote by C_1 the event 'the car is behind door 1'. To complete the algebra, you have

to introduce C_1^c, i.e. 'the car is *not* behind door 1'. It is better to distinguish between door 2 and door 3: hence you also introduce C_2 and C_3, with the meaning $C_i =$ 'the car is behind door i'. Now, $\{C_i\}_{i=1}^3$ is a decomposition of Ω, which allows you to freely speak about the positions of the car. The same procedure can be applied to your choice: let $D_i =$ 'you say that the car is behind door i'. Now C_i and D_j are *logically* independent, that is they belong to different decompositions of Ω, whereas, for instance, C_1 and $C_2 \cup C_3$ have an opposite truth value; if C_1 is true $C_2 \cup C_3$ is false, and if C_1 is false, $C_2 \cup C_3$ is true. Now let the quiz master enter the game, let him/her open one door with no car behind it and let $M_i =$ 'behind door i there is no car'. Based on the discussion in Example 3.6.3, all a priori possible constituents are then the 27 conjunctions $\{D_i \cap C_j \cap M_k; i,j,k = 1,2,3\}$ describing all possible states of affairs after the quiz master opened one door. Indeed, some of them are impossible (empty), and must be discarded. Thanks to the symmetry of this problem one can just consider the case wherein $D_1 =$ 'you say that the car is behind door 1' is true. The following analysis is conditioned on D_1 and the constituents are reduced to the following set of nine elements: $\{C_i \cup M_j; i,j = 1,2,3\}$. These elements are discussed below.

1. $C_i \cap M_1 = \varnothing$, $i = 1,2,3$, as the quiz master never opens a door selected by you; this eliminates 3 elements;
2. $C_i \cap M_i = \varnothing$, $i = 2,3$ as the quiz master does not lie, he/she shows an empty room, and this eliminates 2 more elements.

Therefore conditioned on D_1, the non-empty constituents are four:

1. $C_1 \cap M_2$, $C_1 \cap M_3$, corresponding to a correct guess;
2. $C_2 \cap M_3$, $C_3 \cap M_2$, corresponding to a wrong guess.

Conditioned on D_1, observe that $C_1 = (C_1 \cap M_2) \cup (C_1 \cap M_3)$, whereas $C_2 = C_2 \cap M_3$, and $C_3 = C_3 \cap M_2$. In other words, if $C_2|D_1$ is sure then M_3 is sure, and the same is valid by permuting 2 and 3. In case $C_1|D_1$, corresponding to a win, the quiz master has two possible choices, in the other cases he/she can choose only one door.

2) It is now possible to give probability values to all the non-empty constituents. From Bayes' theorem, one gets

$$\mathbb{P}(C_i|D_1) = \mathbb{P}(C_i \cap D_1)/\mathbb{P}(D_1) = \mathbb{P}(D_1|C_i)\mathbb{P}(C_i)/\mathbb{P}(D_1). \qquad \text{(A3.7)}$$

It is usually assumed that all rooms are equivalent and that this is common knowledge, leading to the uniform probability assignment $\mathbb{P}(C_i) = 1/3$ for $i = 1,2,3$. Another usual assumption is $\mathbb{P}(D_1|C_i) = \mathbb{P}(D_1)$ as no information flows from the chosen room. Therefore, $\mathbb{P}(C_i|D_1) = \mathbb{P}(C_i)$, and the two decompositions $\{C_i\}$ and $\{D_j\}$ are not only *logically*, but also *probabilistically* independent, and $\mathbb{P}(C_i \cap D_j) = \mathbb{P}(C_i)\mathbb{P}(D_j)$. For a complete characterization a probability value must be assumed for $\mathbb{P}(M_i|D_1 \cap C_1)$, that is the choice of the quiz master when the guess is correct. Assuming $\mathbb{P}(M_2|D_1 \cap C_1) = \mathbb{P}(M_3|D_1 \cap C_1) = 1/2$, the four constituents have the following probability:

1.
$$\mathbb{P}(C_1 \cap M_2|D_1) = \mathbb{P}(C_1|D_1)\mathbb{P}(M_2|D_1 \cap C_1) = (1/3)(1/2) = 1/6, \quad \text{(A3.8)}$$
$$\mathbb{P}(C_1 \cap M_3|D_1) = \mathbb{P}(C_1|D_1)\mathbb{P}(M_3|D_1 \cap C_1) = (1/3)(1/2) = 1/6, \quad \text{(A3.9)}$$

corresponding to the right guess, and two door choices for the quiz master; and

2.
$$\mathbb{P}(C_2 \cap M_3|D_1) = \mathbb{P}(C_2|D_1)\mathbb{P}(M_3|D_1 \cap C_2) = (1/3) \cdot 1 = 1/3, \quad \text{(A3.10)}$$
$$\mathbb{P}(C_3 \cap M_2|D_1) = \mathbb{P}(C_3|D_1)\mathbb{P}(M_2|D_1 \cap C_3) = (1/3) \cdot 1 = 1/3, \quad \text{(A3.11)}$$

corresponding to the two wrong guesses, with only a possible action for the quiz master.

Note that the posterior probability of C_1 is equal to the prior:

$$\mathbb{P}(C_1|D_1) = \mathbb{P}(C_1 \cap M_2|D_1) + \mathbb{P}(C_1 \cap M_3|D_1) = 1/3 = \mathbb{P}(C_1). \quad \text{(A3.12)}$$

In other words: *the winning probability does not change as a consequence of the quiz master action!*

3) It is true that, after the action of the quiz master, the possibilities reduce from 3 to 2 and this is perhaps the reason why many people wrongly believe that the quiz master action changes the initial probability of winning from 1/3 to 1/2. However, odds change from $1:1:1$ to $1:0:2$ (assuming the quiz master opens door 2 after your choice of door 1). In other words, if you are allowed to change your initial choice, changing enhances your odds of winning. In fact, assume that, conditioned on D_1, the quiz master opens door 2, meaning that M_2 holds true. Then the space reduces to two constituents, $C_1 \cap M_2$ and $C_3 \cap M_2$, whose union is M_2. Hence the posterior probability of $\{C_i\}_{i=1}^3$ is

$$\mathbb{P}(C_1|M_2 \cap D_1) = \mathbb{P}(C_1 \cap M_2|D_1)/\mathbb{P}(M_2|D_1) = (1/6)/(1/2) = 1/3, \quad \text{(A3.13)}$$

$$\mathbb{P}(C_2|M_2 \cap D_1) = 0, \quad \text{(A3.14)}$$

$$\mathbb{P}(C_3|M_2 \cap D_1) = \mathbb{P}(C_3 \cap M_2|D_1)/\mathbb{P}(M_2|D_1) = (1/3)/(1/2) = 2/3. \quad \text{(A3.15)}$$

As for the meaning of these formulae: given D_1, the quiz master chooses M_2 or M_3 with the same probability, given that C_2 and C_3 are equiprobable, and if

C_1 holds true he/she chooses M_2 or M_3 equiprobably again. Moreover, he/she chooses M_2 every time C_3 holds true, but only half of the times in which C_1 holds true. The fact that $\mathbb{P}(C_3|M_2 \cap D_1) = 2/3 = 2\mathbb{P}(C_1|M_2 \cap D_1)$ indicates that it is convenient to change your choice, if possible.

3.3 Consider the events M = 'you are ill', together with M^c = 'you are not ill' and P = 'you test positive' together with P^c = 'you do not test positive'. In order to answer the question of this exercise, it is necessary to find $\mathbb{P}(M|P)$. From Bayes' rule, one has

$$\mathbb{P}(M|P) = \frac{\mathbb{P}(P|M)\mathbb{P}(M)}{\mathbb{P}(P)}. \tag{A3.16}$$

As a consequence of the theorem of total probability, one further has

$$\mathbb{P}(P) = \mathbb{P}(P|M)\mathbb{P}(M) + \mathbb{P}(P|M^c)\mathbb{P}(M^c), \tag{A3.17}$$

so that (Bayes' theorem)

$$\mathbb{P}(M|P) = \frac{\mathbb{P}(P|M)\mathbb{P}(M)}{\mathbb{P}(P|M)\mathbb{P}(M) + \mathbb{P}(P|M^c)\mathbb{P}(M^c)}. \tag{A3.18}$$

According to recent results by Gigerenzer and coworkers, it seems that the formulation of this problem in terms of conditional probabilities is very difficult to grasp even for the majority of physicians and not only for lay-persons, who confuse $P(M|P)$ (the predictive value of the test) with $\mathbb{P}(P|M)$ (the sensitivity of the test). It is easier to determine the probabilities of the constituents.

Now the data of the problem imply

- $\mathbb{P}(M \cap P) = \mathbb{P}(P|M)\mathbb{P}(M) = (999/1000) \cdot (1/1000) \simeq 1/1000$;
- $\mathbb{P}(M^c \cap P) = \mathbb{P}(P|M^c)\mathbb{P}(M^c) = (5/100) \cdot (999/1000) \simeq 50/1000$;
- $\mathbb{P}(M \cap P^c) = \mathbb{P}(P^c|M)\mathbb{P}(M) = (1/1000) \cdot (1/1000) \simeq 0.001/1000$;
- $\mathbb{P}(M^c \cap P^c) = \mathbb{P}(P^c|M^c)\mathbb{P}(M^c) = (95/100) \cdot (999/1000) \simeq 949/1000$.

Usually, the constituents and their probabilities are represented by the following table in units of $1/1000$.

	P	P^c	
M	1	0	1
M^c	50	949	999
	51	949	1000

The last column of the table represents the marginal probabilities $\mathbb{P}(M)$ and $\mathbb{P}(M^c)$ for the decomposition based on the health conditions. The last row gives the marginal probabilities $\mathbb{P}(P)$ and $\mathbb{P}(P^c)$ for the decomposition based on the test results. Coming back to Equation (A3.18), one finds

$$\mathbb{P}(M|P) = \frac{\mathbb{P}(M \cap P)}{\mathbb{P}(M \cap P) + \mathbb{P}(M^c \cap P)} = \frac{1}{1+50} \approx 2\%, \tag{A3.19}$$

quite different from $\mathbb{P}(P|M) = 99.9\%$. Given the constituent table, any conditional probability can be computed by dividing each entry by the row or column corresponding to the conditioning event.

Until now, the focus was on the test results for a person. Following a method typical of statistical physics, one can build an *ensemble* which represents the individual situation by introducing a large number (1000 in this case) of *copies* of the unique real system. The number of systems with a certain property is set proportional to the probability of that property in the real system. In the particular case of this exercise, the table could represent an ensemble of 1000 persons that have been tested for the disease and classified with frequencies proportional to the probabilities. Once the ensemble is built, if the patient is positive, its probability space is limited to the first column and its probability of being ill is given by the number of ill persons in that column divided by the total number of persons testing positive.

3.4 Taking the parts in order:

1. Let us consider the decomposition of $\{A_j^{(1)}\}_{j=1,\dots,g}$, associated to the description of the first object drawn. The event $A_j^{(1)}$ is the disjoint union of all the sampling sequences (they are $n!/(n_1!\cdots n_j!\cdots n_g!)$) beginning with j. Their number is given by all sampling sequences belonging to $\mathbf{n}_{(j)} = (n_1,\dots,n_j-1,\dots,n_g)$, i.e. $(n-1)!/(n_1!\cdots(n_j-1)!\cdots n_g!)$.

   Therefore, the probability $\mathbb{P}(A_j^{(1)})$ is

$$\mathbb{P}\left(A_j^{(1)}\right) = \frac{(n-1)!}{n_1!\cdots(n_j-1)!\cdots n_g!}\,\frac{n_1!\ldots n_j!\cdots n_g!}{n!} = \frac{n_j}{n}. \tag{A3.20}$$

   The same result holds true for all the other $n-1$ objects.

2. Now consider $A_j^{(1)} \cap A_k^{(2)}$, $k \neq j$, which is the disjoint union of all sampling sequences beginning with j,k. Their number is given by all sampling sequences belonging to $\mathbf{n}_{(j,k)} = (n_1,\dots,n_j-1,\dots,n_k-1,\dots,n_g)$ and they are $(n-2)!/(n_1!\cdots(n_j-1)!\cdots(n_k-1)!\cdots n_g!)$. The probability is then

$$\mathbb{P}\left(A_j^{(1)} \cap A_k^{(2)}\right) = \frac{n_j n_k}{n(n-1)}. \tag{A3.21}$$

Hence, one has

$$\mathbb{P}\left(A_j^{(1)} \cap A_k^{(2)}\right) = \frac{n_j}{n}\frac{n_k}{n-1} > \frac{n_j}{n}\frac{n_k}{n} = \mathbb{P}\left(A_j^{(1)}\right)\mathbb{P}\left(A_k^{(2)}\right), \tag{A3.22}$$

for $k \neq j$. Considering $j = k$, by the same reasoning, one finds

$$\mathbb{P}\left(A_j^{(1)} \cap A_j^{(2)}\right) = \frac{n_j}{n}\frac{n_j - 1}{n-1} < \frac{n_j}{n}\frac{n_j}{n} = \mathbb{P}\left(A_j^{(1)}\right)\mathbb{P}\left(A_j^{(2)}\right), \tag{A3.23}$$

that is the joint probability of the two events of the same type is smaller than the product of the two probabilities. This means that the two events are negatively correlated.

3. The predictive probability at the second step is

$$\mathbb{P}\left(A_k^{(2)} | A_j^{(1)}\right) = \frac{\mathbb{P}\left(A_j^{(1)} \cap A_k^{(2)}\right)}{\mathbb{P}\left(A_j^{(1)}\right)}, \tag{A3.24}$$

and the previous results show that

$$\mathbb{P}\left(A_k^{(2)} | A_j^{(1)}\right) = \frac{n_k - \delta_{k,j}}{n-1}, \tag{A3.25}$$

where the term $\delta_{k,j}$ unifies all the possible cases, including $k = j$. The general result in Equation (3.20) can be obtained by repeated application of the multiplication rule leading to

$$\mathbb{P}\left(A_j^{(m+1)} | A_{\omega_1}^{(1)} \ldots A_{\omega_m}^{(m)}\right) = \frac{n_j - m_j}{n-m}. \tag{A3.26}$$

In the dichotomous case $g = 2$, if $\mathbf{n} = (k, n-k)$, and $\mathbf{m} = (h, m-h)$, Equation (3.20) becomes

$$\mathbb{P}\left(A_1^{(m+1)} | A_{\omega_1}^{(1)} \ldots A_{\omega_m}^{(m)}\right) = \frac{k-h}{n-m}. \tag{A3.27}$$

Considering the sequence with all k 'successes' before the $n-k$ 'failures', and applying Equation (A3.27) as well as Bayes' rule for $m=0,\ldots,n-1$, gives

$$\mathbb{P}(\{\omega\}|(k,n-k)) = \frac{k}{n}\frac{k-1}{n-1}\cdots\frac{1}{n-k+1}\frac{n-k}{n-k}\frac{n-k-1}{n-k-1}\cdots\frac{1}{1}$$
$$= \frac{k!}{n!/(n-k)!} = \binom{n}{k}^{-1} \tag{A3.28}$$

as expected.

3.5 Assume that there are two alternative and exhaustive hypotheses: $H_0 =$ 'the observed motions have a common cause', and $H_1 =$ 'the observed motions are the result of a random assembly of celestial bodies'. The evidence is $E =$ 'all the 43 motions are concordant'. Further assume that before any observation is made $\mathbb{P}(H_0) = \mathbb{P}(H_1) = 1/2$. It means that one could bet $1:1$ on H_0. To obtain $\mathbb{P}(H_0|E)$, one can set $\mathbb{P}(E|H_0) = 1$; $\mathbb{P}(E|H_1)$ can be estimated considering that all motions are independent and the probability of being in one of the two possible directions with respect to a chosen body is 1/2. Therefore, $\mathbb{P}(E|H_1) = 2^{-42}$. Note that the exponent is 42 and not 43 as one is interested in the relative motion of celestial bodies. Now, using Bayes' theorem for both hypotheses and dividing:

$$\frac{\mathbb{P}(H_0|E)}{\mathbb{P}(H_1|E)} = \frac{\mathbb{P}(E|H_0)\mathbb{P}(H_0)}{\mathbb{P}(E|H_1)\mathbb{P}(H_1)} = \frac{\mathbb{P}(E|H_0)}{\mathbb{P}(E|H_1)} = 2^{42}. \tag{A3.29}$$

Now $2^{42} = 2^{2+40} = 4\cdot\left(2^{10}\right)^4 \geq 4\cdot(10^3)^4 = 4\cdot 10^{12}$, where $2^{10} = 1024$ is approximated by 10^3. This application of Bayes' theorem was criticized because the decomposition $\{H_0, H_1\}$ is a rough one; also the interpretation of $\mathbb{P}(H_i)$ was challenged. Also the estimate of $\mathbb{P}(E|H_1)$ is too rough in this case, as the problem is a tridimensional one, whereas, above, the options are reduced to alternatives (that is dichotomous choices).

Moreover, no frequency interpretation is available for $\mathbb{P}(H_i)$, and, as a consequence, for $\mathbb{P}(H_i|E)$. However, an interpretation in terms of betting is possible, as put forward by Laplace himself, as well as a 'rational belief' interpretation. In any case, this is a very nice example of plausible reasoning.

Chapter 4

4.1 An equivalence relation xRy, with $x,y \in A$ satisfies:

1. (identity) any element is in relation with itself, $\forall x \in A$, xRx;
2. (symmetry) $\forall x,y \in A$ if xRy then yRx; and

3. (transitivity) $\forall x, y, z \in A$ if xRy and yRz, then xRz.

One can represent the relation of the problem as a binary function defined on all ordered couples of the elements of the set as follows: $R(\omega_i, \omega_j) = 1$ if the relation holds true (if ω_i and ω_j belong to the inverse image of a given label, otherwise $R(\omega_i, \omega_j) = 0$; this is valid for $i,j = 1, \ldots, \#\Omega$. In other words, one sets

$$R(\omega_i, \omega_j) = \begin{cases} 1 \text{ if } f(\omega_i) = f(\omega_j) = b \\ 0 \text{ if } f(\omega_i) \neq f(\omega_j) \end{cases}. \tag{A4.1}$$

Identity and symmetry are obvious, as $R(\omega_i, \omega_i) = 1$; and if $R(\omega_i, \omega_j) = 1$ then $R(\omega_2, \omega_1) = 1$. Transitivity holds as $R(\omega_i, \omega_j) = 1$ means $f(\omega_i) = f(\omega_j)$, $R(\omega_j, \omega_k) = 1$ means $f(\omega_j) = f(\omega_k)$ leading to $f(\omega_i) = f(\omega_k)$ finally yielding $R(\omega_i, \omega_k) = 1$.

4.2 The result straightforwardly follows from the definition; $\mathbb{E}(a+bX) = \sum_x (a+bx)\mathbb{P}_X(x) = a\sum_x \mathbb{P}_X(x) + b\sum_x x\mathbb{P}_X(x)$; therefore

$$\mathbb{E}(a+bX) = a + b\mathbb{E}(X). \tag{A4.2}$$

4.3 All the three results are direct consequences of the definition. For the first question, one can use the result of Exercise 4.2.

1. One has the following chain of equalities:

$$\mathbb{V}(X) = \mathbb{E}[(X - \mathbb{E}(X))^2] = \mathbb{E}[X^2 - 2X\mathbb{E}(X) + \mathbb{E}^2(X)] = \mathbb{E}(X^2) - \mathbb{E}^2(X). \tag{A4.3}$$

2. The variance is also known as the second centred moment, hence it is not affected by change of the origin: $\mathbb{V}(X+a) = \mathbb{V}(X)$. Indeed, setting $Y = X + a$, $\mathbb{E}[Y] = \mathbb{E}[X] + a$, and $Y - \mathbb{E}[Y] = X - \mathbb{E}[X]$, yielding $\mathbb{V}(Y) = \mathbb{V}(X)$.
3. Note that $\mathbb{V}(bX) = b^2\mathbb{V}(X)$, as $(bX - \mathbb{E}[bX])^2 = b^2(X - \mathbb{E}[X])^2$.

Hence in synthesis:

$$\mathbb{V}(a+bX) = b^2\mathbb{V}(X) = b^2(\mathbb{E}(X^2) - \mathbb{E}^2(X)). \tag{A4.4}$$

4.4 The first three properties can be proved by showing that the first and the right-hand side of the equations always coincide. As for the intersection, $\mathbb{I}_{A\cap B}(\omega)$ is 1 if and only if $\omega \in A$ and $\omega \in B$. In this case both $\mathbb{I}_A(\omega)$ and $\mathbb{I}_B(\omega)$ are 1 and their product is 1. In all the other possible cases, $\mathbb{I}_A(\omega)\mathbb{I}_B(\omega)$ vanishes. For what concerns the union, its indicator function is one if and only if $\omega \in A$ or $\omega \in B$. Now, if $\omega \in A$ and $\omega \notin B$ as well as if $\omega \notin A$ and $\omega \in B$, the sum $\mathbb{I}_A(\omega) + \mathbb{I}_B(\omega) - \mathbb{I}_A(\omega)\mathbb{I}_B(\omega)$ is 1 as either $\mathbb{I}_A(\omega)$ or $\mathbb{I}_B(\omega)$ are 1 and the other terms vanish. If $\omega \in A$ and $\omega \in B$, then again the right-hand side of the equality is 1 as $\mathbb{I}_A(\omega) = 1$ and $\mathbb{I}_B(\omega) = 1$. In the remaining case, $\omega \notin A$ and $\omega \notin B$, all the three terms vanish. Similarly, in the case of the complement, $\mathbb{I}_{A^c}(\omega)$ is 1 if $\omega \notin A$ and is 0 if $\omega \in A$.

The fact that the expected value of the indicator function of an event A coincides with the probability of A is an immediate consequence of the definitions

$$\mathbb{E}(\mathbb{I}_A(\omega)) = \sum_{\omega \in \Omega} \mathbb{I}_A(\omega)\mathbb{P}(\omega) = \sum_{\omega \in A} \mathbb{P}(\omega) = \mathbb{P}(A). \tag{A4.5}$$

Finally, the variance of the indicator function is given by

$$\mathbb{V}(\mathbb{I}_A) = \mathbb{E}[\mathbb{I}_A^2] - \mathbb{P}^2(A), \tag{A4.6}$$

but $\mathbb{I}_A^2 = \mathbb{I}_A$ so that

$$\mathbb{V}(\mathbb{I}_A) = \mathbb{P}(A) - \mathbb{P}^2(A) = \mathbb{P}(A)(1 - \mathbb{P}(A)). \tag{A4.7}$$

4.5 Let $k = 0, 1, \ldots, n$ represent the composition $(k, n-k)$ of the urn. Introducing the partition $\Omega = \cup_{k=0}^{n}\{k\}$, one has $\mathbb{P}(X_2 = x_2|X_1 = x_1) = \sum_k \mathbb{P}(X_2 = x_2, k|X_1 = x_1)$, and $\mathbb{P}(X_2 = x_2, k|X_1 = x_1) = \mathbb{P}(X_2 = x_2|k, X_1 = x_1)\mathbb{P}(k|X_1 = x_1)$ as a consequence of the definition of conditional probability. Note that $\mathbb{P}(X_2 = x_2|X_1 = x_1)$ is the expected value of $\mathbb{P}(X_2 = x_2|k, X_1 = x_1)$, but the weights are not $\mathbb{P}(k)$, rather $\mathbb{P}(k|X_2 = x_1)$, that is the posterior distribution of the urn composition after the observation $X_1 = x_1$. In other words, $\mathbb{P}(X_2 = x_2|X_1 = x_1)$ is the expected value of $\mathbb{P}(X_2 = x_2|k, X_1 = x_1)$, but the weights are updated due to the occurrence of $X_1 = x_1$. Replacing $\mathbb{P}(k|X_1 = x_1) = A\mathbb{P}(x_1|k)\mathbb{P}(k)$, with $A = 1/\mathbb{P}(x_1)$, one recovers $\mathbb{P}(X_2 = x_2|k, X_1 = x_1)\mathbb{P}(k|X_1 = x_1) = A\mathbb{P}(X_2 = x_2|k, X_1 = x_1)\mathbb{P}(X_1 = x_1|k)\mathbb{P}(k) = A\mathbb{P}(X_1 = x_1, X_2 = x_2|k)\mathbb{P}(k)$ leading to (4.73).

Note that $\mathbb{P}(X_2 = x_2|k, X_1 = x_1)$ depends on the chosen sampling model. If sampling is Bernoullian, then one has $\mathbb{P}(X_2 = x_2|k, X_1 = x_1) = \mathbb{P}(x_2|k)$ meaning that the composition of the urn determines the probability whatever $X_1 = x_1$; if sampling is hypergeometric, $X_1 = x_1$ is important. Assume that $x_1 = 1$, and that k is the number of type-1 balls: in the Bernoullian case $\mathbb{P}(X_2 = 1|k, X_1 = 1) = k/n$, whereas in the hypergeometric case $\mathbb{P}(X_2 = 1|k, X_1 = 1) = (k-1)/(n-1)$, and $\mathbb{P}(X_2 = 1|k, X_1 = 2) = k/(n-1)$.

4.6 The reader should take care to avoid confusing Euler's beta function

$$B(a,b) = \frac{\Gamma(a)\Gamma(b)}{\Gamma(a+b)} = \int_0^1 \theta^{a-1}(1-\theta)^{b-1}\mathrm{d}\theta \tag{A4.8}$$

with the beta density

$$\mathrm{beta}(\theta; a, b) = B(a,b)^{-1}\theta^{a-1}(1-\theta)^{b-1}, \tag{A4.9}$$

defined for $\theta \in [0,1]$. Based on the definition of $B(a,b)$, one has that

$$B(a+1,b) = \int_0^1 \theta \cdot \theta^{a-1}(1-\theta)^{b-1}\mathrm{d}\theta, \tag{A4.10}$$

so that the expected value is

$$\begin{aligned}\mathbb{E}(\Theta) &= B(a,b)^{-1}\int_0^1 \theta \cdot \theta^{a-1}(1-\theta)^{b-1}\mathrm{d}\theta \\ &= B(a,b)^{-1}B(a+1,b) = \frac{\Gamma(a+b)}{\Gamma(a)\Gamma(b)}\frac{\Gamma(a+1)\Gamma(b)}{\Gamma(a+b+1)} = \frac{a}{a+b}.\end{aligned} \tag{A4.11}$$

Similarly, the integral for $\mathbb{E}(\Theta^2)$ is given by

$$B(a+2,b) = \int_0^1 \theta^2 \cdot \theta^{a-1}(1-\theta)^{b-1}\mathrm{d}\theta; \tag{A4.12}$$

therefore, one obtains

$$\begin{aligned}\mathbb{E}(\Theta^2) &= B(a,b)^{-1}\int_0^1 \theta^2 \cdot \theta^{a-1}(1-\theta)^{b-1}\mathrm{d}\theta \\ &= \frac{\Gamma(a+b)}{\Gamma(a)\Gamma(b)}\frac{\Gamma(a+2)\Gamma(b)}{\Gamma(a+b+2)} = \frac{a(a+1)}{(a+b+1)(a+b)}.\end{aligned} \tag{A4.13}$$

Finally, one arrives at

$$\begin{aligned}\mathbb{V}(\Theta) &= \mathbb{E}(\Theta^2) - \mathbb{E}^2(\Theta) = \frac{a(a+1)}{(a+b+1)(a+b)} - \left(\frac{a}{a+b}\right)^2 \\ &= \frac{a(a+1)(a+b) - a^2(a+b+1)}{(a+b+1)(a+b)^2} = \frac{ab}{(a+b+1)(a+b)^2}.\end{aligned} \tag{A4.14}$$

Chapter 5

5.1 Assume that the two random variables are defined on the same probability space, and let $P_{X_1,X_2}(x_1,x_2)$ be the joint probability distribution defined in

Section 4.1.7. One has the following chain of equalities

$$\begin{aligned}
\mathbb{E}(X_1+X_2) &= \sum_{x_1,x_2}(x_1+x_2)P_{X_1,X_2}(x_1,x_2) \\
&= \sum_{x_1,x_2}(x_1P_{X_1,X_2}(x_1,x_2)+x_2P_{X_1,X_2}(x_1,x_2)) \\
&= \sum_{x_1,x_2}x_1P_{X_1,X_2}(x_1,x_2)+\sum_{x_1,x_2}x_2P_{X_1,X_2}(x_1,x_2) \\
&= \sum_{x_1}x_1\sum_{x_2}P_{X_1,X_2}(x_1,x_2)\sum_{x_2}x_2\sum_{x_1}P_{X_1,X_2}(x_1,x_2) \\
&= \sum_{x_1}x_1P_{X_1}(x_1)+\sum_{x_2}x_2P_{X_2}(x_2)=\mathbb{E}(X_1)+\mathbb{E}(X_1). \qquad \text{(A5.1)}
\end{aligned}$$

The extension of this result to the sum of n random variables can be straightforwardly obtained by repeated application of the above result.

5.2 Using the properties of the expectation (see Exercise 5.1 above and Exercise 5.2), of the variance (see Exercise 5.3) and of the covariance (Equation (4.18) in Section 4.1.7) one has that

$$\begin{aligned}
\mathbb{V}(X_1+X_2) &= \mathbb{E}[(X_1+X_2-\mathbb{E}(X_1+X_2))^2]=\mathbb{E}[(X_1+X_2)^2]-[\mathbb{E}(X_1+X_2)]^2 \\
&= \mathbb{E}(X_1^2+X_2^2+2X_1X_2)-\mathbb{E}^2(X_1)-\mathbb{E}^2(X_2)-2\mathbb{E}(X_1)\mathbb{E}(X_2) \\
&= \mathbb{E}(X_1^2)-\mathbb{E}^2(X_1)+\mathbb{E}(X_2^2)-\mathbb{E}^2(X_2)+2[\mathbb{E}(X_1X_2)-\mathbb{E}(X_1)\mathbb{E}(X_2)] \\
&= \mathbb{V}(X_1)+\mathbb{V}(X_2)+2\mathbb{C}(X_1,X_2). \qquad \text{(A5.2)}
\end{aligned}$$

The extension of this result to n random variables is straightforward as well. In order to represent the square of the sum of n terms, consider an $n\times n$ square matrix, whose entries are the products $(X_i-\mathbb{E}(X_i))(X_j-\mathbb{E}(X_j))$. The sum of the n diagonal terms is $\sum_{i=1}^{n}(X_i-\mathbb{E}(X_i))^2$, the other $m(m-1)$ terms are $\sum_{i\neq j}(X_i-\mathbb{E}(X_i))(X_j-\mathbb{E}(X_j))$. The expected value of the sum is the sum of the expected values. Therefore, the covariance matrix is a symmetric matrix with variances on the diagonal and covariances out of the diagonal.

5.3 The random variable $I_i^{(k)}I_j^{(k)}$ differs from zero only when $X_i=k$ and $X_j=k$, in which case it takes the value 1; in all the other cases it is 0. Therefore, one immediately gets

$$\mathbb{E}(I_i^{(k)}I_j^{(k)})=1\cdot\mathbb{P}(X_i=k,X_j=k)+0\cdot\dots=\mathbb{P}(X_i=k,X_j=k). \qquad \text{(A5.3)}$$

5.4 There is not a unique solution to this problem and there may be several methods of optimizing the code for the simulation of Pólya processes. The code below uses the predictive probability to create a dichotomous Pólya sequence, calculates the random walk as a function of m and then plots the result. The program is written in *R*, which is a free software environment for statistical computing and graphics. The software can be retrieved from `http://www.r-project.org`.

```
# Program polyarw.R
n <-1000 # number of iterations
alpha0 <- 10 # parameter of category 0 (failure)
alpha1 <- 10 # parameter of category 1 (success)
alpha <- alpha0 + alpha1 # parameter alpha
m0 <- 0 # number of failures
m1 <- 0 # number of successes
X <- rep(c(0), times=n) # defining the vector of jumps
for (m in 0:n-1) { # main cycle
if ( runif(1) < (alpha1+m1)/(alpha+m)){ # predictive probability
# for success
X[m+1] <-1 # upward jump
m1<- m1+1 # update of success number
} # end if
else { # else failure
X[m+1]<--1 # downward jump
m0 <- m0+1 # update of failure number
} # end else
} # end for
S <- cumsum(X) # implementing eq. (5.160)
t <- c(1:n) # time steps
plot(t,S,xlab="m",ylab="S(m)") # plot
```

In Fig. A5.1, a simulation for the case $\alpha_0 = 1, \alpha_1 = 1$ is plotted. Figure A5.2 contains a realization of the Pólya random walk for $\alpha_0 = \alpha_1 = 10$. Finally, Fig. A5.3 shows a simulation of $\alpha_0 = \alpha_1 = 100$. Even if the a priori probabilities do coincide in the three cases $p_0 = p_1 = 1/2$, the behaviour of the random walk is very different. For small α (Fig. A5.1), it is almost deterministic. The initial choice gives a great advantage to the selected category (see the discussion in Section 5.6.2). In Fig. A5.3, by direct visual inspection, one can see that the behaviour of the Pólya random walk for $\alpha_0 = \alpha_1 = 100$ is very close to that of a standard random walk where the variables $X(m)$ are independent and identically distributed with equal probability of success and failure (as in coin tosses). It would not be easy for a statistician to distinguish between the cases presented in Fig. A5.3 and A5.4, the latter being a realization of a standard random walk.

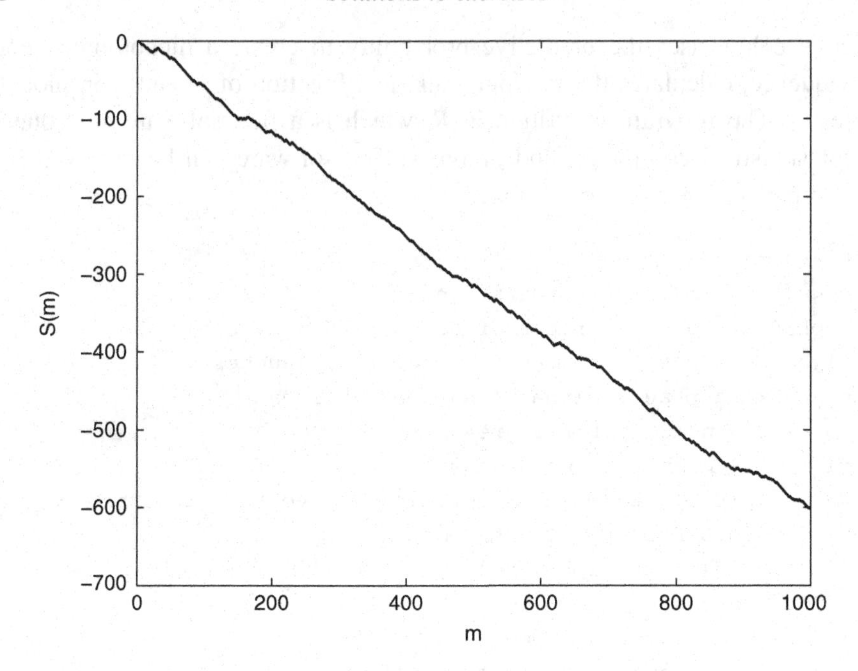

Figure A5.1. Pólya random walk for $\alpha_0 = \alpha_1 = 1$ for a maximum of $m = 1000$ steps. In this particular realization, failures prevail over successes.

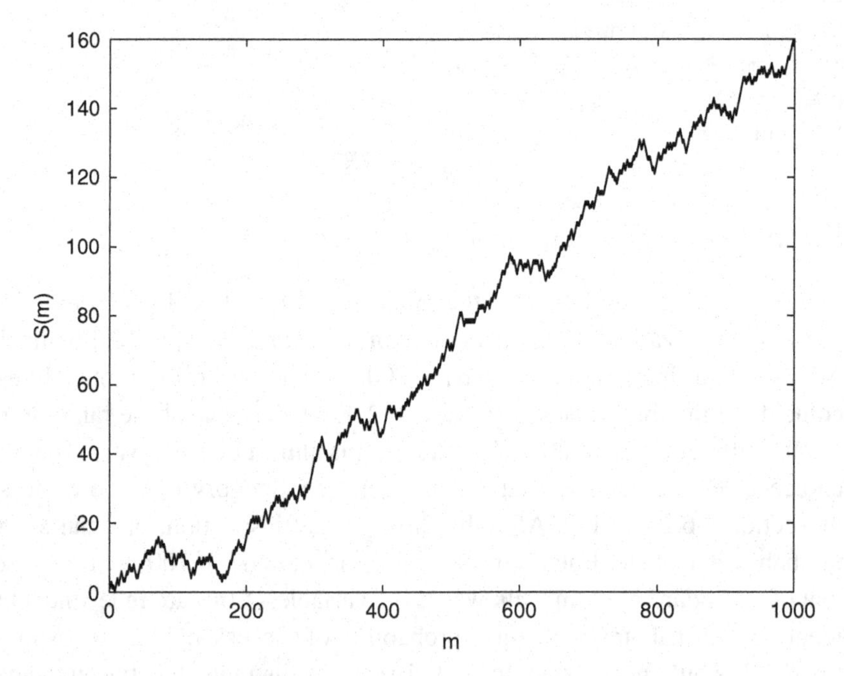

Figure A5.2. Pólya random walk for $\alpha_0 = \alpha_1 = 10$ for a maximum of $m = 1000$ steps. In this particular realization successes prevail over failures, but the behaviour is less 'deterministic' than in Fig. A5.1. The final distance from the origin is much smaller than in the previous case.

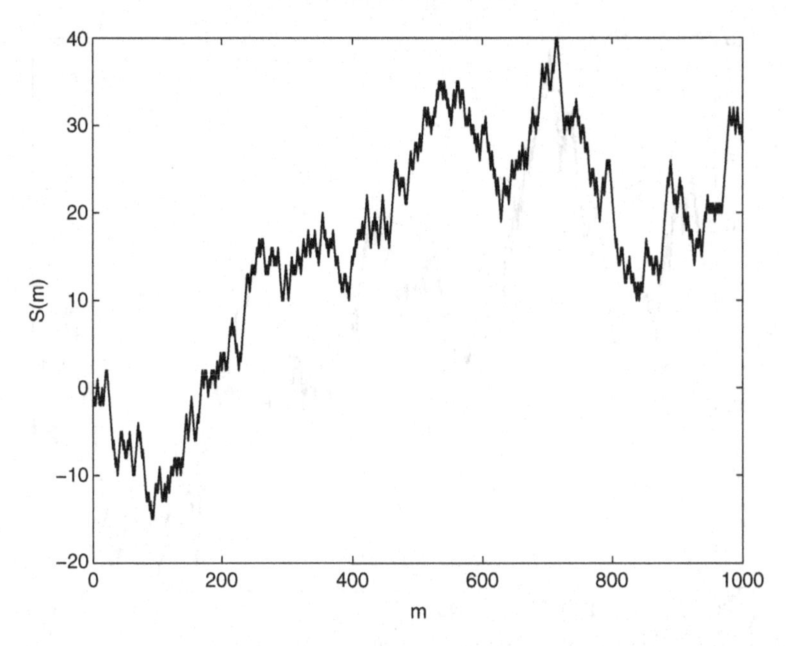

Figure A5.3. Pólya random walk for $\alpha_0 = \alpha_1 = 100$ for a maximum of $m = 1000$ steps. In this particular realization again successes prevail over failures, but the behaviour is very close to the one for a standard random walk (see Fig. A5.4). The final distance from the origin is even smaller.

5.5 In order to get an estimate for the probability of all the possible values of m_1, one has to run many realizations of the Pólya process. The following program does this in a pedestrian way. For the sake of simplicity, the case $n = 10$ with 11 different values of m_1 is taken into account. Without going into the detail of goodness-of-fit tests, visual inspection of Fig. A5.5 shows a good agreement between the two distributions.

```
# Program polyacompbin.R
N <- 10000 # total number of realizations, larger than 2^n
n <- 10 # total number of individuals/observations
alpha0 <- 100 # parameter of category 1 (success)
alpha1 <- 100 # parameter of category 0 (failure)
alpha = alpha0+alpha1
M1 <- rep(c(0), times=n+1) # vector storing MC results
for (i in 1:N){ # cycle on realizations
m0 <- 0 # number of failures
m1 <- 0 # number of successes
for (m in 0:n-1){ # cycle on individuals/observations
if (runif(1) < (alpha1+m1)/(alpha+m)){ # predictive
# probability for success
```

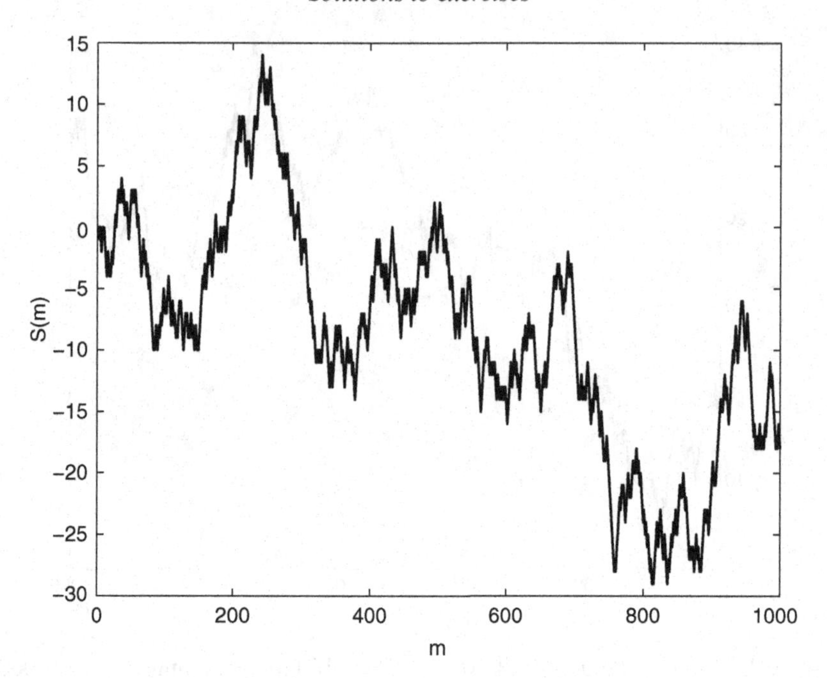

Figure A5.4. Standard symmetric random walk for $p_0 = p_1 = 1/2$ for a maximum of $m = 1000$ steps. It is difficult to distinguish this case from the one in Fig. A5.3.

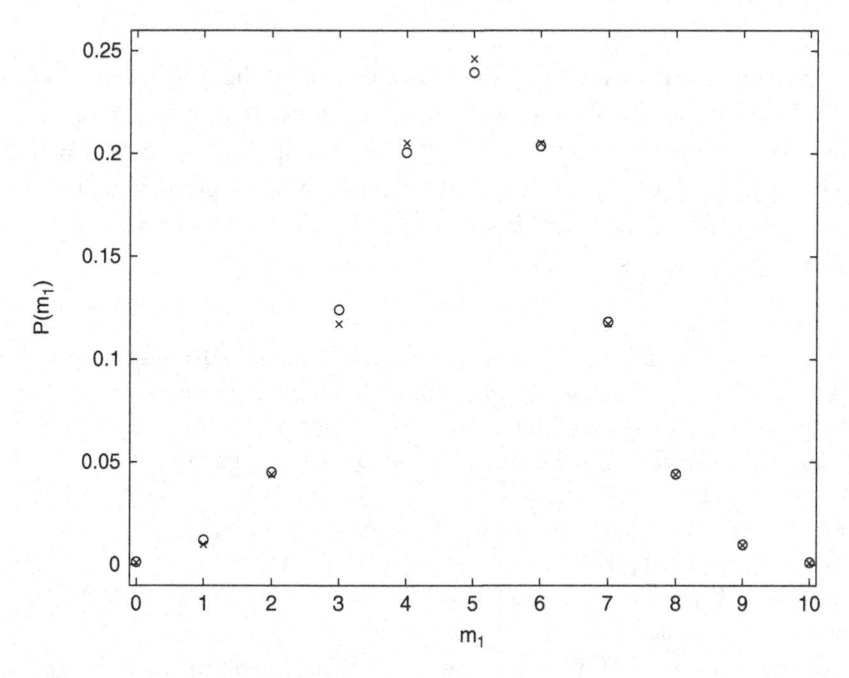

Figure A5.5. Comparison between the simulated relative frequency out of 10000 trials from $\mathbb{P}(m_1) = \text{Polya}(m_1; \alpha_0 = 100, \alpha_1 = 100)$ (open circles) and the symmetric binomial distribution (crosses).

```
m1 <- m1+1 # update of success number
} # end if
else { # else failure
m0 <- m0+1 # update of failure number
} # end else
} # end for (n)
M1[m1+1] <- M1[m1+1] + 1 # storing the number of successes
} # end for (N)
M1[1:11] # displays simulation values
k <- seq (0,10) # creates vector of integers from 0 to 10
N*dbinom(k,10,0.5) # displays binomial prediction
```

5.6 Usually, it is assumed that people born on February 29th in leap years are not taken into account. For the sake of simplicity, this assumption will be made here also.

The first remark is that the birthday problem amounts to allocating the n persons under scrutiny into $g = 365$ categories representing the days of the year. The most complete description is the individual one $(X_1 = x_1, \ldots, X_n = x_n)$ where the event $X_i = x_i$ means that the ith person is born in day x_i and $x_i \in (1, \ldots, 365)$. In this representation, $X_i = 1$ means that the ith person is born on January 1st and $X_i = 365$ means that the birthday is on December 31st. The total number of possible individual descriptions is given by 365^n. Then, there is the frequency description where the vector $(Y_1 = n_1, \ldots Y_{365} = n_{365})$ denotes the intersection of the events $Y_j = n_j$, where n_j is the number of persons with their birthday on the jth day of the year. The constraint $\sum_{j=1}^{365} n_j = n$ holds true and the number of possible frequency vectors is given by $(n+356-1)!/(n!(365-1)!)$. Finally there is the frequency of frequencies (or partition) description in which the vector $(Z_0 = z_0, \ldots, Z_n = z_n)$ is the intersection of the events $Z_k = z_k$, where z_k is the number of days with k birthdays. Now there are two constraints: $\sum_{k=0}^{n} z_k = 365$ and $\sum_{k=0}^{n} k z_k = n$.

The usual assumption in the elementary analyses of the problem is that all the n random variables X_i are independent and identically distributed with

$$\mathbb{P}(X_i = x_i) = \frac{1}{365}, \tag{A5.4}$$

for any x_i (any day of the year). This assumption leads to the uniform distribution over the vectors of individual descriptions

$$\mathbb{P}(\mathbf{X}^{(n)} = \mathbf{x}^{(n)}) = \prod_{i=1}^{n} \mathbb{P}(X_i = x_i) = \frac{1}{365^n}. \tag{A5.5}$$

Now, due to exchangeability, using Equation (4.59) leads to

$$\mathbb{P}(\mathbf{Y}^{(n)} = \mathbf{n}) = \frac{n!}{n_1! \ldots n_{356}!} \mathbb{P}(\mathbf{X}^{(n)} = \mathbf{x}^{(n)}) = \frac{n!}{n_1! \ldots n_{356}!} \frac{1}{365^n}. \quad \text{(A5.6)}$$

Not only are the individual variables exchangeable, but also the frequencies, all the days being on a par. Therefore, a second application of Equation (4.59) yields

$$\mathbb{P}(\mathbf{Z}^{(n)} = \mathbf{z}^{(n)}) = \frac{365!}{z_0! \ldots z_n!} \mathbb{P}(\mathbf{Y}^{(n)} = \mathbf{n}^{(n)}), \quad \text{(A5.7)}$$

as in Equation (5.72). Applying (5.71) also, one gets

$$\mathbb{P}(\mathbf{Z}^{(n)} = \mathbf{z}^{(n)}) = \frac{365!}{z_0! \ldots z_n!} \frac{n!}{(0!^{z_0}) \cdot \ldots \cdot (n!^{z_n})} \frac{1}{365^n}. \quad \text{(A5.8)}$$

Equation (A5.8) can be used to answer all the questions related to birthday coincidences. For instance, the probability of no common birthday in a group of n persons means that $Z_1 = n$ and $Z_0 = 365 - n$, all the other Z_i being 0. In fact, this is satisfied by any frequency vector including n times 1 (for the days of the n different birthdays) and $365 - n$ times 0 for the other days. Therefore, one is led to insert the partition vector $\mathbf{Z}^{(n)} = (365 - n, n, 0, \ldots, 0)$ into Equation (A5.8)

$$\mathbb{P}(365 - n, n, 0, \ldots, 0) = \frac{365!}{(365-n)!n!} n! \frac{1}{365^n} = \frac{365!}{(365-n)!} \frac{1}{365^n}. \quad \text{(A5.9)}$$

The event *at least one common birthday* is the complement of *no common birthday* and its probability is just $1 - \mathbb{P}(365 - n, n, 0, \ldots, 0)$. This answers the initial question.

This solution can be found with more elementary methods, as well. However, Equation (A5.8) can be used to solve other problems. For instance, consider the probability of the event *exactly two birthdays coincide* meaning that all the other birthdays are on different days of the year. This means that the frequency description has $n_i = 2$ for 1 value of i, $n_j = 1$ for $n-2$ values of j and $n_k = 0$ for $365 - n + 1$ values of k because only $n - 1$ days are birthdays. This leads to the partition vector $\mathbf{Z}^{(n)} = (365 - n + 1, n - 2, 1, 0 \ldots, 0)$ whose probability is

$$\mathbb{P}(365 - n + 1, n - 2, 1, 0 \ldots, 0) = \frac{365!}{(365-n+1)!(n-2)!1!} \frac{n!}{2!} \frac{1}{365^n}. \quad \text{(A5.10)}$$

In particular, for $n \geq 23$, it is convenient to bet 1:1 on the presence of a common birthday at least.

Chapter 6

6.1 **Solution of case a)**

Note that $\beta = 0$ is associated to a deterministic system following the cycle $a \to b \to c \to a$ with probability 1 forever. Then the answer to point 1 is yes, as all states are persistent, the chain is periodic (point 2), and the invariant distribution (point 3) is $\boldsymbol{\pi} = (1/3, 1/3, 1/3)$, which is also the limiting frequency distribution (point 4). In fact, considering blocks of three steps, each state occurs with the same frequency, and in the long run the initial steps, whatever they are, do not matter for the relative visiting frequency. That the invariant distribution is $\boldsymbol{\pi} = (1/3, 1/3, 1/3)$ can be shown by building an ensemble of systems (three are sufficient), one starting from $x_0 = a$, one from $x_0 = b$ and one from $x_0 = c$. In the 'ensemble' representation, very common in statistical physics, the probability of a state is given by the number of systems in that state divided by the total number of systems. At the second step, all the systems make a clockwise step, and the number of systems in any state is 1 again. If one is not confident regarding intuition, $\boldsymbol{\pi}$ is the solution of $\boldsymbol{\pi} = \boldsymbol{\pi} \times \mathbb{W}$. The answer to point 5 is negative; indeed, given $x_0 = a$ then $\mathbf{P}^{(3t)} = \delta(x, a)$, $\mathbf{P}^{(3t+1)} = \delta(x, b)$, and $\mathbf{P}^{(3t+2)} = \delta(x, c)$, where $\delta(x, i)$ is the degenerate probability distribution all concentrated on the ith state, i.e. the probability of state i is 1. This simple example illustrates the root of ergodic properties in physics. If the state space is finite and the dynamics is deterministic the motion is periodic on all persistent states. The relative frequency of visits converges to the invariant distribution, therefore, instead of taking the time average over the whole trajectory, one can use the 'ensemble average' on $\boldsymbol{\pi}$, which is necessarily uniform. Building an ensemble proportional to $\boldsymbol{\pi}$ is an artificial device to introduce a distribution that the deterministic dynamics leaves invariant. But one knows that any particular system whose initial conditions are known cannot be described by $\boldsymbol{\pi}$ in the long run. As for point 6, even if the ensembles were described by the invariant $\boldsymbol{\pi} = (1/3, 1/3, 1/3)$, each system of the ensemble moves clockwise for sure, whereas the backward movie is always counterclockwise. Here, there is no reversibility and no detailed balance.

Solution of case b)

Now, consider the case $\beta = 1$. If the chain reaches state a, it stops there forever, and a is said to be an equilibrium state in a physics sense. Given that all states communicate with a, sooner or later the system stops there. In this simple case, after two steps, one is sure that the system is in a.

Then, the limiting and the equilibrium distribution exist in some sense, as, for $t \geq 2$, for any initial state $\mathbb{P}(X_t = j) = \delta(j,a)$, but it is not positive (regular) on all states. States b and c are not persistent (they are called *transient*) and the chain is reducible. There is no reversibility and no detailed balance.

Solution of case c)

In order to fix ideas, let $\beta = 1/2$. The possibility of not immediately leaving a state destroys the periodicity of the chain. In fact, $\mathbb{W}^{(4)}$ is positive. A little thought is enough to understand that, in the long run, state a is visited twice more than states b and c, which, in turn, have the same number of visits. Hence, the limiting visiting frequency distribution is $(1/2, 1/4, 1/4)$. This result can be confirmed by solving the Markov equation $\boldsymbol{\pi} = \boldsymbol{\pi} \times \mathbb{W}$. The chain being aperiodic, it reaches the limiting distribution $\boldsymbol{\pi} = (1/2, 1/4, 1/4)$ after 42 steps within 3-digit precision, whatever the initial state. When the chain becomes stationary, the total balance between entering and exiting probabilities vanishes. For instance $\Delta\pi_b = \pi_a/2 - \pi_b = 0$, that is at each step the mass $\pi_a/2$ enters from state a and compensates the mass π_b which leaves for state c. Detailed balance does not hold: in fact, the backward movie is essentially different from the direct one. Then there is no reversibility and no detailed balance.

6.4 First note that the events $\{|X - \mathbb{E}(X)| \geq \delta\}$ and $\{(X - \mathbb{E}(X))^2 \geq \delta^2\}$ coincide. If V_X denotes the set of allowed values of X, one has that

$$
\begin{aligned}
\frac{\mathbb{V}(X)}{\delta^2} &= \sum_{x \in V_X} \frac{(x - \mathbb{E}(X))^2}{\delta^2} \mathbb{P}(X = x) \\
&= \sum_{(x-\mathbb{E}(X))^2 \geq \delta^2} \frac{(x - \mathbb{E}(X))^2}{\delta^2} \mathbb{P}(X = x) + \sum_{(x-\mathbb{E}(X))^2 < \delta^2} \frac{(x - \mathbb{E}(X))^2}{\delta^2} \mathbb{P}(X = x) \\
&\geq \sum_{(x-\mathbb{E}(X))^2 \geq \delta^2} \frac{(x - \mathbb{E}(X))^2}{\delta^2} \mathbb{P}(X = x) = \sum_{|x-\mathbb{E}(X)| \geq \delta} \frac{(x - \mathbb{E}(X))^2}{\delta^2} \mathbb{P}(X = x) \\
&\geq \sum_{|x-\mathbb{E}(X)| \geq \delta} \mathbb{P}(X = x) = \mathbb{P}(|X - \mathbb{E}(X)| \geq \delta),
\end{aligned} \tag{A6.1}
$$

where the first equality is the definition of variance, whereas the second equality is the division of the sum into complementary terms; the first inequality is due to the fact that all the terms added in the variance are non-negative; the third equality uses the coincidence discussed above; the second inequality is a consequence of the fact that, for the summands, one has $(x - \mathbb{E}(X))/\delta^2 \geq 1$; finally, the last equality is by definition.

6.5 In order to solve this problem, one has to find a normalized left eigenvector of $\mathbb{W}$ with eigenvalue 1. This leads to the following linearly dependent equations

$$\pi_1 = \frac{1}{3}\pi_1 + \frac{3}{4}\pi_2, \tag{A6.2}$$

$$\pi_2 = \frac{2}{3}\pi_1 + \frac{1}{4}\pi_2, \tag{A6.3}$$

corresponding to

$$\pi_1 = \frac{9}{8}\pi_2. \tag{A6.4}$$

Imposing the normalization condition leads to

$$\frac{9}{8}\pi_2 + \pi_2 = 1, \tag{A6.5}$$

so that $\pi_1 = 9/17$ and $\pi_2 = 8/17$ give the invariant probabilities of state 1 and 2, respectively, which are also the equilibrium ones, being the chain irreversible and aperiodic.

6.6 A little algebra shows that

$$\mathbb{W}^{(2)} = \mathbb{W}^2 = \begin{pmatrix} 0.611 & 0.389 \\ 0.438 & 0.563 \end{pmatrix},$$

$$\mathbb{W}^{(5)} = \mathbb{W}^5 = \begin{pmatrix} 0.524 & 0.477 \\ 0.536 & 0.464 \end{pmatrix},$$

$$\mathbb{W}^{(10)} = \mathbb{W}^{10} = \begin{pmatrix} 0.530 & 0.471 \\ 0.529 & 0.471 \end{pmatrix}.$$

These results can be compared with $\pi_1 = 9/17 \simeq 0.529$ and $\pi_2 = 8/17 \simeq 0.471$.

6.7 It turns out that

$$\mathbb{W}^{2k} = \begin{pmatrix} 1 & 0 \\ 0 & 1 \end{pmatrix},$$

and

$$\mathbb{W}^{2k-1} = \begin{pmatrix} 0 & 1 \\ 1 & 0 \end{pmatrix},$$

for any integer $k \geq 1$. Therefore the powers of this matrix do not converge.

Chapter 7

7.1 Let k denote the number of fleas on the left dog and $n-k$ the number of fleas on the right dog. Here, $n = 4$ and k can vary from 0 to 4. If $\mathbf{Y}_0^{(4)} = (0,4)$, then one has $k_0 = 0$, and D_1, D_2 are forced to be moves from left to right, therefore, one has $\mathbf{Y}^{(2)} = (0,2)$ for sure. There is 1 path C_1, C_2 leading to $k = 2$, the paths C_1, C_2 leading to $k = 1$ are 2, and 1 returning to $k = 0$. Therefore, the first row of the binary move matrix $\mathbb{W}_{(2)}$ is $b_{0j} = (1/4, 1/2, 1/4, 0, 0)$. If $k = 1$, that is $\mathbf{Y}_0^{(4)} = (1,3)$, there are three possible transitions D_1, D_2, with probability

$$\begin{cases} \mathbb{P}(D_1 = r, D_2 = r) = (3/4) \cdot (2/3) = 1/2 \\ \mathbb{P}(D_1 = r, D_2 = l) = (1/4) \cdot (3/3) = 1/4, \\ \mathbb{P}(D_1 = l, D_2 = r) = (3/4) \cdot (1/3) = 1/4 \end{cases} \tag{A7.1}$$

where r denotes a move to the right and l a move to the left. Note that $D_1 = r$, $D_2 = r$ brings us to $\mathbf{Y}^{(2)} = (1,1)$, and the uniformity of C_1, C_2 gives probability $(1/4, 1/2, 1/4)$ to the final states $(1,3)$, $(2,2)$, $(3,1)$, respectively. Both $D_1 = r, D_2 = l$ and $D_1 = l, D_2 = r$ bring us to $\mathbf{Y}^{(2)} = (0,2)$, which ends in $(0,4)$, $(1,3)$, $(2,2)$ with probability $(1/4, 1/2, 1/4)$. Then grouping the distinct patterns to each final vector, one finds $b_{10} = (1/2)(1/4) = 1/8$, $b_{11} = (1/2)(1/2) + (1/2)(1/4) = 3/8$, $b_{12} = (1/2)(1/4) + (1/2)(1/2) = 3/8$, $b_{13} = (1/2)(1/4) = 1/8$, $b_{14} = 0$. Note that this very cumbersome calculation is highly simplified by the independent and symmetric creation terms, and by the small size of the population ($n = 4$). Instead of going on with a detailed description of the transitions, the matrix can be completed using the detailed balance Equations (7.113) and (7.114). First, one can check that the previous results are correct. For instance $b_{10}/b_{01} = (1/8)/(1/2) = 1/4$, which is exactly the ratio π_1/π_4, when $\pi_k = \mathrm{Bin}(k|4, 1/2)$. Considering that the symmetry implies that $b_{i,j} = b_{4-i,4-j}$, the third row can be calculated from $b_{2,j} = b_{j,2}\pi_j/\pi_2$. Therefore, one gets $b_{2,0} = b_{0,2}\pi_0/\pi_2 = (1/4)(1/6) = 1/24$, $b_{2,1} = b_{1,2}\pi_1/\pi_2 = (3/8)(4/6) = 1/4$, $b_{2,3} = b_{3,2}\pi_3/\pi_2 = (3/8)(4/6) = 1/4$ and $b_{2,4} = b_{4,2}\pi_4/\pi_2 = (1/4)(1/6) = 1/24$, where $b_{3,2} = b_{1,2}$, $b_{4,2} = b_{0,2}$ and $b_{2,2} = 1 - 1/12 - 1/2 = 5/12$ by normalization. Finally, the matrix $\mathbb{W}_{(2)} = \{b_{ij}\}$ is the following

$$\mathbb{W}_{(2)} = \begin{pmatrix} 1/4 & 1/2 & 1/4 & 0 & 0 \\ 1/8 & 3/8 & 3/8 & 1/8 & 0 \\ 1/24 & 6/24 & 10/24 & 6/24 & 1/24 \\ 0 & 1/8 & 3/8 & 3/8 & 1/8 \\ 0 & 0 & 1/4 & 1/2 & 1/4 \end{pmatrix}, \tag{A7.2}$$

which is different from both the unary matrix (6.30), and from its square $\mathbb{W}^{(2)}$. However, the approach to equilibrium is such that $\mathbb{W}_{(2)}^{(t)}$ is equal to $\mathbb{W}^{(\infty)}$ up to 3-digit precision for $t = 13$ (instead of $t = 30$).

As for the comparison of the two chains, with transition matrices $\mathbb{W}$ and $\mathbb{W}_{(2)}$ corresponding to unary/binary moves ($m = 1$ versus $m = 2$), it is interesting

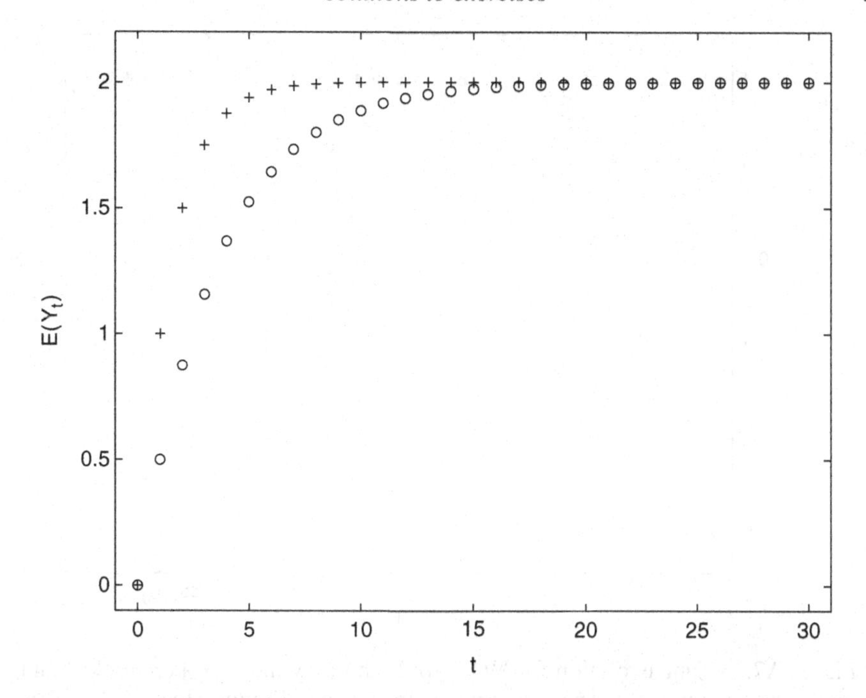

Figure A7.1. Time evolution for the expected value (A7.3) for both unary moves (open circles) and binary moves (crosses). The convergence to the equilibrium value $\mathbb{E}(Y_t) = 2$ is faster for binary moves (see the main text for further explanation).

to compute the evolution of the same initial probability distribution, and to test the corresponding evolution of the expected value (to be compared to Equation (7.62), and to Exercise 6.3). Suppose that the initial state is $k_0 = 0$, that is $\mathbf{P}^{(0)} = (1,0,0,0,0)$. To find $\mathbf{P}^{(t)}$, one just considers the first row of the corresponding t-step matrix, and finds

$$\mathbb{E}(Y_t) = \sum_{k=0}^{4} k w^{(t)}(0,k). \tag{A7.3}$$

The values of the two expected values are presented in Fig. A7.1 as a function of time, for both unary and binary moves. Returning to Equation (7.59), in the case of independent creations ($\alpha \to \infty$) the rate converges to m/n, in formula

$$r = \frac{m\alpha}{n(\alpha + n - m)} \to \frac{m}{n}, \tag{A7.4}$$

so that $r^{(1)} = 1/4$, and $r^{(2)} = 1/2$, and it is a straightforward exercise to verify that Equation (7.62) holds true exactly in both cases.

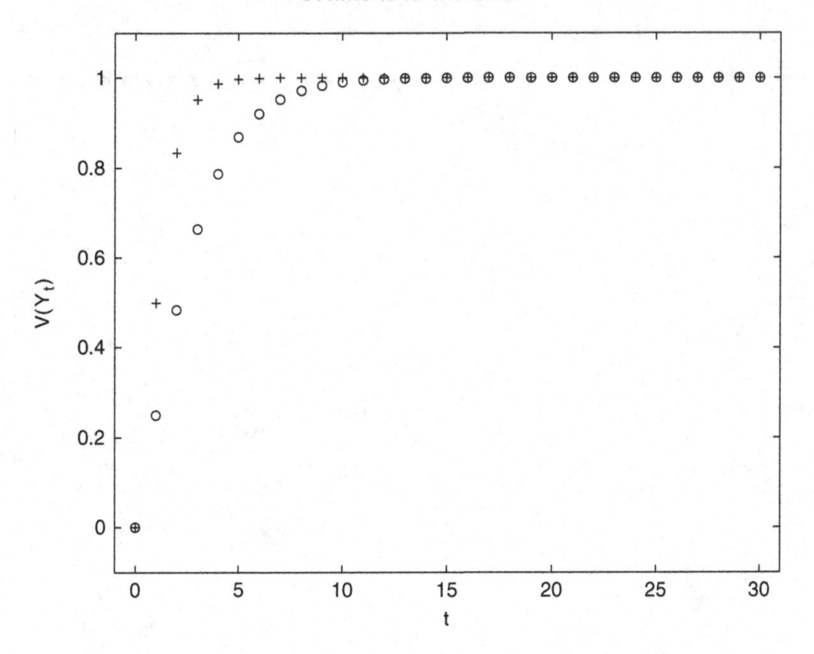

Figure A7.2. Time evolution for $\mathbb{V}(Y_t)$ for both unary moves (open circles) and binary moves (crosses). The convergence to the equilibrium value $\mathbb{V}(Y_t) = 1$ is faster for binary moves (see the main text for further explanation).

Regarding second moments, if the starting distribution is degenerate as before, $\mathbf{P}^{(0)} = (1,0,0,0,0)$, its variance is zero, and it grows to the limiting value $\mathbb{V}(Y_t) = 1$. This is represented in Fig. A7.2. It is apparent that $\mathbb{E}^{(1)}(Y_t)$ and $\mathbb{V}^{(1)}(Y_t)$ have roughly the same rate of approach to equilibrium, and the same happens in the binary case. Note that, if one starts from $k = 2$, that is $\mathbf{P}^{(0)} = (0,0,2,0,0)$, then $\mathbb{E}(Y_0) = 2$ from the beginning, and this value does not change further; but, again, $\mathbb{V}(Y_t)$ must increase up to 1, as the degenerate distribution converges to the binomial distribution.

The covariance function given in Equation (7.73) is valid only at equilibrium, when the two chains become stationary, $\mathbb{E}(Y_t) = 2$ and $\mathbb{V}(Y_t) = 1$. Its main empirical meaning is given by the linear regression

$$\mathbb{E}(Y_{t+1}|Y_t) - \mathbb{E}(Y_{t+1}) = \frac{\mathbb{C}(Y_{t+1}, Y_t)}{\mathbb{V}(Y_t)}(Y_t - \mathbb{E}(Y_t)), \tag{A7.5}$$

discussed in Section 7.5.3. In the case $m = 1$, if one simply knows $\mathbb{E}^{(1)}(Y_t)$ and $\mathbb{E}^{(1)}(Y_{t+1})$, and observes $Y_t = k$, one's final expectation $\mathbb{E}(Y_{t+1}|Y_t = k)$ coincides with the initial one $\mathbb{E}(Y_{t+1})$, to which $(3/4)(k - \mathbb{E}(Y_t))$ must be added, and this holds true also far from equilibrium. The observed deviation

$k - \mathbb{E}(Y_t)$ is weighted by $q = 1 - r$. If $r \to 1$ (for large m, large α and small correlation), the observed deviation has a small weight, and it tends to vanish; if $r \to 0$ (small m, small α, large correlation), the observed deviation has a large weight, and it tends to remain for future steps.

7.2 In Fig. A7.3 and A7.4, the Monte Carlo estimates for the probability distribution are compared to Equation (7.12). Moreover, Fig. A7.5 and A7.6 present a realization of the time-evolution for the occupation variable $k = n_1$ of category 1. As expected from the shape of the distribution, in Kirman's case, when $\alpha_1 = \alpha_2 = 1/2$, the system spends more time close to the extremal values ($k = 0, n$) suddenly jumping between these values.

These results can be generated with the help of the following Monte Carlo simulation.

```
# Program unary.R
n <- 10 # number of objects
g <- 2 # number of categories
T <- 100000 # number of Monte Carlo steps
alpha <- rep(c(0), times = g) # vector of parameters
bn <- rep(c(0), times = g) # occupation vector
probd <- rep(c(0), times = g) # vector of
# hypergeometric probabilities
probc <- rep(c(0), times = g) # vector of Polya probabilities
# creation parameters (leading to symmetric Polya if all equal)
alpha <- rep(c(1/2), times = g)
sumalpha <- sum(alpha) # Polya parameter
# initial occupation vector
bn[1] <- n
A <- bn # vector of results
# Main cycle
for (t in 1:T) {
# destruction (an object is removed from an occupied
# category according to a hypergeometric probability)
probd <- bn/n
cumprobd <- cumsum(probd)
# pointer to category
indexsite <- min(which((cumprobd-runif(1))>0))
bn[indexsite] <- bn[indexsite] - 1
# creation (the object reaccommodates
# according to a Polya probability)
probc <- (alpha + bn)/(sumalpha+n-1)
cumprobc <- cumsum(probc)
```

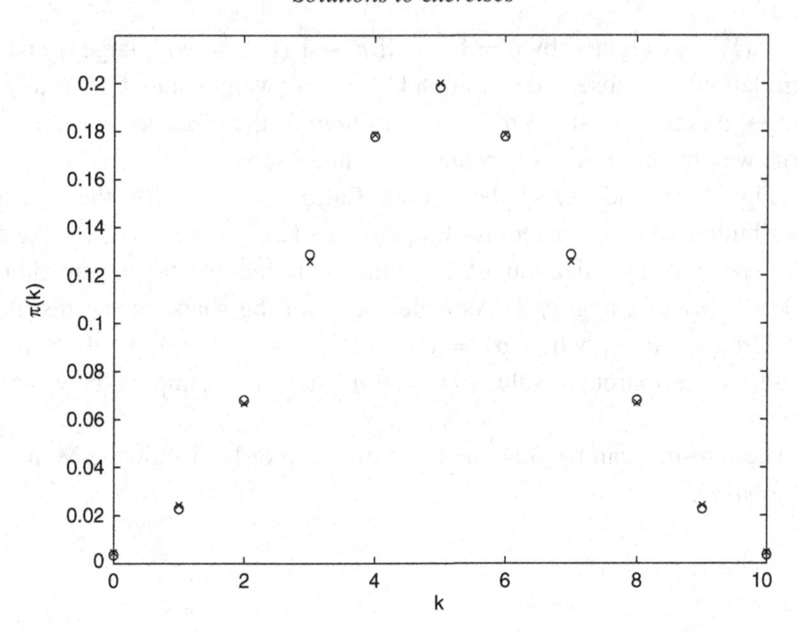

Figure A7.3. Comparison between the Monte Carlo simulation described in this exercise and the Pólya distribution for $g = 2$ with $\alpha_1 = \alpha_2 = 10$ (crosses). In the case represented in this figure, the number of elements is $n = 10$; the circles represent the Monte Carlo probability estimates over $T = 50000$ steps.

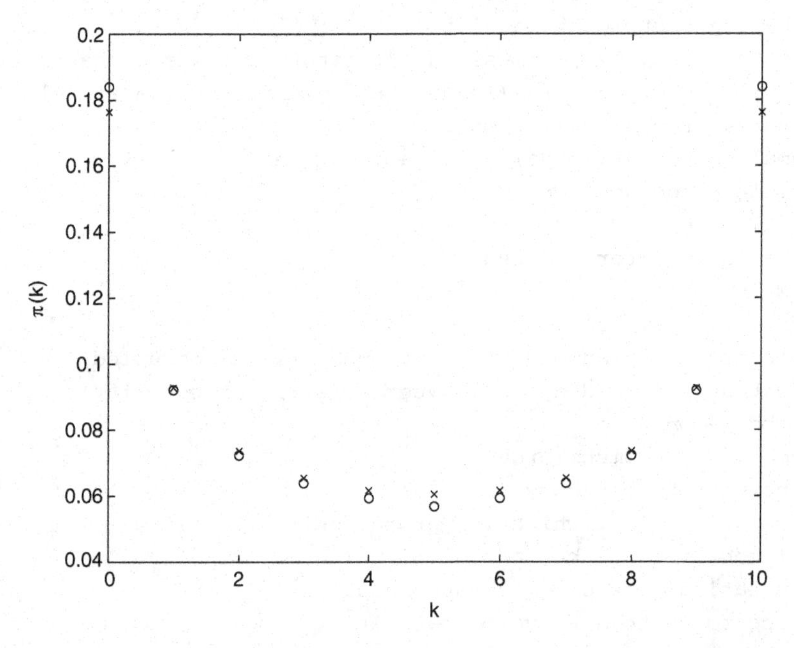

Figure A7.4. Comparison between the Monte Carlo simulation described in this exercise and the Pólya distribution for $g = 2$ with $\alpha_1 = \alpha_2 = 1/2$ (crosses). In the case represented in this figure, the number of elements is $n = 10$; the circles represent the Monte Carlo probability estimates over $T = 100000$ steps.

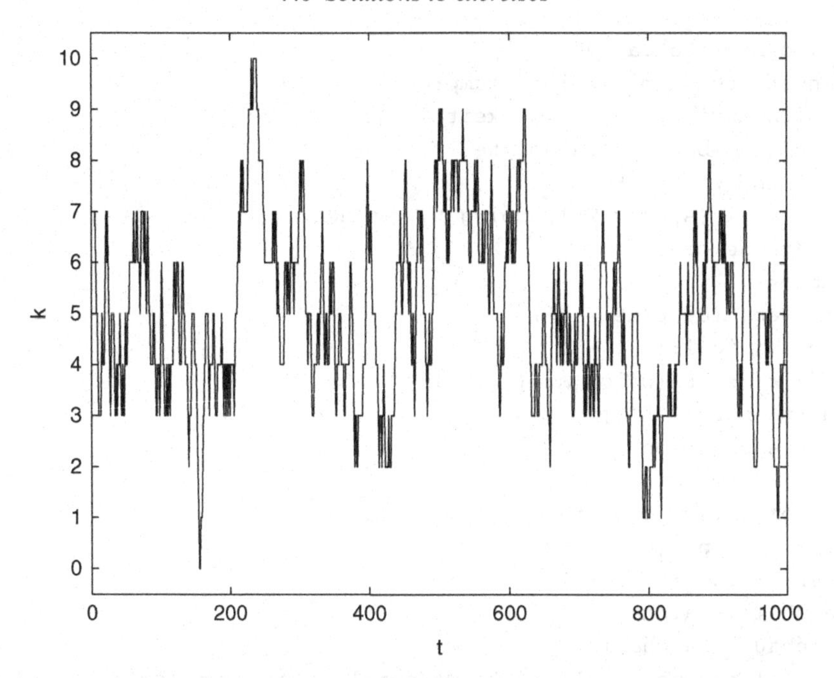

Figure A7.5. Time evolution of $k = n_1$ for $\alpha_1 = \alpha_2 = 10$ for 1000 steps.

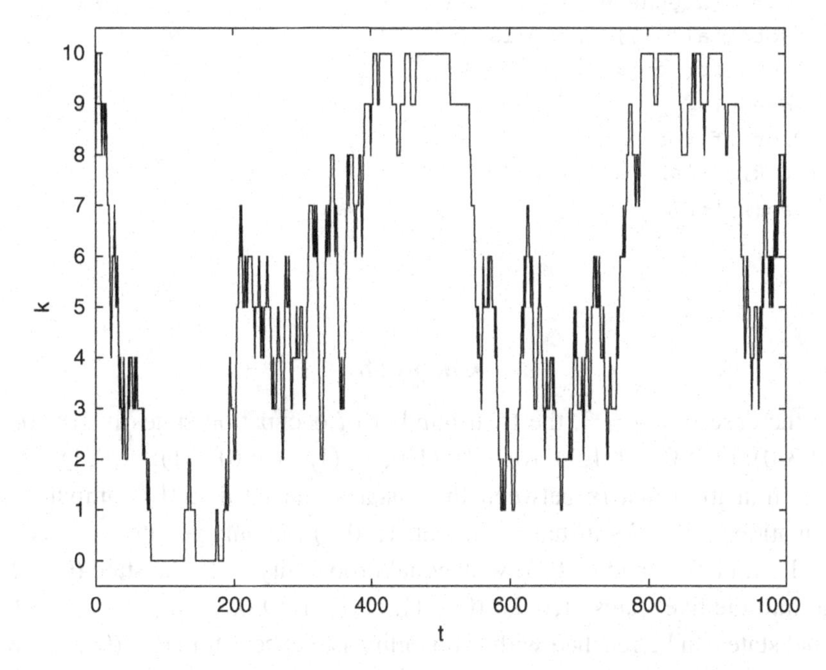

Figure A7.6. Time evolution of $k = n_1$ for $\alpha_1 = \alpha_2 = 1/2$ for 1000 steps.

```
# pointer to category
indexsite <- min(which((cumprobc-runif(1))>0))
bn[indexsite] <- bn[indexsite] + 1
A = c(A, bn) # data update
} # end for on t
A<-matrix(A,nrow=T+1,ncol=g,byrow=TRUE)
# Frequencies
norm<-0
f<-rep(c(0),times=n+1)
for (j in 1:(n+1)) {
f[j]<-length(which(A==(j-1)))
norm<-norm + f[j]
} # end for on j
f<-f/norm
# Comparison with theory (g=2 only)
fth <- rep(c(0),times=n+1)
for (i in 1:(n+1)) {
alpha1 <- alpha[2]
alpha0 <- alpha[1]
fth[i] <- factorial(n)/(gamma(sumalpha+n)/gamma(sumalpha))*
+ (gamma(alpha1+i-1)/gamma(alpha1))*
+ (gamma(alpha0+n-i+1)/gamma(alpha0))/
+ factorial(i-1)/factorial(n-i+1)
} # end for on i
k<-c(0:n)
# Plot of results
plot(k,f,xlab="k",ylab="p(k)")
lines(k,fth)
```

Chapter 8

8.1 In the case $g = n = 3$, the total number of occupation states is 10: $(0,0,3)$; $(0,3,0)$; $(3,0,0)$; $(1,1,1)$; $(0,1,2)$; $(1,0,2)$; $(1,2,0)$; $(0,2,1)$; $(2,0,1)$; $(2,1,0)$. The transition matrix between these states can be directly computed from Equation (8.77). For instance, the state $(0,0,3)$ can only go into the three states $(0,1,2)$, $(1,0,2)$ and $(0,0,3)$ with equal probability $1/3$. The state $(0,1,2)$ can go into the five states $(1,0,2)$, $(0,2,1)$, $(1,1,1)$, $(0,0,3)$ and $(0,1,2)$ and each final state can be reached with probability $1/6$ except for state $(0,1,2)$, which can be reached twice over six times and has probability $1/3$. Repeating these considerations in all the other cases leads to the following 10×10 transition matrix:

$$\mathbb{W} = \begin{pmatrix} 1/3 & 0 & 0 & 0 & 1/3 & 1/3 & 0 & 0 & 0 & 0 \\ 0 & 1/3 & 0 & 0 & 0 & 0 & 1/3 & 1/3 & 0 & 0 \\ 0 & 0 & 1/3 & 0 & 0 & 0 & 0 & 0 & 1/3 & 1/3 \\ 0 & 0 & 0 & 1/3 & 1/9 & 1/9 & 1/9 & 1/9 & 1/9 & 1/9 \\ 1/6 & 0 & 0 & 1/6 & 1/3 & 1/6 & 0 & 1/6 & 0 & 0 \\ 1/6 & 0 & 0 & 1/6 & 1/6 & 1/3 & 0 & 0 & 1/6 & 0 \\ 0 & 1/6 & 0 & 1/6 & 0 & 0 & 1/3 & 1/6 & 0 & 1/6 \\ 0 & 1/6 & 0 & 1/6 & 1/6 & 0 & 1/6 & 1/3 & 0 & 0 \\ 0 & 0 & 1/6 & 1/6 & 0 & 1/6 & 0 & 0 & 1/3 & 1/6 \\ 0 & 0 & 1/6 & 1/6 & 0 & 0 & 1/6 & 0 & 1/6 & 1/3 \end{pmatrix}.$$

Either by inspection of the matrix or by building a graph connecting the states, it can be seen that in the long run any state can lead to any other state, so that this chain is irreducible. Moreover, the chain is aperiodic, as $P(x,x) \neq 0$ for any state x. Therefore this is a finite, irreducible and aperiodic Markov chain. There is a unique invariant distribution which is also an equilibrium distribution. The invariant distribution can be found by looking for the left eigenvector of $\mathbb{W}$ with eigenvalue 1. In *R*, one can use the following sequence of commands.

```
# Program BDYdiag.R
# transition matrix
row1<-c(1/3,0,0,0,1/3,1/3,0,0,0,0)
row2<-c(0,1/3,0,0,0,0,1/3,1/3,0,0)
row3<-c(0,0,1/3,0,0,0,0,0,1/3,1/3)
row4<-c(0,0,0,1/3,1/9,1/9,1/9,1/9,1/9,1/9)
row5<-c(1/6,0,0,1/6,1/3,1/6,0,1/6,0,0)
row6<-c(1/6,0,0,1/6,1/6,1/3,0,0,1/6,0)
row7<-c(0,1/6,0,1/6,0,0,1/3,1/6,0,1/6)
row8<-c(0,1/6,0,1/6,1/6,0,1/6,1/3,0,0)
row9<-c(0,0,1/6,1/6,0,1/6,0,0,1/3,1/6)
row10<-c(0,0,1/6,1/6,0,0,1/6,0,1/6,1/3)
P<-t(matrix(c(row1,row2,row3,row4,row5,
row6,row7,row8,row9,row10),nrow=10,ncol=10))
# diagonalization
ev3<-eigen(t(P))
# eigenvalues (the largest eigenvalue is equal to 1)
ev3$val
# eigenvectors (the first column is the eigenvector
# corresponding to 1)
```

```
ev3$vec
# probability of states (normalization)
p<-abs(ev3$vec[,1])/sum(abs(ev3$vec[,1]))
p
# exact values of the equilibrium distribution
pt<-c(1/18,1/18,1/18,1/6,1/9,1/9,1/9,1/9,1/9,1/9)
pt
# differences
p-pt
p-p%*%P
```

As a final remark one can notice that, as only the eigenvector corresponding to the largest eigenvalue is required, there are efficient numerical methods to compute it.

8.2 There is not a unique way of implementing a Monte Carlo simulation of the Markov chain described by (8.77). Below, a simulation is proposed for the 10-state version discussed in the previous exercise that samples both the transition matrix and the stationary distribution. It is just a tutorial example and more efficient algorithms may be written. Say $(1,1,1)$ is chosen as the initial state: every category is occupied by one object. Then a couple of categories is randomly selected (first the loser and then the winner) and the loser transfers one object to the winner. If the loser has no objects, nothing happens. Note that the loser can coincide with the winner. It would be possible to be a little bit more efficient, by implementing a periodic version of the program. In this case, the simulation would lead to the same values for the stationary probability distribution, but with a different transition matrix in which it is impossible for the winner and the loser to coincide. However, a word of caution is necessary. Sometimes, it is dangerous to use shortcuts of this kind, as one can simulate a different model without immediately realizing it and much time may be necessary to unveil such a bug.

```
# Program BDYMC.R
# Number of objects
n<-3
# Number of categories
g<-3
# Number of Monte Carlo steps
T<-100000
# Initial occupation vector
y<-c(1,1,1)
# Frequencies and transitions
state<-4
```

```
freq<-c(0,0,0,0,0,0,0,0,0,0)
transmat<-matrix(0,nrow=10,ncol=10)
# Loop of Monte Carlo steps
for (i in 1:T) {
# Random selection of the winner
# Generate a uniformly distributed integer between 1 and 3
indexw<-ceiling(3*runif(1))
# Random selection of the loser
# Generate a uniformly distributed integer between 1 and 3
indexl<-ceiling(3*runif(1))
# Verify if the loser has objects
while(y[indexl]==0) indexl<-ceiling(3*runif(1))
# Dynamic step
y[indexl]<-y[indexl]-1
y[indexw]<-y[indexw]+1
# frequencies
if (max(y)==1) newstate<-4
if (max(y)==3) {
if(y[3]==3) newstate<-1
if(y[2]==3) newstate<-2
if(y[1]==3) newstate<-3
# end if
}
if (max(y)==2) {
if(y[1]==2) { if(y[2]==1) newstate<-10
else newstate<-9
# end if
}
if(y[1]==1) { if(y[2]==2) newstate<-7
else newstate<-6
# end if
}
if(y[1]==0) { if(y[2]==1) newstate<-5
else newstate<-8
# end if
}
# end if
}
# Updates
freq[newstate]<-freq[newstate]+1
transmat[newstate,state]<-transmat[newstate,state]+1
state<-newstate
# end for
}
```

```
# Normalization
freq<-freq/T
freq
for (j in 1:10) transmat[j,]<-transmat[j,]/sum(transmat[j,])
transmat
# Differences
pt<-c(1/18,1/18,1/18,1/6,1/9,1/9,1/9,1/9,1/9,1/9)
pt-freq
row1<-c(1/3,0,0,0,1/3,1/3,0,0,0,0)
row2<-c(0,1/3,0,0,0,0,1/3,1/3,0,0)
row3<-c(0,0,1/3,0,0,0,0,0,1/3,1/3)
row4<-c(0,0,0,1/3,1/9,1/9,1/9,1/9,1/9,1/9)
row5<-c(1/6,0,0,1/6,1/3,1/6,0,1/6,0,0)
row6<-c(1/6,0,0,1/6,1/6,1/3,0,0,1/6,0)
row7<-c(0,1/6,0,1/6,0,0,1/3,1/6,0,1/6)
row8<-c(0,1/6,0,1/6,1/6,0,1/6,1/3,0,0)
row9<-c(0,0,1/6,1/6,0,1/6,0,0,1/3,1/6)
row10<-c(0,0,1/6,1/6,0,0,1/6,0,1/6,1/3)
P<-t(matrix(c(row1,row2,row3,row4,row5,
row6,row7,row8,row9,row10),nrow=10,ncol=10))
P-transmat
```

8.3 In Fig. A8.1, the Monte Carlo estimate of n_i^* as the time average of the occupation numbers $n_i(t)$ for $T = 10000$ Monte Carlo steps is compared to the theoretical prediction given by Equation (8.62). This figure can be generated by the following program.

```
# Program binary.R
# This program simulates the Ehrenfest-Brillouin model with
# binary moves and the constraint on energy (or demand).
# The energy (or demand) quantum is e = 1.
n <- 30 # number of particles
c <- 0 # c=1 for BE, c=0 for MB. (c = -1 for FD)
T <- 10000; # number of Monte Carlo steps
# We are only interested in the transitions
# between energy (or productivity) levels!
# Initial occupation vector. We start with all
# particles (workers) in a given cell of a given level.
# This fixes the energy (or production) of the system)
level <- 3
E <- level*n # total initial energy (production)
dmax <- E-(n-1) # the maximum level that can be occupied
q <- 0; # power for degeneracy of levels (0 for the original AYM)
i <- c(1:dmax) # number of levels
g <- i^q; # degeneracy of levels
```

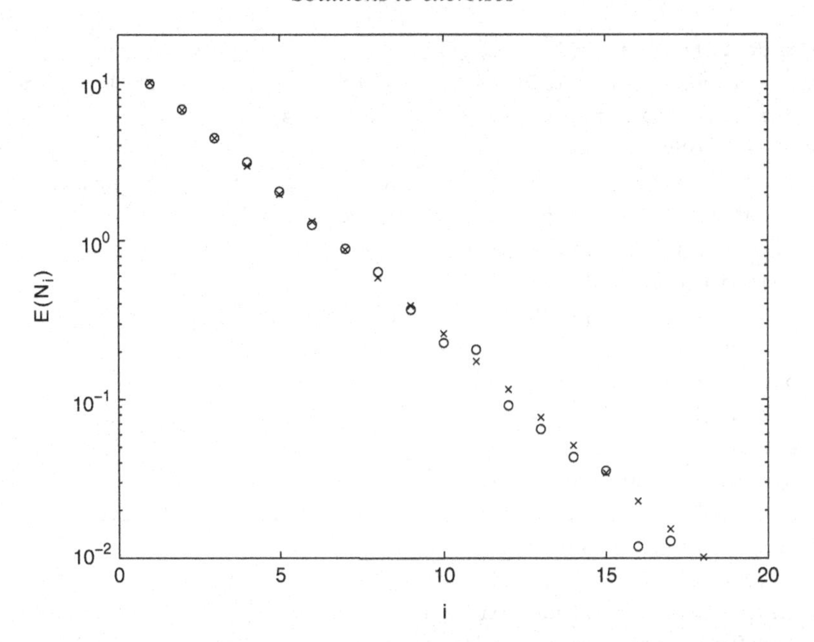

Figure A8.1. Monte Carlo simulation of the Ehrenfest–Brillouin model for $n = 30$ and $T = 10000$ Monte Carlo steps in the Maxwell–Boltzmann case ($c = 0$) and no degeneracy. The circles represent the time average of the level occupation numbers $n_i(t)$, whereas the crosses are a plot of the analytical prediction of n_i^* (Equation (8.62)) for $r = 3$. The agreement between the two curves is very good and it is worth stressing that the crosses *do not represent a fit*!

```
#(number of cells for level) (1 for the original AYM)
N <- rep(c(0), times = dmax) # level occupation vector
EN <- rep(c(0), times = dmax) # time average of N
probd <- rep(c(0), dmax) # vector of hypergeometric probabilities
N[level] <- n
A <- N # vector of results
# Monte Carlo cycle
for (t in 1:T) {
# first destruction (an object is removed from an
# occupied category according to a hypergeometric probability)
probd <- N/n
cumprobd <- cumsum(probd)
indexsite1 <- min(which((cumprobd-runif(1))>0)) # level 1
N[indexsite1] = N[indexsite1] - 1
# second destruction (an object is removed from
# an occupied category according to a hypergeometric probability)
probd <- N/(n-1)
cumprobd <- cumsum(probd)
indexsite2 <- min(which((cumprobd-runif(1))>0)) # level 2
```

```
N[indexsite2] <- N[indexsite2] - 1
en <- indexsite1 + indexsite2 # energy (productivity) of the
# destroyed (removed) particles (workers)
en2 <- floor(en/2)
probc <- rep(c(0), times = en2) # probability of creation
# creation (the two particles reaccommodate
# according to a Polya probability)
for (j in 1:en2) {
probc[j] = 2*(g[j]+c*N[j])*(g[en-j]+c*N[en-j])
if (j==en-j) {
probc[j] <- probc[j]/2
} # end if
} # end for on j
cumprobc <- cumsum(probc/sum(probc))
indexsite3 <- min(which((cumprobc-runif(1))>0)) # level 3
N[indexsite3] <- N[indexsite3]+1
indexsite4 = en - indexsite3 # level 4
N[indexsite4] = N[indexsite4]+1
EN=EN+N/T # the average of N is stored
A = c(A, N); # data update
} # end for on t
A<-matrix(A,nrow=T+1,ncol=dmax,byrow=TRUE)
k=c(1:dmax)
plot(k,log(EN),xlab="i",ylab="N(i)")
ENth = (n*((level-1)/level)^k)/(level-1)
lines(k,log(ENth))
```

Author index

Subject index

Printed in the United States
By Bookmasters